The Writings of CHARLES DE KONINCK

Charles De Koninck, University of Laval, Quebec.

The Writings of
CHARLES DE KONINCK

VOLUME TWO

Edited and Translated by

Ralph McInerny

University of Notre Dame Press
Notre Dame, Indiana

University of Notre Dame Press
Notre Dame, Indiana 46556
www.undpress.nd.edu

Paperback edition published in 2016

Library of Congress Cataloging-in-Publication Data

Koninck, Charles de, d. 1965.
[Works. English. 2008]
The writings of Charles De Koninck / edited and translated by Ralph McInerny ; with a biography by Thomas De Koninck ; and an introduction by Leslie Armour.
p. cm.
Includes bibliographical references and index.
ISBN-13: 978-0-268-02597-7 (cloth : alk. paper) — ISBN-10: 0-268-02597-5 (cloth : alk. paper) — ISBN-13: 978-0-268-02623-3 (pbk : alk. paper) — ISBN-10: 0-268-02623-8 (pbk : alk. paper)
1. Koninck, Charles de, d. 1965. 2. Science—Philosophy.
I. McInerny, Ralph M. II. Title.
Q175.K664 2008
149'.91—dc22

2008010161

∞ *The paper in this book meets the guidelines for permanence and durability of the Committee on Production Guidelines for Book Longevity of the Council on Library Resources.*

Contents

Preface

This second volume of the *Writings of Charles De Koninck* begins with two short works published in 1943, *Ego Sapientia: The Wisdom That Is Mary*, and *The Primacy of the Common Good against the Personalists*. The former is De Koninck's first work in Mariology and is self-effacing in that it is made up in great part of citations from the Fathers and the saints on the role of Mary. And there is the pronounced influence of Marie Grignion de Montfort, then Blessed, since Saint, whose teaching on true devotion to the Blessed Virgin is often invoked. The serenity and edifying character of this work provides vivid stylistic contrast with *The Primacy.*

The lengthy preface to the latter work by Cardinal Villeneuve sets the tone for what follows. What follows is not a single work, but two essays, that on the primacy of the common good and another called "The Principle of the New Order." They are of course related: the second might be said to put historical clothing on the argument of the first, somewhat as Kierkegaard's *Postscript* pays off in historical coin on the argument of the *Philosophical Fragments.* The personalism criticized in the *Primacy* is shown to have been exemplified in the history of western thought, beginning with Renaissance Humanism and culminating—that is, reaching its nadir—in Marxism and Fascism.

The Primacy of the Common Good stirred up a surprising reaction, something manifest in the reviews of it by Yves Simon and Father Eschmann. While praising the book, Simon has a number of critical things to say. Among them is the danger that the Personalists attacked might be thought to include in their number Jacques Maritain. Any such assumption, he says, would be libelous. That is, Simon holds that the positive doctrine of De Koninck's little treatise is not at odds with that of Maritain and that Maritain would share his criticisms of the "personalists." Father Eschmann, on the other hand, rejects what De Koninck has to say of the primacy of the common good and takes the position he is attacking to be that of Jacques Maritain.

His essay is called "In Defense of Jacques Maritain," but it is characterized by the almost total absence of Maritain. Very far along in the essay, Father Eschmann finally turns to Maritain, but only to give the reader his own brief paraphrase of what he takes Martain's personalism to be. Maritain is almost as absent from this defense as he is from De Koninck's little book.[1]

It seemed well to include in this volume both Simon's and Eschmann's reviews. Without Eschmann's review before him, no reader of De Koninck's response, *In Defence of St. Thomas,* could possibly understand the length and tone of the work. That there were personal undercurrents at work, that relations between Eschmann and Cardinal Villeneuve were thought to motivate the Dominican's attempted put-down of De Koninck, is true enough. De Koninck certainly rose in defense of the man who had written the preface to his book and notes that Eschmann attributes to De Koninck himself (laudatory) things the cardinal had said of him in his preface. No more need be known in order to appreciate the two defenses, one allegedly of Maritain, the other of St. Thomas. What galvanized De Koninck was the suggestion that he had misrepresented the thought of Thomas Aquinas. Since he felt that it was Eschmann who misrepresented that thought, a long and detailed refutation was called for. Many of the lacunae Simon noted in the *Primacy* are filled in by the lengthy *Defence.*

Ralph McInerny, Notre Dame

The originals of the texts included in this volume are:

1. *Ego Sapientia . . . La Sagesse qui est Marie,* Laval-Fides, 1943, 176 pp.
2. *De la primauté du bien commun contre les personnalistes; le principe de l'ordre nouveau.* Quebec, 1943, xxiii–195 pp.
3. Yves Simon, "On the Common Good." Review of 2 in *The Review of Politics* 6, no. 4 (1944): 530–33.
4. I. Thomas Eschmann, "In Defense of Jacques Maritain," *The Modern Schoolman* 22, no. 4 (May 1945): 183–208.
5. "In Defence of Saint Thomas: A Reply to Father Eschmann's Attack on the Primacy of the Common Good," *Laval théologique et philosophique* 1, no. 2 (1945): 8–109.

6. "La dialectique des limites comme critique de la raison," *Laval théologique et philosophique* 1, no. 1 (1945): 177–85
7. "Notes sur le marxisme," *Laval théologique et philosophique* 1, no. 1 (1945): 192–99.
8. "Cette parole est dure," *Revue Dominicaine* 51, no. 1 (1945): 65–73.
9. "Between Heaven and Earth," *Laval théologique et philosophique* 2, no. 2 (1946): 131–33.
10. "Concept, Process, and Reality," *Laval théologique et philosophique* 2, no. 2 (1946): 141–46.

NOTE

1. In *The Person and the Common Good,* Maritain alludes to Eschmann's review and thanks him for it. He does not mention the review by Simon. One cannot conclude from this that Maritain accepts Eschmann's identification of his position with that attacked in *The Primacy.* For all that, it is a mystifying footnote.

EGO SAPIENTIA

The Wisdom That Is Mary

Charles De Koninck

1943

To my wife Zoé
and to my daughters
Godelieve and Marie-Charlotte

I confess my own ignorance and do not hide my own pusillanimity. And yet, nothing gives me greater joy, but also nothing frightens me so much, as to speak of the glory of the Virgin Mary.

—Saint Bernard, *In Assumpt.*, sermo 4

The excellence of the glorious Virgin is such that every language is impotent to recount and praise it: Scripture is powerless, the prophets are powerless, and the parabolic images are also. That is why the Holy Spirit, speaking by the mouth of the Prophets, praises her not only with words but also by figures and parabolic images: because no parabolic image suffices perfectly to express her excellence, the similitudes and metaphors have been multiplied with a view to better celebrate her praise.

—St. Bonaventure, *De Nativitate B.V.M*, sermo 3

Preface

How may we truly apply to the Blessed Mary all that is said of Wisdom in the Sapiential Books? In answering this question we are attempting no innovation as will be evident from the constant use made of the Doctors of the Church. Rather, in composing this study, whose main purpose is to assemble a certain amount of pertinent quotations relating to the Mother of God as Wisdom, we have been prompted by the miseries of our times which indicate more than ever our need to keep our eyes steadily fixed upon one of the most outstanding manifestations of the Wisdom and Mercy of God.

Those texts which in their mystical sense the liturgy applies to the Mother of God will be utilized merely to illustrate conclusions deduced from the literal sense of other passages of Scripture. Such an illustration, nevertheless, supported as it is by the liturgy, has a truly illuminating quality.[1]

How to express my acknowledgment to all my friends who, without always realizing it, collaborated in this collection? It was Jacques de Monleon who first helped me understand the role of mercy; it was an American Jew who told me of the writings of the Blessed de Montfort; I do not forget the part played by Father Maurice Dionne nor that of Father Alphonse-Marie Parent who undertook the thankless task of correcting my manuscript and the proofs; I think too of all those I cannot name. If this little work has any value, it is uniquely due to a merciful Providence which arranged the fortuitous encounters and confided this work to the most unworthy of the servants of His Mother.

PART ONE

Ego Sapientia

1. Ego Sapientia

Mary must shine forth
more than ever in these latter times,
in mercy, in strength, and in grace.
—St. Grignion de Montfort

The words which the Church places in the mouth of the Blessed Virgin are not, "I, the wise," nor "I, the wisest of all creatures," but "*I, wisdom—Ego Sapientia.*"[2] Of a very good person we may say that he is goodness itself, but this attribution is to be understood in a purely metaphorical or parabolical sense. There are only two cases in which one may predicate an abstract term of a concrete term in an essential proposition: when it is a question of God or of the transcendentals. "Abstract things," says John of St. Thomas, "cannot be truly predicated of concrete things, nor concrete things of abstract things, because of the mode of signifying . . . although sometimes they are really identical, as in divine things Divinity and God, Paternity and the Father."[3] How then can the Blessed Virgin claim the sovereign affinity to God which such a mode of attribution implies?

2. Ut orietur lumen indeficiens

What is proper to wisdom? The adage says, "*Sapientis est ordinare*—It is for the wise man to order." How are we to understand the term "to order"? Two things are included in the notion of order: distinction and principle. Principle is that from which something proceeds in any way whatsoever. Principle implies proceeding. Proceeding or procession is a movement from a principle, movement which can be understood in the broad sense of any action, the action of thinking as well as of physical motion. Accordingly insofar as the principle is a principle of place, a principle of time, or a principle of nature, order will be divided into local order, temporal order, and the order of nature. Of these three orders the last is the most profound, since it implies the notion of origination, inasmuch as nature is "that from which is first born the thing which is born—*ex qua pullulat pullulans primo.*" Under another aspect order is divided into universal and particular order according to whether the principle is absolutely first or first in a given genus only.

What order is in question in the adage: "It is for the wise man to order?" It belongs to the wise man to set things in order, says St. Thomas, "because wisdom is the highest perfection of reason, the proper role of which is to know order."[4] Since order implies principle, and principle implies relation, the intellect alone can grasp order as order. "Since the intellect (unlike the will) draws things to itself, and proceeds by passing from one to the other, it can compare and formally grasp the relation of one thing to another; the intellect therefore possesses within itself the primary root and cause necessary for ordering things—comparing them among themselves and establishing a relation of one to the other."[5] However, the mere knowledge of an order is not, as such, sapiential. Simple apprehension can attain order, and every science involves a certain order. Wisdom alone will be the highest perfection of reason insofar as it implies an order proceeding from a principle which is wholly first. The verb "to order" expresses this originative primacy. "It is not to be ordered," says Aristotle, "but to order, which belongs to a wise man."[6] That is why wisdom is radical. It not only shows the interlocking of one thing with another, but it grasps things in their primary root, wherein all the things that proceed therefrom are, in a certain way, pre-contained; and it grasps this root under its proper formality of origin. If this root were not at the same time origin, the absolutely first principle would be in dependence upon that of which it is the first principle; the multiple would then have, as such, the nature of a first principle.

Wisdom may be predicated substantially only of a thing which in its being and operation is of the nature of the first principle from which all things proceed by way of origination. It would not suffice for it to attain the primary root solely according to knowledge because then it would be wise only, but it must substantially possess the nature of a first principle, and know itself as such.

In order for the Blessed Virgin to be called Wisdom, she must be first principle in this sense. She must be herself first principle, not merely according to intellect and will, but also according to her substance and being. And who is first principle according to his very being except God? To be truly a first principle would not Mary have to be such a first principle even in her relation to God, would she not have to be so close to God that she would somehow participate even in His nature of first principle, be as the root of the universal order, even, in a way, that from which God Himself proceeds in a certain manner, the origin and genetrix of God?

3. Ecce virgo concipiet

Fear not, Mary, for thou has found grace with God. Behold thou shalt conceive in thy womb, and shalt bring forth a son: and thou shalt call his name Jesus. He shall be great, and shall be called the Son of the most High; and the Lord God shall give unto him the throne of David his father; and he shall reign in the house of Jacob forever.[7] The Virgin gives birth to the God-man, the branch of Jesse has flowered.[8] She is henceforth truly the *mother of God*,[9] she who has engendered God.

Generation means vital origin and assimilation. It is the procession of a living thing from within a living thing conjoined as a principle of life which assimilates the product of generation to its proper nature by virtue of this very procession. Generation consists therefore in expressing a likeness propagative of the nature of the generator. The generator draws that which is generated from its own substance while forming it. If the Blessed Virgin is truly a generator, this definition of generation must fully apply to her. Let us here note that although in the act of conception the mother is merely a passive principle which, while properly a nature, does not of itself imply an active and expressive assimilation, nevertheless, considered in her relation to the one engendered, the mother is properly an active principle which vitally assimilates the one engendered. An assimilative action takes place formally in the production of the passive principle of conception, a production which results from the active generative power of the woman, in view of the one engendered. For this reason, the mother participates actively in the vital assimilation of the one engendered. She is properly a genetrix.[10]

Birth regards primarily and principally the being of the hypostasis and person. Hence, since the Blessed Virgin is the mother of Christ according to the hypostasis, she is truly the mother of God and of the man, of the God-man.[11] In relation to that in Him who is born, the Blessed Virgin is properly cause and origin of God, *causa Dei et origo Dei.*[12]

Being the cause of the cause of all things, the mother of God is consequently the mother of all things. "She is the mother of all things," says St. Albert, "and God the Father is the origin of all things: but whatever is per se the origin and cause of the cause is per se origin and cause of that which is caused. But she is the mother of Him who is the cause and origin of all things: therefore she is per se the mother of all things."[13] Is she not under this aspect an absolutely universal cause? Is there any work of God which is not to be related to her as to its principle?

Insofar as she is the substantial principle of Him who made her—*genuisti qui te fecit*—she fulfills by her divine maternity an essential condition of the appellation "Wisdom." Since she is truly mother of the Son, and the Son is incarnate Wisdom, she is the mother of Wisdom engendered entitatively both of the eternal Father and the temporal mother. "She is the mother," says Cornelius a Lapide, "of the eternal Wisdom incarnated in her. Just as the Son is Wisdom engendered and incarnate, so she is the Wisdom which engenders and incarnates."[14]

4. Fiat!

Nevertheless, Wisdom implies knowledge, a procession according to knowledge. In order that the Blessed Virgin be truly Wisdom, she must, even in relation to God, in addition to her divine maternity according to the flesh, attain to the nature of a first principle according to intellect. That is what she declares in her *Fiat—may it be done unto me according to thy word.* The *Fiat* of Mary is the echo of the *Fiat* of Genesis, the word whence proceeds the new order to which the ancient had been ordered. *For behold I create new heavens and a new earth: and the former things will no longer be remembered, and will no longer come to mind.*[15] "God who made all things," says St. Anselm, "is Himself made from Mary (*ipse se ex Maria fecit*) and thus all that He made He has made again."[16]

My heart hath uttered a good word. "When the angel had spoken," says St. Augustine, "Mary, full of faith and conceiving Christ in spirit before conceiving Him in her womb, said: Behold the handmaid of the Lord, be it done unto me according to thy word."[17] "The Blessed Virgin," adds St. Albert, "would not have engendered Christ in the flesh if she had not first of all conceived and preserved the Word in the ear of her heart (*aure cordis*), bearing Him so to speak in the womb of her heart (*in cordis utero*)."[18]

The *Fiat* of Mary—on this word, principle of the Eternal Word by whom all things are made, depends the entire new order. Judge Mary to be wisdom, and the whole universe will be remade. "Hasten, O Virgin," cries St. Bernard, "to give your answer. O my Sovereign, pronounce the word which the earth, hell, and heaven await. . . . Say but the word, and receive the Word; give your word and receive the divine Word: pronounce a passing word and embrace the eternal Word."[19] *Be it done unto me according to thy word. Let him kiss me with the kiss of his mouth.*[20] Let the Holy Spirit take my flesh and unite it intimately to the Son of God. Let the Word that is the Light become flesh.

M. Olier, in the most express way, tells us that in her *Fiat* the Blessed Virgin imitated the procession of the Son in God according to knowledge.

> . . . Just as [the eternal Father] engenders His Word through all eternity by His knowledge, by a return upon and vision of Himself, so He wills that Mary, the supremely perfect and holy image of His virginal fecundity, should engender the Word with knowledge; and for that reason He decrees that she shall give her consent to the generation of the word in flesh in an express and solemn way presupposing knowledge and reason. Whereas other mothers will not know the one who is to be born of them, He wishes that Mary should know previously what manner of son she shall conceive: an angel will make known to her that this son will be the Son of the Most High, both God and man, the Redeemer of the world, and that His reign will endure forever.[21]

5. In columna nubis

This God, whose mother she is, is God the Redeemer who as Redeemer is the final and consequently absolutely first cause of the entire universe, for

Christ was never efficaciously willed as the end of all things except as the Redeemer.[22] As Mother of the Redeemer, Mary is inseparably united to this final cause as co-principle. *The Lord possessed me in the beginning of his ways, before he made any thing from the beginning. I was set up from eternity, and of old before the earth was made.*[23] Mother of Emmanuel, that is to say, of "the Mighty with us," she is the first one predestined among all pure creatures. "She came from God in the beginning," says St. Albert, "because from all eternity she was predestined to become the mother of the Son of God."[24]

The mother is inconceivable without the Son, nor is the Son and the Redeemer Son conceivable without the mother. She proceeds from Him who made her in order that He might proceed from her. It is as principle that she proceeds from the Principle: her procession from the Principle is ordained in the procession of this same Principle, and she envelops the Principle in her procession from Him, she is held by Him in His procession from her. *I came out of the mouth of the most High, the firstborn of all creatures.*[25] In coming out of the mouth of the most High she is herself the mouth which proffers the Word. *Let him kiss me with the kiss of his mouth.*[26] She proceeds from the true light, from Him who is the unfailing light. *Ego feci ut in caelis oriretur lumen indeficiens.*[27] The Son who is in the bosom of the Father, precontains all things, including the Virgin, and causes Himself to be contained in the womb of the Virgin. *He whom the whole universe cannot contain has enclosed Himself in your womb, becoming man.*[28] The Son and the mother thus constitute from the very beginning a kind of circular motion wherein the principle is the term, and the term, principle, a motion which is the symbol of Wisdom *which reaches from end to end.*[29] This circular motion of Wisdom, which is more mobile than mobile things,[30] is like unto play: *playing before him at all times.*[31]

6. A mari abundavit cogitatio ejus

Being truly the mother of God,[32] the Blessed Virgin is bound to the hypostatic order in the most intimate way possible for a pure creature. "Hence," says St. Albert, "since birth primarily and principally has respect to the being of the hypostasis and the person, and secondarily to the nature, the Blessed Virgin is called the mother of Christ according to the hypostasis, which hypostasis is

God and Man, and this is why she is the mother of God and of the man—although she is not consubstantial with God except with respect to His human nature, since consubstantiality taken in itself means nothing other than *convenientia* in substance. Birth, then, belongs primarily and of itself to the person, and to the nature by consequence and secondarily."[33] She alone among all pure creatures thus occupies the very summit: *I dwelt in the highest places, and my throne is in the pillar of a cloud,* wherein the incarnate Wisdom is hidden.[34] Starting out from the summit of heaven, her course terminates upon the same summit. *A summo caelo egressio ejus: et occursus ejus usque ad summum ejus—His going forth is from one end of the heavens, and his circuit ends at the other.*[35] She alone has compassed the circuit of heaven, of all mere creatures she alone is the Wisdom which has penetrated into the bottom of the deep. *Gyrum caeli circuivi sola, et profundum abyssi penetravi—I alone have compassed the circuit of heaven, and have penetrated to the bottom of the deep.*[36]

It would be impossible for a pure creature to be raised any higher. By the grace of her maternity, she exhausts, so to speak, the very possibility of a higher elevation.

> The plenitude of the Blessed Virgin deprives her of all emptiness. As long as a vessel can receive anything, it retains some emptiness. That is why every creature includes a certain emptiness, because it can also always receive a greater grace. But she alone is full of grace, because she could have no greater grace. She would have to be herself united to the divinity in order to conceive a grace greater than that according to which that is drawn from her which is united to her. Unless she were herself God, it is impossible to conceive a greater grace than that of being the mother of God.[37]

7. Gratia plena, in Sion firmata

In order to be praised and glorified in Mary, God was not content to express Himself in her maternity alone, in which Mary herself does not accomplish in the fullness of her being a complete return to the principle. "Thus, maternal parenthood," says St. Augustine, "would have been of no advantage to Mary if she had not experienced more joy by bearing Christ in her heart than in her flesh."[38] She was full of grace even before her consent to maternity. The angel called her full of grace before the Holy Spirit had come upon her.

> The Holy Spirit descended upon Mary in order that she might be the mother of God, and in order that she might thus attain to the hypostatic order since she was already full of grace. Because she belongs to the hypostatic order which of itself carries with it a higher form of sanctity, her maternal dignity demands sanctity by fittingness and connaturality. If by God's absolute power there had been maternity without sanctity, then the Holy Spirit would not have decended upon her after the manner of a mission in the absolute sense, because the Holy Spirit would not have dwelt in her, but would have descended upon her in the relative sense according to a mission.[39]

Fullness of grace in Mary thus becomes the root of her consent to maternity, of the most free and liberal act that a pure creature can accomplish, of the most radical human act, upon which all the works of God are made to depend. *For her thoughts are more vast than the sea, and her counsel deeper than the great ocean—A mari enim abundavit cogitatio ejus, et consilium ejus ab abysso magno.*[40] Chosen in the beginning of all the works of Divine Wisdom, the strength and sweetness of the power of premotion caused to spring up in her a vast determination wherein she is established and establishes herself as first principle. *There is none that can resist thy will, if thou determine to save Israel—Non est qui possit tuae resistere voluntati, si decrevis salvare Israel.*[41] Because she herself becomes a sapiential principle, it is fitting that in her quality as Wisdom she be imbued with immutability.[42] *And so I was established in Sion—Et sic in Sion firmata sum.*[43] "Confirmation in good was fitting for the Blessed Virgin," says St. Thomas, "because she was the mother of divine Wisdom, in which there is nothing defiled, as it is said in Chapter 7 of the Book of Wisdom."[44]

Just as our liberty is so much the more our own since it is received universally both with regard to act and modality (*Deus est qui operatur in nobis et velle, et perficiere—For it is God who of his good pleasure works in you both the will and the performance*[45]), so the fact of being first principle, as with the Blessed Virgin since it is entirely received according to this properly divine modality, is all the more truly her own. There is thus established between the grace of maternity and her sanctification a certain circular motion which it has pleased God to arouse within her. It is God, the origin of all things, who gives her the power of giving herself as origin of God. "Behold all things are subject to the command of God, even the Virgin, behold all things are subject to the Virgin, even God."[46] By her free consent to the maternity which properly comes from her nature, God gives the Blessed Virgin the means to raise herself further to the dignity of her maternity which both fittingly (*congrue*) and connaturally demands sanctity.

8. Mitte radices

In this exhaustive superabundance of grace and glory expressed in her, the Blessed Virgin accomplishes the return to the principle under the very aspect of principle of all grace and glory. To her in her quality of Wisdom it has been confided to place in the elect the principle of their conversion to God, to place in them the divine roots. *Then the creator of all things commanded me and said to me: and he that made me, rested in my tabernacle, and he said to me: Let thy dwelling be in Jacob, and thy inheritance in Israel, and take root in my elect—Tunc praecepit, et dixit mihi Creator omnium: et qui creavit me, requievit in tabernaculo meo, et dixit mihi: in Jacob inhabita, et in Israel haereditare, et in electis meis mitte radices.*[47] In this Wisdom dwells *all the grace of the way and of the truth*, in her *all hope of life and of virtue.*[48]

9. Appropinquavit ad mare

A House built by Wisdom—*Sapientia aedificavit sibi domum.*[49] "Mary is the sanctuary and the place of repose of the Holy Trinity, where God is present more magnificently and divinely than in any place in the universe, not excepting His dwelling above the Cherubim and Seraphim."[50] This indwelling is so full and so complete that as Wisdom the Blessed Virgin is the *brightness of eternal light, and the unspotted mirror of God's majesty, and the image of his goodness—Candor lucis aeternae, speculum sine macula Dei majestatis, et imago bonitatis illius.*[51] This image is so perfect that it in turn fulfills the function of root and exemplar for all creatures. This image was the sapiential exemplar that God followed in the composition of all things. *I was with him forming all things—Cum eo eram cuncta componens.*[52] By this she is united to the consubstantial image of the Father, to the incarnate Wisdom, to the Word by which all things were made, and without whom was made nothing that was made. Being the image of Goodness, she imitates the original in universal diffusion of goodness, and she gives to things their first impetus and motion: as spouse of the Holy Spirit who is compared to the waters, and who *moves over the waters*[53] and, herself a spirit of Wisdom, she too can say: *I, Wisdom, have poured out rivers. I, like a brook out of a river of a mighty water; I like a channel of a river and like an aqueduct, came out of paradise—Ego sapientia effudi flumina, ego quasi tramis aquae immensae de fluvio.*[54] Her diffusion is so universal that she reaches God and imitates the manner in which God Himself is found in every diffusion of His goodness: *And behold my brook became a great river, and my river came near to the sea—Et ecce facta est mihi tramis abundans, et fluvius meus appropinquavit ad mare.*[55] A mirror of unspotted purity of the majesty of God, this created Wisdom is in its effusion of graces like a formal sign: no limitation is imposed upon her mediation. That is why she is called *subtilis.* "She reacheth everywhere by reason of her purity. She is a vapor of the power of God, and a certain pure emanation of the glory of the almighty God, and therefore no defiled thing cometh into her—*attingit autem ubique et capit propter suam munditiam, vapor est enim virtutis Dei et emanatio quaedam est claritatis omnipotentis Dei sincera et ideo nihil inquinatum in illa incurrit.*"[56] This same tabernacle of the Holy Spirit, this dwelling built by Wisdom, this holy city, this new Jerusalem, this new heaven which renews the earth, becomes the tabernacle of God with men.[57]

10. Omnia innovat

Order is implied by wisdom. Wisdom is at once one and manifold, steadfast and mobile.[58] Wisdom may be predicated of the principle of the sapiential order insofar as this principle is the root, and pre-contains the order, of which it is the principle. Together with her Son at the very origin of the universe, she is in a way the root of the universal order: *Ego sum radix—I am the root.*[59] That which God principally desires in the universe is the good of order. This order is better in proportion as its principle, which is interior to the universe, is the more profoundly rooted in God. But Mary is the purely created principle of this order, the purely created principle which is nearest to God and the most perfect that can be conceived. As a principle of the sapiential order, she participates in the unity and the unicity of this principle, she is at once an *emanation* and an *indwelling*,[60] her power extends to all things which take from her their constant renewal. We conceive vital emanation as a constant renewing from within, and in their relation to the first principle things receive being in an ever-new procession. Whatever being they might have of themselves would be nothingness. *One is my dove, my perfect one. And being one, she can do all things, and remaining in herself the same, she reneweth all things—Una est columba mea, perfecta mea. Et cum sit una, omnia potest: et in se permanens omnia innovat.*[61] Daughter of the eternal Father, mother of the Son, spouse of the Holy Spirit, she is rooted in the order of the Trinity, and she links up the order of the universe in a radically new way to the order which is in God according to the processions. *Thy neck is as a tower of ivory—Collum tuum sicut turris eburnea.*[62]

11. Imago bonitatis illius

As the principle whence comes the good of the universe, as *Regina et Domina* of all things, she is a good separated from the universal order, a good which is properly universal, a good which in its indivisible and superabundant unity is the good of all things. This good is better than the good which exists as a form in the order of the parts of the universe, it is anterior to it and is its principle, as the leader is the principle of the order in an army. Her good does not even imply a material dependency upon the things which are ordained or upon the form that is their order. Being wisdom, *all her glory is from within—omnis gloria ejus filiae regis ab intus.*[63] Because she carries with her the notion of the properly universal common good, because she is for us the principle of every spiritual good, it is not enough to love the Blessed Virgin as one loves oneself, nor to love her as much as oneself. Just as it is necessary to love Christ more than oneself, so too it is necessary to love the Blessed Virgin more than oneself.

> Each loves himself, after God, more than his neighbor. One must love others as oneself, hence one's own self is so to speak the primary exemplar of those one must love; oneself as participating in the divine glory

> and others as associated in this participation. I except, however, the Lord Christ, even as man, and the Blessed Virgin the mother, because they take on the character for us of a principle diffusive of grace and blessedness. Christ as man is the head (*caput*) of glory, and the Blessed Virgin is the mother of this head, and she is the neck through which grace descends from this head down to us, and for this reason we should love them more than ourselves.[64]

12. Circumdata varietate

When from another point of view, we consider the Blessed Virgin as interior to the universe, we can compare her to the intrinsic good of the universe, a good which consists in the form which is none other than the order of its parts. This form is comparable to the visage and the face. In this form consists the highest dignity of pure creation, that is to say, that which by God's will is the most desired for itself and most perfectly ordered to Him. Considered as a separate good of the universe, the Blessed Virgin is more worthy than the order of the universe whose transcendental principle she is. On the other hand, when we consider her as interior to the universe as a part, the dignity of the universe is greater than that of the Blessed Virgin considered, not absolutely, but formally insofar as she is a part, a consideration which in her case is secondary.

Nevertheless, it should be noted that even under this aspect she remains the intrinsic root of the dignity inherent in the universe as a form and at the same time she has the greater share in this dignity. The dignity of the whole depends materially upon the dignity of the parts and upon the relation that these parts have to one another. But the excellence of the inferior parts is contained in a more eminent way in the superior parts themselves. The superior parts are to the inferior parts as form: the splendor that the inferior parts derive from the subordination to the superior is greater than that which they themselves have of themselves absolutely. This derived splendor is foremost in the ordination of the parts to the whole. Considered from the aspect of the parts, the excellence of the principal part is of the nature of a form for all the subordinated parts. Consequently the Blessed Virgin is, as a part, the form and the purely created principal cause of the dignity which ordains them most proximately and most perfectly to the dignity of the whole. As the prior and principal part, she draws all the other parts after her toward the dignity of the whole. *Draw me, we will run after thee—Trahe me: post te curremus.*[65]

Among all the purely created parts of the universe, she participates in a greater degree in the order of the universe, and she is invested in a greater way with its splendor and variety. Her splendor is most comparable to that of the whole (*decora sicut Jerusalem—comely as Jerusalem*[66]), she is surrounded with the variety of all the other parts—*circumdata varietate.*[67] She is that purely created part of the universe thanks to which it can be indued with a great dignity. *A spring rose out of the earth, watering all the face of the earth—Fons autem ascendebat de terra, et irrigabat omnem faciem terrae.*[68] "Face of the earth, that is to say, dignity of the earth," St. Augustine comments, "is a name very rightly (*rectissime*) applied to the Virgin Mary, Mother of God, she whom the Holy Spirit, who is called in the Gospel by the name of fountain and water, waters so that out of the clay there might be formed the man who is placed in paradise to dress and keep it, that is, in the will of the Father, in order to accomplish it and keep it."[69]

13. Quasi rota in medio rotae

If she is already so beautiful and so worthy of praise as a part of the universe, *even without that which is hid within—absque eo quod intrinsecus latet,*[70] how much more so is she as a separate principle and good. Under the latter aspect she is absolutely anterior to her character as part and a principle in relation to this character as a part. Her position as a part is ordained to her position as a separate principle. She is born within to be a separated principle, she is born in the universe to become the mother of all things. That which is without proceeds from that which is within, and that which is within proceeds thence in order to proceed without. As a separate principle she is more within the universe than when considered as a part within it. *Intrinsecus ejus per circuitum—as the appearance of fire round about.*[71] There is thus established a circular motion between her dignity as a separated principle and her dignity as the noblest part of pure creation, a circular motion which embraces the very order of the parts of the universe. The order and dignity inherent in the universe are thus all the more intimately linked to the separated principle insofar as this principle is at the same time the principal intrinsic part of the universe. This circular motion imitates in a way the circular motion between the incarnate Wisdom and the mother of this Wisdom, which in turn imitates more deeply still the circular motion between the Father and the perfect and consubstantial image of the Father, *as it were a wheel within a wheel—quasi sit rota in medio rotae.*[72]

14. De fructu suo cognoscitur

The fact that her Son infinitely surpasses her in privileges and dignity manifests the sovereign dignity of the mother. It is her Son, *the fruit of her womb,*[73] the word whom she has drawn from her heart, who surpasses her infinitely.

> We grant that her Son surpasses her in all privileges: but this, far from diminishing the praise of the mother, exalts it in that she has not only engendered a Son equal to her, but a Son infinitely better than herself. Under this aspect the goodness of the mother is rendered, in a sense, infinite. Each tree is known by its fruit: hence, if the goodness of the fruit makes the tree good, the infinite goodness of the fruit manifests an infinite goodness of the tree.[74]

15. Mons in vertice montium

In his admirable commentary on the Sapiential Books, Cornelius a Lapide expresses in the most formal way the primary reason for this appellation of Wisdom from the point of view of the end, which is the cause of causes. How can the Blessed Virgin be made to say the things which Wisdom says of itself: *I brought it about that in the heavens there should rise light that never faileth—ego in caelis feci ut oriretur lumen indeficiens*—and *I have poured out rivers—ego sapientia effudi flumina?*[75] To this

> I answer, first, that in the mystical sense these words should be understood as follows: I brought it about that in the heavens, that is, in the Churches, Christ should be born, He who is the sun of justice; I have

brought it about that in the Church there should rise the light of faith. Furthermore, the Virgin, comparable to a sea of graces, pours out her rivers upon the Church and the faithful. Secondly, according to the literal sense, one should read: I have been the cause for God to create the light, the heavens, the sea, the rivers and all the universe. The creation of God is ordained, as to its end, to the justification and glorification of the Saints, accomplished by Christ through the Blessed Virgin; for the order of nature was created and instituted for the order of grace. Thus, then, it is because the Blessed Virgin was the mother of Christ that she subsequently becomes the mediatrix of all the other graces instituted by Christ; whence for the same reason, she was the final cause of the creation of the universe. Indeed, the end of the universe is Christ such that His mother and the saints, that is to say, this universe was created in order for the Saints to enjoy grace and glory through the intermediary of Christ and the Blessed Virgin. That is why the final cause of the creation of the universe was the predestination of Christ, the Blessed Virgin, and the Saints. Although Christ and the Blessed Virgin are parts of the universe and consequently posterior to it in the order of material causality, nevertheless they are anterior to it in the order of final causality. There is thus a certain reciprocal dependency between the creation of the universe and the birth of Christ and the Blessed Virgin. God did not wish Christ and the Blessed Virgin to be born except in this world; nor did He wish that this universe exist without Christ and the Blessed Virgin; on the contrary, it was for them that He created it. He wished that the whole universe, no less than the order of grace, should be referred and ordained to Christ and the Blessed Virgin as to its complement and end. Christ and the Blessed Virgin are thus the final cause of the creation of the universe and at the same time they are its formal, that is, exemplary cause, namely the idea. In effect, the order of grace in which Christ with the Blessed Virgin occupies the first place is the idea and exemplar according to which God created and disposed the order of nature and of all the universe.[76]

16. Quae est ista?

Are we not amazed at the supereminent *beauty* of this pure creature, of this purely created Wisdom? Is she not called "Mother most admirable," that is, Mother, principle, whose cause is unfathomable? Mary, the most astounding of all mere creatures, allows us, in a manner most proportioned to us, to surmise the incomprehensibility of Wisdom which precedes all things. *Who hath searched out the wisdom of God that goeth before all things?—Sapientiam Dei praecedentem omnia quis investigavit?*[77] To what purely created work could she be compared? *There was no such work made in any kingdom—Non est factum tale opus in universis regnis.*[78] Is she not so astounding that, according to St. Bernard, even the princes of the celestial court are filled with astonishment: *Who is this that cometh up from the desert, leaning upon her beloved? Who is she that cometh forth as the morning rising, fair as the moon, bright as the sun, terrible as an army set in array?—quae est ista quae ascendit de deserto; quae est ista quae progreditur quasi aurora consurgens pulchra ut luna electa ut sol terribilis ut acies ordinata?*[79]

PART TWO

Nigra sum, sed formosa

17. Universae viae Domini misericordia et veritas

Why does the Holy Virgin tell us "I am black, but beautiful—*Nigra sum sed formosa*"? What connection could there be between the attribution of wisdom and of blackness that signifies a state of inferiority, as can be seen in the term 'but'? Far from excluding one another, would not these two qualifications have a link of dependence between them? Would there not be a very intimate connection between that blackness the mother of God attributes to herself and her note as first principle?

To gauge the nature of this link, we must go back to the primary motive and to the universal way of God's communication without—*ad extra.* But this motive is nothing other than the divine goodness insofar as it is diffusive of itself. The root of the primary way of this diffusing and of this manifestation outside is mercy. *All the ways of the Lord are mercy and truth—Universae viae Domini misericordia et veritas.*[80] That is why St. Bernard calls the mercy of God *causalissima causarum—of causes the one that is most cause.*[81] Mercy is the first root, even of justice. "But the work of divine justice," St. Thomas writes,

> always presupposes the work of mercy and is grounded in it. Nothing is owed to the creature save because of something preexisting in him, or pre-considered in him: again, if it is owed to the creature, this will be because of something prior. And since we cannot be involved in an infinite regress, we must come to something which depends solely on the good of the divine will, which is the ultimate end. For example, it is as if we should say that to have a hand is owed to man because of his rational soul; and to have a rational soul, in order that he be man; and he is a

> man because of the divine goodness. So it is that mercy shows up in any work of God, as its first root. And its power is saved in everything consequent upon it, and even operates more strongly in it, as the primary cause has a stronger effect than does the secondary cause.[82]

Mercy, having the meaning of absolute universal root, extends from one end of the universe to the other. Even the sovereign dignity of the Incarnation is willed only with a view to the manifestation of the divine glory by way of mercy and justice. Any dignity other than God's own is only a relative absolute. "The end for the sake of which the glory of God should be made manifest is by way of mercy and justice; hence because the Incarnation can accomplish this effect of manifesting mercy and justice in man's redemption, the motive for willing the Incarnation was not the dignity of the Incarnation taken absolutely, but the Incarnation as bringing about such an effect."[83] Since it has the note of the perfectly universal root, "mercy is apparent even in the damnation of the reprobate, not as completely relaxing it, but in a way alleviating it, never punishing beyond what is due."[84]

The concept of mercy includes a quite eminent perfection: it is the virtue of the superior precisely as superior.

> I reply that it should be said that a virtue can be the highest in two ways: in one way, taken in itself; in another way, by comparison to the one having it.—In itself indeed mercy is highest, for it pertains to mercy that it flows to the other, and, what is more, that it makes up for the defects of the other; and this belongs most to the superior. Hence to be merciful is said to be proper to God, and in it His omnipotence is especially made manifest.
>
> But with respect to the one having it, mercy is not the greatest, unless he who has it is greatest, having nothing above himself but everything else beneath him. For one who has others above himself it is higher and better to be joined to the superior than to make up the defects of inferiors. Therefore, as for man who has God superior to him, charity, by which he is united to God, is more powerful than mercy through which he supplies what is lacking in his neighbor. But of all the virtues which pertain to the neighbor, mercy is most powerful, as its act too is more powerful, for to supply the lack of another is, as such, the act of one superior and better.[85]

18. Miserationes ejus super omnia opera ejus

If mercy is fulfilled in the elevation of the inferior, this elevation will be the more merciful and revealing of the divine goodness and omnipotence when it raises up that which is most inferior. In other words, we can judge the measure in which God has willed to manifest Himself by the degree of merciful raising up that He has chosen to realize.

If the divine mercy is already manifest in creation,[86] it shines forth even more insofar as it raises others above their defects (*inquantum defectus aliorum sublevat*). *The Lord's mercies are above all his works—Miserationes ejus super omnia opera ejus.*[87] But among all defects, properly speaking evil is the greatest. It is evil, insofar as it has the note of misery, which will be the motive of the plenitude of mercy, of mercy victorious over evil:[88] "evil is the motive of this mercy."[89]

19. Angeli fortudine, et virtute cum sint majores . . .

In order to gauge the height and depth of what God has chosen to make manifest outside Himself, we must see the baseness of the nature He has elevated above all other creatures. It is indeed in this that merciful omnipotence most properly shines forth. Let us consider first off the hierarchy of created things in the perfection that belongs to them by nature.

At the summit of creation envisaged from a purely natural point of view are found the angels, pure spirits, beings very perfect with respect to both substance and operation. Their essence being simple, each of them is of itself a complete and individual species, subsisting outside every common natural genus. Each of them exhausts a degree of being. Radically hierarchized, each of the angels occupies in this hierarchy an absolutely determined place. Even the lowest pure spirit constitutes by himself a universe incommensurably more perfect than the cosmos and humanity combined.

The cosmos and its most perfect interior term, humanity, are only a remote echo of the spiritual universe—*quaedam resonantia*.[90] One can show this by considering in a dialectical manner the angelic hierarchy in the sense of its inferior limit. In proportion as the angels are distant from Pure Act, the simplicity of their essence diminishes. The limit of this departure from the note of simplicity is an essence composed of matter, form, and privation.

While pure spirits are immutable in their substance and absolutely necessary, in the sense that they do not contain in themselves any principle of non-being,[91] essences which include privation entail so to speak their own negation. At this level, the species, diffused in individuals, is only maintained by their generation and corruption. It is to matter, insofar as it is deprived of form, that the existence of chance and disorder here below must be attributed, privation which expresses our distance from the first principle who is in Himself always uniform—*semper eodem modo se habente.*[92] And chance only doubles the fortuitous. We live at the confines of the universe where we are diffused both according to substance and to quantity and according to temporal duration.

Our days and our places are uncertain. Everything here below is variable and decrepit, and it is only by a great effort that we sometimes succeed in impressing on things a momentary direction. It is only by a familiarity that blinds us and a sort of animal resignation that we have become unconscious of the immense confusion in which we live and to which only violence seems able to awaken us. Our substance is truly at the confines of being.

Envisaged in its natural condition, the intelligence of separated substances is always in act. They judge without composition and division; they know the notions of things, one in another, without discourse; intuitively they grasp in a quasi-circular motion the essence from whom they emanate and in the light of which they see. Because the angel is too perfect to be subject to other creatures in his knowing, God infused in him in the morning of his existence intelligible species representative of the universe He had chosen to form, species anterior to the things themselves. Imitating God who knows all things in one unique universal species, the pure spirits, in proportion as they approach Him, know this universe by means of an always smaller number of species. But when we regard the angelic hierarchy in the direction of its distance from the first intelligence, the intuition of essence impoverishes along with the imperfection of that essence and of the intelligence which emanates from it. In order to know other things, this intelligence has need of more and more ideas, its activity is more and more fragmented; the discrete time constituted by the sequence of always increasing thoughts and desires becomes more and more atomized, the present is diffuse, scattered in an ever more distant past and future. The intelligence is more and more remote from itself and from the other things that it knows. At the limit of this declension there arises an intelligence turned outside himself, in pure potency, similar to prime matter, a blank slate, a non-intuitive intelligence which can only be awakened to its proper act by means of a sensible singular, intelligible only potentially. *Ratio oritur in umbra intelligentiae—human reason emerges in the shadow of intelligence.*[93] It can know only in dependence upon a species representing something other than itself. In order to know things in their proper nature, it requires a number of intelligible species equal to the number of the natures it knows; it puts itself in dependence on the senses which need as many species as there are single forms known. On this level, knowledge requires not only a great number of internal and external senses, but a doubling of the intellectual faculty into an intellect which precedes knowledge by penetrating into the penumbra of the sensible world and illumines objects so that they might be assimilable, and another intellect which properly knows and names things. Our intellect can only live in the shadows. The necessity of the shadows of the sensible world has its origin in the weakness of our intelligence. By its nature, our rational life is the least perfect intellectual life it is possible to conceive.

The union of the intellectual nature and sensible nature makes man subject to a certain contrariety. Our sensible nature bears us toward the sensible and private good, our intellectual nature has for its object the universal and the good under the very notion of good, which is found principally in the common good. But, in us, the sensitive life is primary; we can only attain the acts of reason by passing through the senses which, in this respect, play the role of principles. So long as a man is not rectified by the cardinal virtues, he is principally drawn toward the sensible good[94] against the good of intelligence.[95] Men for the most part succumb to this attraction, and that for two connected reasons. The good, in fact, demands perfect integrity; evil, on the contrary, results from no matter what defect.[96] So long as a man has not acquired the virtues which determine him *ad unum* to the correct integrity conformed to reason, his action is uncertain and easily turns from the true good. Hence the adage: *malum ut in pluribus in specie humana.*[97] Most men follow the inclination to the sensible good and allow themselves to be led by it against the order of reason.

Moreover, considering ourselves in our natural condition compared with pure spirits who are always in act, immutable, and incapable of error or fault in the natural order, we are already black enough: in substance, because of matter and privation; in knowledge, because of the nocturnal potentiality of intellect and the opacity of sense; in the order of action, because of the contrariety of our composed nature.

That is the order of things envisaged in their natures and the place we have in that order. If we run great risks, we nonetheless have every reason to rejoice in this existence that the divine mercy has deigned to confer on us. "Who has not received this mercy from God," asks St. Augustine, "first, to exist, to be set apart from the brute animals, to be a rational animal that can know God and, further, to enjoy this light, this air, the rain, fruits, the seasons, the charms of the earth, health of body, the affection of friends, and the welfare of his home?"[98]

20. Orietur in tenebris lux tua. Et tenebrae tuae erunt sicut meridies

Nonetheless, out of pure liberality, God has chosen to manifest Himself in a manner incomparably more profound by elevating the created intelligence to an end which surpasses infinitely the active nature of this intelligence, to a supernatural life, which has for its term the vision of God as He is in Himself. But the ways in which God can realize this return to Himself under the very note of His deity are still many, some more profound and revealing of His mercy than others.

Elevation to the life of God can come about immediately and without any intermediary condition other than acceptance of the glory promised, as was the case with the angels. But this elevation can also be accomplished in a much more striking manner, namely by the visible sending of a divine person in hypostatic union with a created nature. Descending thus into His creation in order to elevate it from within to the properly divine order, God would already manifest the mercy of His omnipotence in an infinitely greater measure than in the creation alone of intellectual creatures so perfect in themselves or in their immediate elevation.

But this same hypostatic union can in turn be accomplished in different ways, one more merciful than the other, and consequently more profound, insofar as what is elevated is lower. It could be accomplished in the assumption of an angelic nature. Since this nature is the most perfect and most worthy of all created intellectual natures, would it not be better disposed to this elevation? And was it not this apparent fittingness which deceived the prince of darkness?

The hypostatic union can be realized in a more admirable manner in the assumption of the lower nature that is human nature, the least worthy of all intellectual natures. The divine wisdom and power confounds the most powerful spirits.

The assumption of human nature can itself be accomplished in two ways, either immediately, without prior condition, as would be the case if God formed immediately the nature assumed, or by assuming human nature by way of birth, God thus putting Himself in dependence on man and proceeding because of that into the universe itself by way of origination. And the nature itself from which He is born becomes in that way properly the origin of God. Let us quickly note that this most radical communication would in no wise be possible in assuming an angelic nature. God could not proceed from an angelic nature, for this nature is, on the one hand, too perfect to engender in the manner of natural beings, and, on the other, too imperfect to engender as God does. *Perfecta imperfecte, imperfecta perfecte.* It is then thanks to the potentiality of matter, indeed of matter as deprived of form, and thus to the privation that is the weakest, that the Son of God can proceed from within His creation, thus imitating in the most profound way His generation from the eternal Father. *Infixus sum in limo profundi: et non est substantia—I am plunged into the mire of the deep, and there is nowhere to set my foot.*[99] Happy imperfection of matter that permits such a formation!

This same Son arises from two extremes of the universe, reuniting our baseness with His supreme grandeur—*ima summis.* Here is the degree of communication and merciful elevation that it has pleased God to bring about. *Ecce virgo concipiet, et pariet filium: et vocabitur nomen ejus Emmanuel—Behold a virgin shall conceive and bring forth a son, and his name shall be called Emmanuel.*[100] God raises Himself up and causes Himself to be engendered from the most remote limits of His creation. *Aperiatur terra, et germinet Salvatorem—Let the earth be opened and bud forth a saviour.*[101]

Here in the substantial order the *nigra sum, sed formosa* can already be applied. Mary is beautiful because of the divine maternity, but, on the side of the creature herself, this maternity is possible only thanks to the darkness of potentiality and privation. Therefore, it is thanks to that darkness that God can Himself proceed from a created principle and that a mere creature can be called wisdom. Note this intimate rapprochement of God that permits maternity in virtue of this very passivity in conception. God could not proceed here below from a principle which is active in its fecundation. Such a principle would have to take on the note of a passive principle.[102] It is only in the passive principle of generation, the principle that has the note of malleable matter, that the fecundity of Pure Act can find its echo in an entitative and substantial mode. *Imperfecta perfecte.* Only the woman can have along with God the notion of first principle in the origination of God. If a man could have been the father of God, not only would generation be less perfect, paternity would be possible only insofar as it would imitate maternity: it is the maternity of the woman, and not the Paternity of God, which would be the original of it.

21. Abyssus abyssum vocat

Nigra sum, sed formosa. This proposition expresses at once the two extreme virtues in the realm of spirit: humility, the most fundamental virtue for man, the least perfect intellectual creature possible, the weakest of all; and mercy, the virtue proper to the Almighty. *Sapientia illius erupuerunt abyssi—His wisdom has opened the depths,*[103] one on the other. The abyss of plenitude calls to the abyss of emptiness. *Abyssus abyssum invocat—deep calls to deep.*[104] It was because of her humility that the Blessed Virgin was agreeable to God. *Quia respexit humilitatem ancillae suae: ecce ex hoc beatam me dicent omnes generationes—Because he has cast his eyes on the lowliness of his servant, behold, because of this all generations will call me blessed.*[105]

The humility of the Most Blessed Virgin can in no wise be compared to the act of humility of the most perfect angel before God. His natural condition being most superior, he has that much more reason to humble himself, since he has received from God all that he is.

Notice that there is a divergence among the commentators on Mary's Canticle as to the meaning of the term *humilitas.* Some see it as expressing her natural condition, others understand by it the virtue of humility. The Greek text of the Magnificat seems to favor the first interpretation, for *tapeinosis* signifies abasement, whereas humility properly so called is signified by the term *tapeinophrosune.*[106] Still, the opinions are not mutually exclusive. Before

God Mary recognized the lowliness of her condition, and it is in this that her act of the virtue of humility consists. She did not see the lowliness in which she found herself as a condition contrary to her dignity, as a humiliation afflicted on her and from which she would have praised the Lord for delivering her. Here is where the act of humility of the Handmaid of the Lord attains the most sublime: it attains to the two extremities of the universe. The angel does not have in himself a substance which would permit so profound an act *which extends from one end to the other.*[107]

22. Quia respexit humilitatem ancillae suae

Humility alone among the virtues is unaware of itself, and one who boasts of his humility is proud. The total effacement of the self, it is in this that its character, at once radical and universal, consists. But do we not find in the Canticle of Mary a teaching and exaltation of humility? We must notice that the Blessed Virgin does not glorify humility in an absolute manner, as if the exaltation of humility was a debt founded absolutely on this virtue. The Blessed Virgin submits herself entirely to the omnipotence and mercy of the Lord: her soul magnifies the Lord, and her spirit is transported with joy to God her Savior, to Him who is powerful and whose name is holy and whose mercy extends from age to age.[108] She does not exalt humility in herself but in God. For He has now given Himself to her who is "the immaculate way,"[109] "the path that God has made through the desert,"[110] "the holy way which is for us the straight way,"[111] to comprehend the hidden and properly divine ways of mercy. God has let her know that she is "at the beginning of all His ways,"[112] *Viam sapientiae monstrabo tibi duxi te per semitas aequitatis—I will show you the path of wisdom.*[113] She who understood so perfectly that "all the ways of the Lord are mercy—*omnes semitae Domini misericordia et veritas,*"[114] how could she have exalted her own humility? It is for the Most High "to exalt the humble,"[115] and the humble exalt humility only in the mighty one.

Although it is not the greatest of virtues, humility is nonetheless for us the condition of every other virtue; it makes us malleable to higher perfections. It has, with respect to the other virtues, a maternal role. The person in whom this mother-virtue was most profound has been chosen as the mother of God. "The Blessed Virgin," Cajetan writes, "recalls that the Lord regarded her humility as a universal virtue which was most broadly and deeply open (*patula*) to receive the celestial influence of the divine largess."[116] "*He has looked on the lowliness of his servant,*" says John of St. Thomas, "that is, to spread a plenitude of grace as great as that received by the Blessed Virgin, God looked only to the depth of her humility by which she was made capable of receiving, as into the deepest concavity, the immense grandeur of grace."[117] Humility is the truly liberating virtue and is the very principle of the dignity to which God has deigned to call us. "Humility is like a certain disposition to man's free access to spiritual and divine goods."[118] By her act of humility, Mary was entirely emptied of herself, she was liberated from herself in a total conversion to God.

23. Humiliavit semetipsum

Behold thou shalt conceive in thy womb. In her perfect humility, founded on a right understanding of the human condition, Mary comprehended the humiliation to which God chose to submit Himself. "Moreover," says St. Bernard,

> of all the infirmities or of all the human injuries that the divine goodness has taken on for us, I think that the first in the order of time and the greatest with respect to abasement was when the infinite majesty allowed Himself to be conceived in the womb of a woman and to remain there for the space of nine months. Indeed, at what time has God ever so despoiled Himself, or when has He ever been seen to turn so completely away from Himself? All that time, this wisdom spoke not a word, this power did nothing apparent, this majesty enclosed and hidden did not manifest itself by any visible sign. God did not seem this weak on the cross where that which was weak in Him showed itself to be stronger than that which is most powerful among men: when, dying, He glorified the thief and, expiring, He inspired the centurion; when His hours of suffering not only made all the created elements suffer with Him, but also consigned the strong set against Him to the suffering of eternal woe. But in the womb He—He Who Is—was as though He were not, and the Eternal Word was silent.[119]

And yet in this silence is hidden the most powerful manifestation of the Word: by this silence in the womb of His mother, the Word imitated at the same time in a most striking way His silent procession from the bosom of the Father.

24. Ubi humilitas, ibi sapientia

Ubi humilitas, ibi sapientia—Where there is humility, there is wisdom.[120] St. Albert says,

> This proposition is in theology one known per se: therefore, the greater the humility, the greater the wisdom, and where there is perfect humility, there is perfect wisdom. But, in the Blessed Virgin, humility was incommensurable; therefore her wisdom is incommensurable. The minor is made evident by this passage of the Gospel, "He who humbles himself will be exalted" (Matthew 23:12). This proposition too is per se nota, self-evident: therefore, she who is exalted above other creatures in an incommensurable manner is thereby incommensurably more humble than they; exalted above all the choirs of the angels to the fourth hierarchy, the Blessed Virgin, according to St. Jerome, was raised incommensurably above all other creatures; therefore, she was the most humble among men and angels, and thereby surpasses them all in wisdom.[121]

Humility touches the same root as mercy. Mercy, indeed, looks to the inferior as such. But "God resists the proud and gives His grace to the humble—*Deus superbis resistit, humilibus autem dat gratiam.*"[122] Mercy only gives its largess to the inferior that recognizes itself as such, and the lower it is, the more reason for humbling itself it has. But this humility will be fecund only if it is rooted in a knowledge whereby one sees at the same time how little we are and how great the power of Him who is the Lord. This very great humility of the Blessed Virgin has to be based on faith in the omnipotence of God. *Et beata, quae credidisti, quoniam perficientur ea, quae dicta sunt tibi a Domino—Happy she who has believed,* cries Saint Elizabeth, *since there will be accomplished the things which were told you by the Lord.*[123]

25. Dominus tecum

"The Lord is with you,"[124] that is, the Almighty, He who is purely and simply Lord. He before whom Mary humbles herself is at the same time one who can do the most astonishing of things. "[B]ecause it is faith which best disposes to consent to marvelous things, and above all faith in omnipotence: and because he who believes and is convinced that God can do all things, admits that He can also change and command natures. So much so that, in what is the most unexpected of unexpected things (*omnium novorum novissimum*) that is here announced to the Blessed Virgin, it is wholly fitting that the name Lord is used since it designates in an absolute manner the omnipotence of God."[125] The handmaid's faith in omnipotence must be so much the greater that it acts to raise a nature most humble in its natural condition. "O Virgin," cries St. Bernard, "sublime branch, you raise yourself to the holiest height, to Him who is seated on the throne, to the majesty of the Lord Himself! And why be astonished when you hid in the heights the roots of humility."[126] "O humility, by which the woman became the mother of God, by which God descended from heaven to earth, by which souls have been transported from hell to heaven (Genesis 28:12). Behold the ladder that God proposes to you and by which one mounts from earth to heaven. For it is this ladder that our fathers have climbed to the heavens, and it is by means of it that we too as well must climb, since otherwise we shall not rise."[127]

Only the abyss of humility can envelop the Infinite without limiting it and be in the world an unshakable foundation for the Changeless. St. Bonaventure asks himself,

> If [wisdom] had been conceived from all eternity, how can it, after so many centuries, take birth from the Virgin Mary? If indeed it was eternal, it was immoveable, therefore it could not be contained, it could not be circumscribed. So how, without being circumscribed, was it enclosed in the womb of a young woman? How, being without possible limit, could it be conceived by this fragile and delicate child? And yet it is such and so great a Wisdom that the Virgin conceived, according to angelic testimony. At the beginning of St. Luke, the angel says to the virgin, "Behold you will conceive in your womb and give birth to a son, etc." and goes on to describe this son, "He will be great," that is, by His infinity, "and will be

> called the son of the most high: because of the immutability of His essence, "and His reign will be everlasting," because it cannot be limited. The divine Majesty by this conception has been humbled in an astonishing way, and the virginal humility exalted in a marvelous manner.[128]

That is why St. Bernard cries, "Admire these two things, then tell me which ought to astonish the more, the most benevolent favor of the Son, or the most excellent dignity of the Mother! On one side or the other, one is stunned; on one side and the other, one touches on a miracle; and that God should submit Himself to a woman is a humility without equal; and that a woman command God, that is a sublimity without equal."[129]

26. Felix culpa!

Nigra sum, sed formosa. In fact, mercy is manifested even beyond the assumption alone of human nature by way of birth. Man, whom God had established in a state of original justice infinitely higher than what would have been befitting by nature, succumbed to the temptation to be himself the origin of the dignity to which God had deigned to raise him. *Et homo cum in honore esset, non intellexit: comparatus est jumentis insipientibus, et similis factus est illis—And man, when he was in splendor, did not understand; he became comparable to the stupid beasts, and has become like them.*[130] By original sin, this human nature became liable to suffering. We were born into a state of misery properly speaking. *Ecce enim in iniquitatibus conceptus sum: et in peccatis concepit me mater mea—Behold in iniquity was I conceived and in sin did my mother conceive me.*[131]

Sin is not just any defect: it is that which is at the farthest remove from God. Properly speaking, evil is not a simple privation, it is opposed to the good as a contrary. Consequently, the mercy which will face down evil, which will be victorious over it, will also be, in a sense, the greatest possible. The manifestation of the divine omnipotence will be, here, in the universe itself, like a return on itself: it will be the fullness of mercy. Evil (*malum poenae*) is ordered to the greatest manifestation of mercy that could be conceived. *O felix culpa quae talem ac tantum meruit habere Redemptorem!—O happy fault that has won for us such and so great a Redeemer!*[132]

If according to the ordinary power of God only man could be redeemed, would not that follow from the imperfection of our intelligence, which was also the root of the contrariety of the two natures? The fallen angel, on the contrary, was at once obstinate and confirmed in evil. And this because the angelic intelligence is so perfect that it grasps without composition or division and without discourse whatever we know by understanding principles, by knowledge most difficult to acquire. He seizes his object in an immutable manner, and the adhesion of his will, too, is fixed and immutable. That is why man is more open to mercy because of his very imperfection. Man's free will remains flexible as much after choice as before; that of the angel, on the contrary, flexible before choice, becomes, after that choice, immutably fixed.[133]

27. Quid mihi et tibi est, mulier?

The merciful one takes on himself the misery of another as if he made it his own. But that can come about in two ways. One can take on the misery of another by affective union. So it is that we suffer the evil that befalls a friend as if we ourselves were afflicted by it. But one can also take on the misery of another by a real union, by undergoing this misery in the same way as it afflicts the one who is the object of compassion. Thus a man might expose himself to illness with a view to easing or curing his neighbor's illness. But that presupposes a proximity, a similitude of nature such that it permits one thus to take on, in a physical manner, the misery of another.[134] In this way a real union in misery is accomplished. But God has assumed human nature along with its passibility, thus taking on himself our misery in the manner that it affects us, that is, physically; assuming evil (*malum poenae*) in this way—a darkness far more profound than any that comes from our nature, the most profound that God could assume. *Although He was by nature God, He did not consider being equal to God a thing to be clung to, but emptied Himself, taking the nature of a slave, and being like unto men* (Phil. 2:6).[135]

The principle from which Christ received this passibility through which He accomplished His redemptive passion, and thanks to which God has become our brother in misery, is once more the Blessed Virgin. As Our Lord seemed to insinuate at the wedding at Cana, the mother of mercy will be manifest in the very passion of Christ. "Woman, what is this to me and thee?

My hour has not yet come."[136] As if to say: that thanks to which I perform a miracle, that you have not engendered, you have not engendered my divinity: but because you have engendered my infirmity, I will acknowledge you when this infirmity is hung upon the cross.[137] By doing this, God has placed the Blessed Virgin at the very beginning of His work of mercy, in which both the darkness, communicative of passibility, and the beauty, *formositas,* instrument of redemptive grace, are present.

28. Et macula non est in te

St. Albert asks why the geneaology of Our Lady contains not only good ancestors, but the bad as well. Certainly because the comparison makes one of the extremes stand out the more—*like a lily among thorns.*[138] The genealogy mentions bad ancestors "in order that the wisdom of God might appear more merciful. Indeed, there is an origination by which good comes from good, and another by which evil comes from evil. In the first way, *God saw all the things he had made and they were very good* (Genesis 1:31). In the second way, *Pride is the principle of every sin.* But there is a third way, according to which evil comes from good, as the woman with whom sin began. And there is a fourth way as well, by which good comes from evil, and this is proper to God alone because His wisdom conquers evil and He attains all things in their beginning and their end—*attingens a fine usque ad finem.*"[139] Is it not a sovereignly merciful privilege that Mary was conceived without the stain of original sin? *Et macula non est in te—There is no blemish in thee.*[140] This should "increase in sinners their confidence, since their mediatrix unites the two extremes in one and the same relationship, namely, just as she is both mother and daughter of God, so she is both our mother and our sister and thus, by nature, inclined to have pity on the sinner."[141] The condition which she herself would have known if she had not been preserved from it, draws her closer to us, and this closeness is more profound and efficacious in that Mary is clothed herself in the greatness of this mercy that preserved her. In this respect, "the Blessed Virgin was dark, not in herself, but in her father Adam who sinned and by sin contaminated all his posterity—with the exception of the Blessed Virgin. Thus she is called dark by extrinsic denomination, as the daughter of a sinner, but in herself she is beautiful, by the fullness of grace that is in her."[142]

29. Discite a me, quia mitis sum, et humilis corde

By communicating to her son human nature with its passibility, Mary is the source of the humble condition of Christ. But her gracious humility is at the same time at the source of the humility of her son, of this wisdom which once spoke, "Learn of me, for I am meek and humble of heart." "She knew," says Cornelius a Lapide,

> that her Son was going to restore the world at the price of His great humility, and that He had to abase His deity to the point of taking on human flesh, more, even to the scourge of the whip, the cross, death. She had then to adapt herself to the future condition of her Son, and even in a way precede Him and prepare His way; above all because, just as proud mothers inculcate pride in their sons, so humble mothers inculcate in theirs a sweet and submissive spirit. That is why Canisius says in Book Four of his *Mariology,* chapter 8, "The mother in no way degenerates her Son; on the contrary, the Son reproduced rather the character and nature of His mother." In fact, children tend to be closer to their mother than to their father. As St. Ambrose says, "Having to give both to the sweet and humble Christ, Mary had to prefer humility." She knew that the head of the devil, that is, his pride, would be crushed by her humility, according to Genesis, "She will crush thy head." So too St. Idelphonse in his second sermon on the Assumption says, "That is why the humble Christ has come from the humble Virgin in order that from the depths of such humility He might bring the victory of salvation."[143]

30. Et tuam ipsius animam pertransivit gladius

And a sword will pierce your soul in order that there might be revealed thoughts hidden in a great many.[144] By her participation in the redemptive passion, of which along with her Son she is a first principle "to suppress the misery of another as if that misery were her own," Mary remains, and in the most profound sense, black but beautiful: black in compassion and sorrow, beautiful in the ineffable merit of compassion.

I am black, but beautiful. Indeed, O most sweet Virgin, you have been, in the night of your most sorrowful compassion, your sadness and affliction, in the whole redemptive passion of your well-beloved Son, you have, I say, been hidden, filled, wounded, even more, pierced by the sword of sorrow, and knew yourself the sorrow of a second birthing. And if the omnipotence of your Son had not preserved you, your heart would have broken with sorrow, and you would soon have died also; but your Son saved you for His Church for the spiritual growth of believers. And yet, in the midst of such tribulation and sorrow, of such shame and such a burden, you remained beautiful, since such compassion and affliction were for you the cause of ineffable merit, by which you gained the power and efficacity to help us all. Thus, if the Apostle could say, "What was lacking in the passion of Christ, I have accomplished in my flesh, for the body which is the church," with how much more reason can that be said of the most holy mother of Christ? O happiest of queens, your beauty, the charm of your sorrow, the splendor of your patience, your most profound humility, and the holiness of your charity shone forth in this, that in the whole dolorous and ignominious passion of your only and well-beloved Son, you did not feel the slightest urge to indignation, aversion, and impatience toward the persecutors and cruel and criminal torturers of your most precious Son, they who regarded you as vile, iniquitous, deformed, "like the tabernacle of Cedar," that is, as the unfortunate mother of the most impious seducer, when your soul was as beautiful as the tent of Solomon, adorned with a celestial beauty like the splendor of the true Peaceful One who laid out the firmament like a tent.[145]

Just as the Son, dying not miserably but with mercy, had only disdain for the unworthy and unseemly grief given Him, so too His most happy Mother, sharing by love in the death of her Son, and in a sense dying in Him because He was bone of her bone, flesh of her flesh, said to us, "Why do you weep for me as if I were a miserable woman and mother of a miserable man? At this hour, I am black, because with my scorned Son I must also be scorned, and with Him who is considered a leper I too must be thought a leper. He is, as the prophet says, my sun, become now like an outlaw in your eyes, and in whom there appears no appearance of beauty; it is right that I should be conformed to Him, and like Him in the sad and somber aspect of one accused; like the tabernacle of Cedar, like a sinner among sinners, as Honorius says."[146]

Note that all our grace is essentially redemptive. But, as Christ is our head insofar as He communicates to us the grace merited by His passion—whereas with respect to the grace and substantial glory of the angels, He is the head only by His authority, as John of St. Thomas says,[147] so the Most Holy Virgin is our queen not only because of her dignity, as she is queen as well of the angels, but even more is our mother with respect to the generation of redemptive grace. *Mater divinae gratiae.*

31. Mater misericordiae

When we call the Blessed Virgin mother of mercy we do not mean only the mercy that is in her by way of an accident and inherence, but we also understand her maternity as the essential origin of mercy. "According to the universal usage of the Church," says St. Albert, "the Blessed Virgin is called, and is in fact, mother of mercy, something that can be said of no other creature. Sometimes men are called men of mercy, that is, men made humble by mercy, through which they have some relation to mercy, whether in a principal or accidental manner, but the relation she has to mercy is one of essential origin, in the manner of generation (*per modum matris*). This essential conformity surpasses immeasurably the modes of inherence and accident. The Blessed Virgin thus disproportionately surpasses all created persons in mercy."[148]

32. Regina misericordiae

The Queen of Mercy, she is so profoundly rooted in the *causalissima causarum,* that she has the note of "first root," and consequently her empire over the work of God is absolutely universal. Just as God is merciful even with regard to those who are confirmed in evil, so Mary is queen not only of the blessed angels but also of those who are in the eternal Gehenna. "All those who are under God's rule," says St. Albert, "come under His mercy, but not all who are under God's rule share in His glory, His grace, and His justice; thus only mercy embraces His entire realm and she who rules over the kingdom of God is above all other titles the Queen of Mercy."[149]

Mary is Queen of Mercy not only because she is the cause of all the mercy that God chooses to manifest, but, as we have seen, one can attribute mercy to her as a substantial predicate. "If one forms this proposition in an intransitive manner," says St. Albert, "the meaning would be: She is the queen of mercy, that is, of mercy itself, and then she is truly called queen of mercy, and Esther, who is a figure of the Blessed Virgin, is also called Edissa, which signifies mercy. So too Isaiah: And a throne will be prepared in mercy; but the proper place for a throne is the proper place of a kingdom, and mercy is the proper place of the kingdom. But in the womb of the Blessed Virgin, a womb prepared by the Holy Spirit, all the divinity and all the humanity of Christ were placed and established, and it was thus the proper place of the kingdom: therefore she has been mercy itself and at the same time has been, not without reason, queen of mercy, because, possessing it perfectly herself, she has always governed well, for nothing has ever gone out from her that has not been directed."[150] That is why Queen of Mercy is the most proper title of the Blessed Virgin according to her dignity.[151]

33. Nigra sum, sed formosa

Given the immensity of the mercy that the Almighty has chosen to manifest, it is only fitting that the universal royalty of Christ and of His mother was manifested in the Passion. "*Pilate then said to him, You are a king? Jesus replied: Thou has said it, I am a king.*"[152] It is the same Christ who said, "*I am a worm and no man, the opprobrium of men and the castoff of the people,*" and "*I am king, king of kings and lord of lords.*"[153] It is in the Passion that shines forth in all its profundity and extent the meaning of *Nigra sum, sed formosa.*

Queen of mercy, the Blessed Virgin is so profoundly rooted in the divine omnipotence, that in the issue, in the procession of this power, she participates so to speak in the incomprehensibility of that same power. *Sol in aspectu annuncians in exitu, vas admirabilis opus excelsi—Coming forth from God she announces the sun in all her glory, this wonderful vase and work of the Most High.*[154] Was she not herself at the outset troubled by the proximity of

God as announced to her by Gabriel. *She was troubled by these words.*[155] If the most powerful of the blessed angels tremble and humble themselves before the power that raised them high above the dignity due them by nature,[156] how much more profound will be the astonishment and humility of the Blessed Virgin called to sovereign dignity. *Totam habet potestatem—She possesses all power.* This astonishment, this imperfect knowledge of the cause, will remain with us forever. *Admirabilis ero—I will be astonished.*[157] *In plenitudine sancta admirabitur—She will astound the assembly of the saints.*[158]

34. Nonne stultum fecit Deus sapientiam hujus mundi?

Because she proceeds so admirably from the incomprehensible abyss of the divine wisdom and omnipotence, is it to be wondered at that the world finds so hard every word that exalts the grandeur and glory of Mary? Has not God made foolish the wisdom of this world? Indeed, the wisdom of this world is foolishness to God.[159] How can this mere creature, so weak in her nature, be clothed with all the power that God has deigned to manifest? The foolishness of God is wiser than the wisdom of men, and the weakness of God is stronger than the strength of men.[160] Is not the Blessed Virgin for us, in her darkness and in her beauty, the touchstone of divine wisdom? *Cunctas haereses sola interemisti—You have rid us of all heresies.*[161]

> All true and predestined children of God have God for their father and Mary for their mother, and he who does not have Mary for his mother does not have God for his father. That is why reprobates, heretics, and schismatics, who hate or look askance or with indifference at the Blessed Virgin Mary, do not have God for their father, although they glorify themselves in this, because they do not have Mary for their mother; if they had her for their mother, they would love and honor her like true and good children naturally love their mother who gave them life
>
> The most infallible and indubitable sign to distinguish a heretic, a man of bad doctrine, a reprobate, from the predestined is that the heretic and the reprobate regard with indifference the Blessed Virgin Mary, and try by word and example to diminish her cult and the love of her, openly or tacitly, and sometimes with good pretexts. Alas. God the Father has not told Mary to make her dwelling with them because they are Esaus.[162]

35. Terribilis ut castrorum acies ordinata

This purely created wisdom called *Mater timoris, et agnitionis—Mother of filial fear and the beginning of knowledge*[163] is for us the beginning of wisdom. *Initium sapientiae timor Domini.*[164] But she who has received everything in humility, becomes as well, by her wisdom, by her practical wisdom, her prudence, and by her power, the terrible enemy of the creature that God created most sublime and powerful in nature and who by his pride became the prince of all evil. For him who is the very head of all evils, she who has received the fullness of power is *terrible as an army drawn up for battle.*[165] "God has made and formed but one enmity," says St. Grignion de Montfort, "but irreconciliable, which will endure and grow even to the end."[166] And in this enmity, it is the most humble Virgin, the sweetest of mothers, who will conquer. *She will crush your head.*

36. Non serviam!

In order the better to see the greatness and power of the Woman, let us take the measure of him who is the *head and king of all the sons of pride.*[167] The Lord Himself has called him *Prince of the ways of God—Ipse est principium viarum Dei.*[168] Indeed, at the beginning he was established as the first and most sublime of all creatures, not only according to nature, but also according to grace, for the separate intelligences receive a grace proportional to the perfection of their nature. *You were the seal of perfection, full of wisdom and beauty. You were perfect in your ways from the day you were created.*[169] It was he who bore the name of light—*Lucifer.*[170] In all creation, *there was no power comparable to his, he has been created to fear nothing.*[171] He was like *the cedar of Lebanon, beautiful of branch, of deep shadow, great height, having its top in the clouds.*[172] He who was already so sublime in his nature, *the waters made grow, the abyss made him expand, causing its waters to flow around where it was planted and sending its runnels to all the trees of the field.* Prince who bore the light, he could illumine every intelligence in his realm and they would see in him the most perfect image of their God. *That is why his height rose higher than the trees of the field.* He could reach from one end of the universe to the other. *He was beautiful in his greatness, by the length of his branches, for his roots plunged into abundant waters. No other tree in the garden of God equaled him in beauty.*

But then the most sublime of intelligences is astonished by a new plan of God, a new plan which was also the most ancient. *Ecce ego facio nova—Behold I am going to make a new marvel, it is ready to hatch, do you not recognize it? I will make a road,* not in the heavens, but *in the desert, and flowers in arid land . . . to water my people, my chosen, the people I have formed for myself.*[173]

Grace, those abundant waters that will make the cedar of Lebanon grow, does not have for its end to increase the excellence of the angelic nature. Principle only of merit, it is ordered to a new life, to a supernatural life where one can do nothing by the powers that belong to us by nature. A purely gratuitous benefit, it must be recognized as such. Without this practical recognition, the grace that is the principle of merit cannot lead to the grace of glory. Compared to the order of nature, the order of grace is radically new. In the former, each intellectual creature conducts himself by himself to his singular good and toward the natural common good. In the new order, this same creature must put himself under a power quite external to his nature; he must allow himself to be drawn beyond himself. In the order of nature, the angel has a right to natural beatitude in virtue of his very creation; in the order of grace it is not owed him. Moreover, grace orders so perfectly to the most universal good that the least worthy creature from a natural standpoint can participate in it in a manner more abundant than the most perfect creature: grace is not tied to the order of natures.[174]

Because he was already raised to the supernatural order by grace the principle of merit, the angel could establish a comparison between the two orders. He thus saw his condition of nature in a light he would never have known if he had not been raised up. And he cast a new look on his great dignity and the singularity that was his with respect to his natural good. Whatever belonged to him in that order belonged to him as a matter of right. In the new light he makes the discovery of himself. Has not the first of angels the note of first principle, *the principle of the ways of God,* and is it not in this that he is most like Him? In the new order I will be dethroned and I will have to share with those inferior to me as with equals and even as superiors; I would lose my singularity and my dignity would be bypassed;[175] love would no longer be my right. In this deliberate turn to the self there rose up in him a disordered desire for his own excellence. And he said to himself: *I am a god, I sit on the throne of God in the midst of the seas.*[176] *And his heart was lifted up because of his beauty.*[177] *I will mount into the heavens, above the stars of God I will lift up my throne; I will sit on the mountain of the assembly, in the depths of septentrion; I will climb to the tops of the clouds, I will be like the Most High.*[178] I am the principle of the ways of God, and in this order I would always be the principle of all His ways, and it would be contrary to the dignity in which God Himself established me to serve anything that is beneath me. In the new order, He by whom all things have been made will not unite with the most splendid and worthy nature, but to that which is most distant from me. And his neck stiffened and he cried: *Non serviam—I will not serve.*[179]

37. Michael?!

But an intelligence very inferior to that of the first of the angels rebelled against the decree of the *prince of the ways of God,* the head of the angelic hierarchies, and he cried out with a cry that is also his name: Michael—Who is like God? In that cry is stated the absolute transcendence of Him who is above all ways. This cry, the echo of humility, intoned the *new canticle,*[180] and it merited for Michael the leadership of the heavenly host. This was, in the order of time, the first appearance of the *new city* where *mercy is above all works.*[181] He who had been the prince of the ways of God *perverted his wisdom because of his splendor.*[182] *He fell from heaven like a thunderbolt.*[183] *It is for this that the Lord Yahweh spoke: Because he was lifted to the heights, because his head had reached almost to the clouds, his heart was filled with pride at his elevation, I have left him to one who is most powerful among the people who will treat him as he pleases.*[184] He who did not wish to serve the creature will be vanquished by the servitude of God *who abased Himself, making Himself obedient unto death, even the death of the cross. And that is why God has lifted Him up and given Him a name above every name.*[185] He who made resound his *Non serviam!* and who denied, not directly the negation he is before God, but the negation that involved his submission to grace and to a nature in itself most inferior to his own, he was vanquished by the negation of the things that are, by the *via negationis* that Jesus Himself lived: *He chose the things that are not in order to destroy the things that are—et ignobilia mundi, et contemptibilia elegit Deus, et ea quae non sunt, ut ea quae sunt destrueret.*[186] Everything that will truly be is from now on linked to that which is not.

Behemoth was the beginning of the ways of God, but the one who made him, that is the one by whom all things were made, *turned his sword against him.*[187] That which from the beginning was the instrument of revolt against God will be at the same time *the sword that pierces the soul* of the most humble of all mere creatures whom God has conceived *before making anything else,* and she will vanquish in her sorrow him *who sinned from the beginning*[188] and who is *the original of all those who rally to him.*[189]

38. Novissimi primi, et primi novissimi

In the new order, the order of the fullness of mercy, *behold the last shall be first and the first shall be last.*[190] "Thus it is right," St. Bernard says, "that Mary who was made the last of all when she was the first, was lifted from the lowest to the highest rank; it is right that she became mistress of all, it is fitting that she was elevated above the angels themselves, after having been placed with ineffable sweetness above the widows and penitent sinners, even above the place from which the seven demons were driven out."[191] In this new hierarchy, the creature most humble in natural condition is borne to the top of the universal order. *Quoniam excelsus Dominus humilia respicit: et alta a longe cognoscit—The Lord being most high, regards the humble and knows the proud from afar.*[192] The very imperfection of human nature has made it open to a greater elevation. In its dignity as the image of the Most Holy Trinity, it can be elevated to the order of grace. It is true that the angels are more worthy by nature, and "the image of creation," that which is in every intellectual creature prior to any elevation and which in us is the basis of our being raised, is more perfect in them. But it remains that man is more malleable to supernatural dignity, to the image of grace and glory, not because of any natural dignity hidden in his infirmity, but because of the incommensurable mercy of God. *That which the world takes for foolish, is that which God has chosen to confound the wise; and that which the world holds to be weak, is what God has chosen to confound the strong.*[193] An excess of power most perfectly joins in an excess of weakness. But it is true that the formlessness of human nature gives it a quite particular affinity with its creator, there is only the excess of divine power going beyond in mercy which can realize this affinity. *Wisdom extends from one end to the other.* It is in the land of mercy that the abyss of fullness calls to the abyss of emptiness, even of misery.

39. Attendite a falsis prophetis

But this affinity will be for man the source of a temptation which will increase in the course of time. Is it not quite fitting that the demon who sinned despite the perfection and infallibility of his speculative knowledge, *who because of his splendor perverted his wisdom,* attacks the new order in awaking in man so feeble in his speculative intelligence a most ignoble pride which will have for an object the very darkness of his nature, under the pretext that it has been assumed by the Son of God. He will press man to seize upon himself, to turn on himself, to identify the power of the weak with the weakness of Him and to see in his nature a right to supreme love and a demand, whereas in truth the power of the weak is nothing other than the merciful omnipotence of God. Man will seek in the awareness of his singularity and in his very weakness a good in itself and for itself in virtue of which he will be judged worthy of a love, even to the abasement of God. Man will thus turn the nature into which God has been incarnated against God, attacked in his condition of humiliation. Thus will the demon accomplish the promise made in the wickedness of his heart: *Sedebo in monte testamenti—I will sit myself upon the mountain* that is Christ.[194] For *the son of perdition will seat himself in the sanctuary of God and present himself as if he were God.*[195] The chief of this simulation of the new order will be announced by men who will cloak themselves in the Incarnation: *For false Christs and false prophets will rise up and they will perform great prodigies and extraordinary things in order to seduce, if this be possible, even the elect.*[196] They will be the more insidious because *they will have the appearance of piety.*[197] Those who openly rise up against God will not be the most dangerous, but those who will come in the name of Christ: *Take care that you are not led astray, for many will come in my name.*[198] They will not want to believe that in becoming man, the majesty of God was made nothing (*exinanivit*) and that it is the things that are not (*ea quae non sunt*) which have been chosen. "It is an intolerable impudence," says St. Bernard, "for a miserable work of the earth (*vermiculus*), to inflate and puff up himself when the majesty of God itself was made nothing."[199] *Existimasti inique quod ero tui similis: arguam te, et statuam contra faciem tuam—You have thought, wicked one, that I will be like you: I will chastise you and make you see yourself in all your ugliness.*[200]

One will be able to recognize false prophets by this sign: they will preach *the great prostitute who sits upon the deep water,* the woman who, she too, is called a great city—*the great city which has dominion over the kings of the earth.*[201]

40. Cantate canticum novum

Let us oppose to "those old canticles of Lucifer, canticles of pride, canticles of detraction, canticles of doubt, canticles of lying and canticles of excuse," the new canticle that sings the mother of Christ. "Raised up," an author says,

> and more than it is possible to say or to think, placed above the world and all the heavens, she is not raised above herself, but sings the canticle of humility, which is also the canticle of charity, because charity is not puffed up. She draws from her whole inebriated heart a word good and soothing, the new canticle which virgins who come after her should repeat. And what does she say. *My soul doth magnify the Lord.* See how this canticle is opposed to that which the angel sings at the brink of the abyss. The one begins high, so it does not descend, but falls into the deepest abyss. Mary begins with what is low that she might be elevated on high. She magnifies the Lord, she does not magnify herself, although she was elevated in an incomparable manner, fulfilling what was written: the greater you are, the more you will be brought low in all things. Thus, she merits to be placed above the choirs of all the angels. Satan raised himself above the Lord, and thus was justly plunged beneath everything that is. The spirit of the humble Virgin Mary trembles in Jesus her Lord and thus receives more abundantly than her companions the unction of the oil of joy. The foolish angel exalted himself, and thus was justly condemned to perpetual woe. Mary gloried in that her humility had been regarded, but she will have a recompense in the judgment made by holy souls: the angel gloried himself by the sign of his power, and thus he incurred the chastisement of an eternal contempt.[202]

41. Infirma elegit, et ea quae non sunt

God has chosen to realize the limit of mercy: He has chosen the things that are not and has exalted the humble. One can see in this how much those who would wish man to be by nature at least equal to the angels and the woman in everything the equal of man diminish the true scale that God has deigned to give His work of predilection, where the Woman is Queen of the Angels. It is an injury to the sublime humility of the Servant of God. *Quae cum audisset, turbata est in sermone ejus, et cogitabat qualis esset ista salutatio—When she had heard, she was disturbed by these words* [of the angel Gabriel], *wondering what kind of salutation it was.*[203] "She thought to herself," Cornelius comments, "I see myself unworthy of all grace; how then can the angel call me full of grace? Pauperess, I pass my life with the poor virgins; how then can the angel say that the Lord is with me? I think of myself as the least and most vile[204] of all women: how then can the angels announce, You are blessed among women."[205]

42. Civitas Dei

Given the growing misery of this world and the grief to which our Holy Mother the Church is subjected, it will be agreed that it is opportune to repeat these truths. The intellectual and moral misery of the contemporary world reveals more than ever the darkness of man, a darkness one ignores more and more. *Et nescis quia tu es miser, et miserabilis, et pauper, et caecus, et nudus—Do you not know that you are unhappy, miserable, poor, blind, and naked?*[206] Never before has humanity been more deprived of the purely human virtues of wisdom and prudence. Never has it been more mistaken about the nature it exalts. Men retain but the name of science; research becomes more and more an end in itself; we become like those men of the last times, *who are always learning but without ever being able to arrive at knowledge of the truth.*[207]

As the encyclical *Divini Redemptoris* says, never more fundamentally than in our time has the work of Redemption been attacked. And this attack is radical. It touches the work of Redemption at its root—mercy. Insidiously, it masks itself as charity. It turns to the crowd that drew forth the mercy of the Savior: *I have compassion on this crowd!*[208] "*Fear, posterity of Israel, for he has not ignored or rejected the cry of the poor.*"[209] Some attack this mercy by pushing the masses into extreme necessity where they themselves lose the sense of mercy. "Those who have already attained the extreme of misery, do not fear to suffer more and that is why they are without pity."[210] Others, instead of preaching and establishing justice, seek to instill pride in the miserable, the false power of the weak. But no one is less worthy of mercy than one who is miserable and proud.

But the Church teaches us that along with this growing darkness, the divine mercy will manifest itself more and more in the course of time, and that quite particularly in the more and more explicit revelation of the mysteries of Mary, Mother of Mercy. "The Lord," says Psalm 58, "will rule in Jacob and in the whole earth; they will be overcome with thirst and will suffer hunger like dogs and they will go around the city looking for something to eat." "This city," adds St. Grignion de Montfort, "that men will find at the end of the world in order to convert and give them hunger for justice, is the most Holy Virgin, who is called by the Holy Spirit, the village and city of God."[211]

IN LAUDEM DEIPARAE VIRGINIS
QUAE NON TANTUM OPPORTUNE,
SED ETIAM IMPORTUNE
INTERPELLAT PRO NOBIS.[212]

NOTES

Here is the original text of two passages cited at the beginning of this little book:

> "Fateor, inquit Bernardus, imperitiam meam, pusillanimatatem propriam non abscondo. Non est quidem quod me magis delectet, sed nec quod terreat magis, quam de gloria Virginis Matris habere sermonem" (*In Assumpt. B.V.M.*, sermo 4, n. 5). "Hoc igitur in principio supponamus, quod quidquid laudis dicitur de beata Maria, non hyperbolice dicitur, sed defective, juxta verbum beati Hieronymi: 'Quidquid humanis dici potest verbis, minus est a laude Dei'" (St. Bonaventure, *De Assumptione B.V.M.*, sermo 3, t. 9, p. 693b).
>
> "Tanta est excellentia Virginis gloriosae, ut si ab ejus narratione et laude deficiant omnes linguae, deficiant Scripturae, deficiant, prophetiae et similitudines parabolicae. Unde et Spiritus Sanctus per ora Prophetarum commendat eam non solum verbis, verum etiam figuris et similitudinibus parabolicis; et quia nulla similitudo parabolica perfecte sufficit ad ejus excellentiam exprimendam, ideo ad ipsius laudem multiformes introducuntur similitudines et metaphorae" (ibid., sermo 2, p. 708a).

1. Cf. Cornelius a Lapide, *Commentaria in Ecclesiasticum*, 24:1, 2 (ed. Crampon, Vivès), t. 9, p. 617b. "Mystice, apte Ecclesia in Officiis divinis B. Virginis haec

omnia (quidquid frustra obstrepat Lutherus, et occlament haeretici) accipi de B. Virgine, idque justissimis et gravissimis de causis."

2. Ecclus. 24:40 (Sirac).

3. John of St. Thomas, *Cursus philosophicus* (ed. Reiser), t. 1, II p., q. 5, a. 4, 364b: ". . . abstracta non praedicantur de concretis nec concreta de abstractis ex vi modo significandi . . . licet aliquando identice verificentur, ut in divinis est idem Deitas et Deus, Paternitas et Pater, et in transcendentibus idem est unitas et unum, entitas et ens."

4. St. Thomas, *In I Ethicorum* (ed. Pirotta), lectio 1, n. 1: "Sapientis est ordinare, quia sapientia est potissima perfectio rationis, cujus proprium est cognoscere ordinem."

5. John of St. Thomas, *Cursus theologicus* (ed. Vivès), t. 7, d. 21, a. 1, p. 744b: "Intellectus autem qui trahit res ad se, et ex una procedit in alterum, potest comparare, et attingere formaliter habitudinem unius ad alterum: habet ergo intellectus in se primam radicem et primam rationem ordinandi res; sicut et comparandi et instituendi habitudinem unius ad alteram."

6. Aristotle, *Metaphysics I*, c. 1, 982a15. *Ou gar dein epitattesthai ton sophon all' epitattein. . . .*

7. Luke 1:28–33.

8. Num. 17:8.

9. Luke 1:43.

10. St. Thomas, *Summa theologiae*, IIIa, q. 32, a. 4. Cajetan's commentary: "Adverte hic quod aliud est loqui de ipso actu conceptionis: et aliud de mensura ipsius actus. In littera non dicitur quod Beata Virgo nihil active operata est in mensura actus conceptionis: sed dicitur quod nihil active operata est in ipso actu conceptionis. Nam Beata Virgo in illo instanti conceptioinis materiam ministravit; ac per hoc active operata est; quia materiam ministrare agere procul dubio est. Ita quod in certo tempore praecedente instans conceptionis, Beata Virgo active praeparavit materiam, ut ita in secunda conclusione dicitur: et in instanti terminante tempus ministravit, ut in prima conclusione dicitur, quasi ad terminum suae actionis perveniens. Ad ipsum autem conceptionis actum non concurrit active, sed passive suscipiendo in materia ministrata actionem agentis Spiritus Sancti loco seminis (n. VI). Feminarum enim potentiae ad partem vegetativam spectantes activae procul dubio sunt: sed res facta ab activa potentia feminae materiale est, non activum generationis principium. Ita quod mas et femina conveniunt in hoc quod utrisque potentiae sunt activae, sed different in re facta per illas potentias: nam factum a mare semen est activum generationis principium. Factum vero a femina, quidquid sit illud, menstruum aut semen, materiale est generationis principium (n. VIII). Dic ergo quod potentia generativa feminae est activa, non generationis, sed materiae proximae et propriae ipsius generationis et geniti . . . quia mater est causa effectiva materiae proximae filii. Ideo filius assimilatur matri. Sed falluntur arguentes a negata efficientia respectu actus conceptionis seu generationis in utero, ad inferendum negari omnem efficentiam respectu filii" (n. X).

11. St. Thomas, *Summa theologiae*, IIIa, q. 35, a. 4. "Concipi autem et nasci personae attribuitur et hypostasi secundum naturam illa in qua concipitur et nascitur. Cum igitur in ipso principio conceptionis fuerit humana natura assumpta a divina

persona, sicut praedictum est (q. 33, a. 3), consequens est quod vere posset dici Deum esse conceptum et natum de Virgine. Ex hoc autem dicitur aliqua mulier alicujus mater, quod eum concipit et genuit. Unde consequens est quod B. Virgo vere dicatur mater Dei. Solum enim sic negari posset B. Virginem esse matrem Dei, si vel humanitas prius fuisset subjecta conceptioni et nativitati, quam homo ille fuisset Filius Dei, sicut Photinus posuit; vel humanitas non fuisset assumpta in unitatem personae, vel hypostasis Verbi Dei, sicut posuit Nestorius. Utrumque autem horum est erroneum. Unde Haereticum est negare B. Virginem esse matrem Dei." See John of St. Thomas, *Cursus philosophicus,* t. 1, p. 569a45: "Et ita sicut actio generativa ordinario illud esse proprium et naturale animae communicat corpori, et hac ratione dicitur formalissime mater hominis et personae subsistentis in humanitate, ita generatio Virginis esse et subsistentiam Verbi, quam invenit in anima assumpta, communicavit corpori, et ita formalissime dicitur mater illius hominis seu Personae Divinae ut subsistentis in illa humanitate constante corpore et anima."

12. Albertus Magnus, *Mariale, sive quaestiones super Evangelium,* q. 141, *Opera Omnia* (ed. Borgnet), t. 37, p. 200a. ". . . mater Dei est causa Dei et origo Dei secundum illud quod natum est." M.-M. Desmarais, O.P., *S. Albert le Grand, Docteur de la Mediation Mariale* (Paris-Ottawa, 1935) has been our guide in the Mariology of St. Albert.

13. St. Albert, *Mariale,* q. 145, p. 206a: "Ipsa est mater omnium, et Deus Pater est origo omnium: quidquid autem per se origo et causa causae est, per se est origo et causa causati: sed ipsa est mater ejus qui est causa et origo omnium: ergo ipsa per se est mater omnium."

14. Cornelius a Lapide, *In Ecclesiasticum,* 24:1, 2, t. 9, p. 617b: "Ipsa est mater aeternae Sapientiae in se incarnatae. Sicut ergo filius ejus est Sapientia genita et incarnata: sic ipsa est sapientiae gignens et incarnans."

15. Isa. 66:17.

16. St. Anselm, in Migne, *Patrologia Latina,* t. 158, col. 956, oratio 52.

17. St. Augustine, *Patrologia Latina,* 7, col. 1074, sermo 215, n. 4.

18. St. Albert, t. 38, p. 158b, *In Lucam,* 11:27.

19. St. Bernard, *De Laudibus Virginis Mariae,* homilia 4, in *Opera Omnia* (ed. Carpenter, Vivès), t. 2, p. 619.

20. Luke 1:38, Canticle 1:1.

21. *Vie intérieure de la Très Sainte Vierge,* a work collected from the writings of M. Olier (Paris, 1875), 5–6.

22. John of St. Thomas, *Cursus theologicus,* t. 8, d. 3, a. 2, n. 52, p. 108b.

23. Prov. 8:22–23.

24. St. Albert, *Sermones de Sanctis,* 38, t. 13, p. 563a.

25. Ecclus. 24:5.

26. Canticle 1:1.

27. Ecclus. 24:6.

28. Mass, *Salve sancta parens,* Gradual.

29. Ws. 8:1.

30. Ibid., 2:24.

31. Prov. 8:30.
32. St. Thomas, *Summa theologiae,* IIIa, q. 35, a. 4.
33. St. Albert, *In III Sententiarum,* d. 4, a. 5, ad 2 (t. 28, p. 85b).
34. Cornelius a Lapide, *In Ecclesiasticum,* 34:7 (t. 9, p. 623).
35. Ps. 28:7.
36. Ecclus. 24:8.
37. St. Albert, *Mariale,* q. 34, par. 4 (t. 37, p. 73a): "Plenitudo beatissimae Virginis privat vacuitatem: quamdiu enim vas potest recipere aliquid, tamdiu habet aliquid de vacuitate. Unde omnis creatura alia habet aliquid vacuitatis: quia majorem gratiam potest recipere. Ipsa autem sola gratia plena, quia majorem gratiam non potuit habere: nisi enim ipsa divinitati unireretur, major gratia non potuit intelligi, quam quod de ipsa accipieretur quod uniretur: hoc enim est, nisi ipsa (esset) Deus, non posset major gratia intelligi, quam quod ipsa esse Dei mater."
38. St. Augustine, *De sancta Virginitate,* c. 3 (t. 9, col. 398). "Sic et materna propinquitas nihil Mariae profuisset, nisi felicius Christum corde quam carne gestasset."
39. "Spiritus Sanctus supervenit in Virginem ut esset Mater Dei et attingeret sic ordinem hypostaticum, supponendo illam esse plenam gratia: sic etiam prius Angelus eam gratia plenam nominavit, quam diceret Spiritum Sanctum superventurum in illam, Et sic dignitas maternitatis congrue et connaturaliter exigit sanctitatem, qui pertinet ad ordinem hypostaticum qui de se altiorem formam sanctitatis importat. Quod si de potentia absoluta esse maternitas sine sanctitatem tunc non supervenerit in eam Spiritus Sanctus per modum missionis simlpliciter, qui non inhabitaret in ea, sed secundum missionem secundum quid" (John of St. Thomas, *Cursus theologicus,* t. 4, d. 17, a, 2, p. 465).
40. Ecclus. 24:39.
41. Esther 13:9.
42. Ws. 7:23.
43. Ecclus. 24:15.
44. "Confirmatio in bono beatae Virgini competebat, quia mater erat divinae sapientiae, in quam nihil inquinatum incurrit, ut dicitur Sap. 7" (St. Thomas, *Q.D. de veritate,* q. 24, a. 9, ad 2).
45. Phil. 2:13.
46. "Ecce imperio Dei omnia subjiciuntur et Virgo; ecce imperio Virginis omnia subjiciuntur et Deus" (St. Louis Marie Grignion de Montfort, *La vraie dévotion à la Sainte Vierge,* n. 76 [Canadian edition, 1940], p. 61).
47. Ecclus. 24:12–13.
48. Ibid., 24:25.
49. Prov. 9:1.
50. St. Grignion de Montfort, *La vraie dévotion,* n. 5, p. 3.
51. Ws. 7:26.
52. Prov. 8:30.
53. Gen. 1:2.
54. Ecclus. 24:40–41.

55. Ibid., 24:43.
56. Ws. 7:24–25.
57. Apoc. 21:2–3.
58. Ws. 7:22–23.
59. Mass of Mary the Mediatrix.
60. Ws. 7:25, 27.
61. Canticle 6:8 and Ws. 7:27.
62. Canticle 7:4.
63. Ps. 44:14.
64. "Post Deum autem, unusquisque magis diligit se quam proximum, debet enim diligere alios sicut seipsum, unde ipsemet est quasi exemplar primum et diligendorum, quia se ut participem gloriae divinae alios ut socios in participando. Excipio tamen Christum Dominum, etiam ut hominem, et Beatissimam Virginem matrem, eo quod participant quamdam rationem communicantis nobis gratiam et beatitudinem, est enim Christus ut homo caput gloriae et Beatissima Virgo mater capitis, et collum per quod derivatur gratia, et ideo magis debemus ipsos diligere quam nos" (John of St. Thomas, *Cursus theologicus,* t. 7, q. 23, p. 403).
65. Canticle 1:3.
66. Ibid., 6:3.
67. Ps. 44:10.
68. Gen. 2:6.
69. "Facies terrae, id est, dignitas terrae, Mater Dominis Virgo Maria rectissime accipitur, quam irrigavit Spiritus Sanctus, qui fontis et acquae nomine in Evangelio significatur, et quasi de limo tali homo fieret, qui constitutus est in paradiso, ut operaretur et custodiret, id est, in voluntate Patris, ut eam impleret atque servaret" (*De Genesi contra Manichaeos,* lib. II, c. 24, t. 3, col. 216).
70. Canticle 4:1 and 3.
71. Ezek. 1:2.
72. Ibid., 1:16.
73. Luke 1:42.
74. "Concedimus etiam quod a Filio suo in omnibus privilegiis suis antecellitur: nec in hoc laus matris suae diminuitur, sed exaltatur in eo quod non tantum genuit sibi filium coaequalem, sed infinitum meliorem. Quod etiam ex hac parte quodammodo infinitat bonitatem matris: omnis enim arbor ex fructu cognoscitur proprio; unde si bonitas fructus bonificat arborem, infinita bonitas in fructu adhuc ostendit infinitatem in arbore bonitatem" (St. Albert, *Mariale,* q. 197, p. 287b).
75. Ecclus. 24:6 and 40.
76. Cornelius a Lapide, *In Ecclesiasticum,* 24:1, 2 (t. 9, p. 618a). "Dices: quomodo enim veritate aptari possunt B. Virgini illa, quae Sapientia dicit de se, vers. 6: *Ego feci ut in coelo oriretur lumen indeficiens;* vers. 40, *effudi flumina?* Respondeo, primo, mystice, q.d. Ego feci ut in coelis, id est in Ecclesiis, oriretur Christus qui est sol justitiae, feci ut in iis oriretur lux fidei. Rursum, ipsa, quasi mare gratiarum earum flumina in Ecclesiam et fideles effundit. Secundo, ad literam, hoc sensu, q.d. Ego fui causa, cur

Deus creavit lucem, coelos, mare, flumina totumque universum. Hujus enim creatio ordinata fuit ad justificationem et glorificationem Sanctorum, factam a Christo per B. Virginem, tamquam ad suum finem; ordo enim naturae creatus et institutus est propter ordinem gratiae. Quia ergo B. Virgo fuit mater Christi, ac consequenter fuit medium nostrae redemptionis, ac totius ordinis gratiarum a Christo instituti; hinc pariter fuit causa finalis creationis universi; universi enim finis est Christus, ejusque mater et Sancti; ut scilicet, Sancti in universo hoc per Christum et B. Virginem, gratia et gloria donentur. Quare creationis universi causa finalis fuit praedestinatio Christi, B. Virginis et Sanctorum. Licet enim universi partes quaedam sint Christus et B. Virgo, ideoque eo posteriores in genere causae materialis; tamen in genere causae finalis sunt priores. Quare inter creationem universi et nativitatem Christi et B. Virginis, est mutua quaedam contra-dependentia; nec enim Deus nasci voluit Christum et B. Virginem, nisi in universo hoc; nec vicissim voluit universum hoc existere sine Christo et Beata Virgine, imo propter illos illud creavit. Totum enim universum ad Christum et Beatam Virginem, ordinemque gratiarum, velut ad sui complementum et finem referri et ordinari voluit. Christus ergo et B. Virgo sunt causa finalis, ob quam creatum est universum, ac proinde ejusdem sunt causa formalis, puta exemplaris. Scilicet, idea. Ordo enim gratiarum, in quo primus est Christus et B. Virgo est idea et exemplar juxta quod Deus creavit et disposuit ordinem naturae totiusque universi."

77. Ecclus. 1:3.

78. 3 Kings 10:20.

79. Canticle 8:5 and 6:9.

80. Ps. 24:10.

81. *In antiphonam Salve Regina,* sermo 1, n. 3 (t. 7, p. 43a).

82. "Opus autem divinae justitiae semper praesupponi opus misericordiae, et in eo fundatur. Creaturae enim non debetur aliquid, nisi propter aliquid in eo praeexistens, vel praeconsideratum: et rursus, si illud creaturae debetur, hoc erit propter aliquid prius. Et cum non sit procedere in infinitum, oportet devenire ad aliquod quod ex sola bonitate divinae voluntatis dependeat, quae est ultimus finis. Utpote si dicamus quod habere manus debitum est homini proper animam rationalem; anima vero rationalem habere, ad hoc quod sit homo; hominem vero esse, propter divinam bonitatem. Et sic in quolibet opere Dei apparet misericordia, quantum ad primam radicem ejus. Cujus virtus salvatur in omnibus consequentibus; et etiam vehementius in eis operatur, sicut causa primaria vehementius influit quam causa secunda" (St. Thomas, *Summa theologiae,* Ia, q. 21, a. 4, c.).

83. John of St. Thomas, *Cursus theologicus,* t. 7, d. 3, a. 3, n. 13 (p. 114a): "Finis cujus gratia fuit gloria Dei manifestanda per viam miseriordiae et justitiae, unde quia incarnatio valebat ad hunc effectum manifestandi misericordiam et justitiam in redemptione hominum motivum volendi incarnationem fuit non ipsa incarnationis dignitas absolute, sed ipsa incarnatio ut valens ad talem effectum."

84. "Et tamen in damnatione reproborum apparet misericordia, non quidem taliter relaxans, sed aliqualiter allevians, dum punit citra condignum" (St. Thomas, *Summa theologiae,* q. 21, a. 4, ad 1).

85. Ibid., IIaIIae, q. 30, a. 4, c.: "Respondeo dicendum quod aliqua virtus potest esse maxima dupliciter: uno modo secundum se; alio modo per comparationem ad habentem.—Secundum se quidem misericordia maxima est; pertinet enim ad misericordiam quod alii effundat; et quod plus est, quod defectus aliorum sublevet; et hoc est maxime superioris. Unde et miserei ponitur proprium Deo: et in hoc maxime dicitur eius omnipotentia manifestari.

"Sed quoad habentem, misericordia non est maxima, nisi ille qui habet sit maximus, qui nullum supra se habeat, sed omnes sub se. Ei enim qui supra se aliquem habet majus est et melius conjungi superiori quam supplere defectum inferioris. Et ideo quantum ad hominem qui habet Deum superiorem, caritas, per quam Deo unitur, est potior quam misericordia, per quam defectus proximorum supplet. Sed inter omnes virtutes quae ad proximum pertinent potissima est misericordia, sicut etiam est potioris actus: nam supplere defectum alterius, in quantum hujusmodi, est superioris et melioris."

86. "It is necessary that mercy and truth be found in every work of God; if, however, mercy is understood as an action that puts aside some defect, although not every defect can properly be called a misery, but only a defect in rational nature, which is capable of happiness, for misery is the opposite of happiness. . . . Although nothing, in the nature of things, is presupposed to creation, nonetheless something is presupposed in the knowledge of God. In this respect, even creation takes on the note of justice, insofar as the being of things is produced in a way agreeable to divine wisdom and goodness. It has as well, in a way, the note of mercy, insofar as things pass from non-being to being" (ibid., Ia, q. 21, a. 4, c. and ad 4.

87. Ps. 144:9.

88. "It is the characteristic of fault (*culpa*) that it is voluntary. And in this respect, it is not worthy of pity, but rather of chastisement. But because fault can, in a certain manner, be a punishment, that is insofar as it involves something contrary to the will of the one who sins, in this respect it can have the note of misery. And it is for this reason that we have pity and compassion for those who sin. For as St. Gregory says in a homily, 'true justice is not a matter of disdain,' that is to say, of sinners, 'but a matter of compassion.' And in Matthew it is written 'But in seeing this multitude of men (Jesus) was moved by compassion for them, because they had been harassed and buffeted, like a flock without a shepherd'" (St. Thomas, *Summa theologiae,* IaIIae, q. 36, a. 1, ad 1).

89. Ibid., IIaIIae, q. 30, a. 1. ". . . sed contra: motivum ad misericordiam est malum" (IIaIIae, q. 30, a. 4). And, in the ad 1 of this same article, the text already quoted: "de ratione culpae est quod sit voluntaria. Et quantum ad hoc non habet rationem miserabilis, sed magis rationem puniendi. Sed quia culpa potest esse aliquomodo poena, in quantum scilicet habet aliquid annexum quod est contra voluntatem peccantis, secundum hoc potest habere rationem miserabilis. Et secundum hoc miseremur et compatimur peccantibus: sicut Gregorius dicit in quadam homilia, quod *vera justitia non habet dedignationem,* scilicet ad peccatores, *sed compassionem.* Et dicit Matthew 9, *Videns Jesus turbas misertus est eis, quia erant vexati, et jacentes sicut oves non habentes pastorem*" (IIaIIae, q. 30, a. 1, ad 1).

90. St. Thomas, *In de divinis nominibus,* c. 7, lectio 2.

91. St. Thomas, *II Summa contra gentes,* c. 30.

92. "Est ergo summa solutionis, quod ordo duo requirit, scilicet ordinatorum distinctionem et communicantiam distinctorum ad totum. Quantum autem ad secundum est quidem ordo indeficienter in aliquibus, quae sunt suprema et proxima primo principio, sicut substantiae separatae et corpora caelestia, in quibus nihil casualiter accidit et praeter naturam: in aliquibus autem deficit, scilicet corporibus, in quibus interdum aliquid accidit casualiter praeter naturam. Et hoc propter remotionem a primo principio semper eodem modo se habente" (St. Thomas, *In XII Metaphysicorum,* lectio 12, n. 2637).

93. St. Thomas, *In II Sententarium,* d. 3. q. 1, a. 6, c.

94. ". . . homo est ex duabus contrariis naturis, quarum una retrahitur ab alio a suo corpore" (St. Thomas, *In Matth.,* 25:15).

95. St. Thomas, *In II Ethicorum,* lectio 7, nn. 319–21.

96. ". . . in natura humana bonum videtur esse in paucioribus, et hujusmodi ratio potest assignari dupliciter. Una est propter corruptionem humanae naturae ex peccato originali. . . . Alia ratio sumi potest ex ipsa natura conditionis humanae" (St. Thomas, *In I Sententiarum,* d. 39, q. 2, a. 2, c.). On this question, see John of St. Thomas, *Cursus theologicus,* t. 6, q. 109, dd. 19 and 20).

97. ". . . in homine est duplex natura, scilicet rationalis et sensitiva. Et quia per operationem sensus homo pervenit ad actus rationis, ideo plures sequuntur inclinationes naturae sensitivae quam ordinem rationis. Plures enim sunt qui assequuntur principium rei, quam qui ad consummationem perveniunt. Ex hoc autem vitia et peccata in hominibus proveniunt, quod sequuntur inclinationes naturae sensitivae contra ordinem rationis" (St. Thomas, *Summa theologiae,* IaIIae, q. 71, a. 2, ad 3).

98. St. Augustine, *In Psalm* 35, verse 6 (t. 5, col. 346).

99. Ps. 68:3.

100. Isa. 7:14.

101. Ibid., 45:8.

102. St. Albert, *Mariale,* c. 141 (t. 37, p. 200): "Quod non posset homo esse pater Dei, sic probatur: Dicit Philosophus, quod masculus est, qui generat in alio: ergo ille qui esset pater Dei, aut generaret in alio, aut in seipso. Si in alio, ergo duo: unus generans in alio, et unus vel una ab altero generans: ergo esset ibi masculus et femina. Si autem ille pater generaret in se: ergo non esset masculus, ergo non esset pater.—Item quae est ratio quare diceretur pater quam mater? . . . Esset nec pater nec mater."

103. Prov. 3:20.

104. Ps. 41:8.

105. Luke 1:48.

106. Cornelius a Lapide, *In Lucam,* 1:48, at the word 'humilitatem,' t. 16, p. 36; *In Proverbia Salomonis,* 8:13, t. 5, p. 210b.

107. Ws. 8:1.

108. Luke 1:46–50.

109. Ps. 17:33.

110. Isa. 63:19.

111. Ibid., 35:8.

112. Prov. 8:22.

113. Ibid., 4:11.

114. Ps. 24:10.

115. Luke 1:52.

116. Cajetan, *In IIamIIae,* q. 161, a. 5, n. xxv: "Unde beata Virgo *humilitatem suam Dominum respexisse,* tanquam universalem virtutem, quia ad supernum suscipiendum divinae largitatis influxum latissime ac profundissime patula erat."

117. John of St. Thomas, *Cursus theologicus,* t. 8, d. 19, a. 6, n. 27, p. 700a: "Respexit humilitatem ancillae, quia videlicet . . . ad tantam plenitudinem gratiae infundendam, quanta fuit in B. Virgine nihil aliud respexit Deus, quam profunditatem humilitatis ejus, qua reddita est capacissima ad suscpiendam quasi in concavitate maxima immensam gratiae magnitudinem."

118. St. Thomas, *Summa theologiae,* IIaIIae, q. 161, a. 5, ad 4: "Humilitas est quasi quaedam dispositio ad liberum accessum hominis in spiritualia et divina bona."

119. St. Bernard, *In Annunciatione Domini,* sermo III, n. 4, t. 7, p. 452a: "Omnium tamen humanarum infirmitatum vel injuriarum quas pro nobis pertulit divina dignatio, sicut tempore primum, sic etiam humilitatem fere maximum existimo, quod in utero concipi, in utero novem mensium tempore majestas illa incircumscripta passa est contineri. Ubi enim sic se exinanivit, aut quando ita penitus a semetipso defecisse visus fuit? Tanto tempore nihil illa sapientia loquitur, nihil illa virtus manifestum operatur: nullo signo visibili majestas quae clausa latet, proditur. Non sic in cruce visus est infirmus, ubi quod infirmum ipsius fuit, statim apparuit fortius omnibus hominibus: quando et moriens glorificat latronem, et expirans inspirat Centurionem: quando horarius dolor passionis ei non solum compati fecit elementa creaturarum, sed etiam contrarias fortitudines subigit aeternorum passioni dolorum. In utero autem sic est quasi non sit: sic omnipotens virtus vacat, quasi nihil possit, et Verbum aeternum sub silentio se premit."

120. Prov. 11:2.

121. St. Albert, *Mariale,* q. 63, p. 119a: "*Ubi humilitas, ibi sapientia.* Prov. XI, 2. Haec propositio est per se in theologia: ergo ubi major humilitas, ibi major est sapientia: ubi summa, ibi summa. Sed in beatissima Virgine improportionabilis fuit humilitas: ergo improportionabilis sapientia. Minor patet in Evangelio ubi dicitur: *Qui se humiliat exaltabitur.* Quae similiter est per se: ergo qui improportionabiliter aliis exaltatur, etiam aliis improportionabiliter demonstratur humilis: sed beatissima Virgo omnes choros Angelorum in quartam hierarchiam secundum Hieronymum exaltata, improportionabiliter aliis ascendit: ergo omnium hominum et Angelorum humillima fuit: ergo et omnes in sapientiae transcendit."

122. James 4:6; Prov. 3:34.

123. Luke 1:45.

124. Ibid., 1:28.

125. St. Albert, *Mariale,* q. 165, p. 247b: "quia ad consentiendum in mirabilia maxime disponit fides, et maxime fides de omnipotentia: quia qui credit et advertit

Deum omnia posse facere, acquiescit ipsum posse naturas mutare, et naturis imperare. Unde cum hic omnium novorum novissimum nuntietur beatissimae Virgini. . . ."

126. St. Bernard, *In Adventu Domini,* sermo II, t. 2, p. 571b: "O Virgo, virga sublimis, in quam sublime vertice sanctum erigis! Usque ad sedentem in throno, usque ad Dominum majestatis. Neque enim mirum quoniam in altum mittis radices humilitatis."

127. St. Bernard, *Tractatus de statu virtutum,* Ia, c. 13, t. 6, p. 314a: "O humilitas, per quam femina mater Dei effecta est, per quam Deus de coelo descendit ad terras, per quam animae de inferno ad coelum translatae sunt. Haec est scala proposita vobis a Deo, per quam ascenditur de terris ad coelum. Per hanc ascenderent patres nostri, per hanc et nos ascendere oportet, alioquin non ascendemus."

128. St. Bonaventure, *De Annuntiatione B.V.M.,* sermo II, *Opera Omnia* (Quaracchi). t. 9, p. 664a: "Si ab aeterno fuit concepta, quomodo concipi potuit in fine saecularum a Virgine Maria? Si enim eterna erat, ergo immutabilis, ergo incomprehensibilis, ergo interminabilis. Quomodo ergo Interminabilis concipi potuit a juvencula? Quomodo Incomprehensibilis a parvula? Quomodo Immutabilis a fragili et tenella? Et tamen talem et tantum concepit, secund testimonium angelicum, Lucae primo (v. 31 et seq.), inquit Angelus ad Virginem, *Ecce, concipies in utero et paries Filium, etc.,* et post describit eum: *Hic erit Magnus,* scilicet in comprehensione; *et Filius Altissimi vocabitur,* suae essentiae immutabilitate; *et regni ejus non erit finis,* sua interminabilitate. Unde in hoc conceptu est divina Majestas mirabiliter humiliata, et virginalis humilitas mirabiliter exaltata."

129. St. Bernard, *Super Missus est,* hom. I, n. 7, t. 2, p. 591b: "Mirare ergo utrumlibet, et elige quid amplius mireris, sive Filii begnissimam dignationem, sive Matris excellentissimam dignitatem. Utrinque stupor, utrinque miraculum. Et quod Deus feminae obtemperet, humilitas absque exemplo; et quod Deo femina principetur, sublimitas sine socio." See too hom. II, n. 9, p. 597b: "Porro ibi agnoscitur longitudo brevis, latitudo angusta, altitudo subdita, profunditas plena. Ibi agnoscitur lux non lucens, verbum infans, aqua sitiens, panis esuriens. Videas, si attendas, potentiam regi, sapientiam instrui, virtutem sustentari; Deum denique lactantem, sed angelos reficientem; vagientem, sed miseros consolantem. Videas, si attendas, tristari laetitiam, pavere fiduciam, salutem pati, vita mori, fortitudinem infirmari. Sed, quod non minus mirandum est, ipsa ibi cernitur tristitia laetificans, pavor confortans, passio salvans, mors vivificans, infirmitas roborans. Cui jam illud quoque non occurrat, quod quaerebam? Numquid non facile tibi inter haec feminam agnoscere virum circumdantem, cum Mariam vides virum approbatum a Deo Jesum suo utero circumplecententem."

130. Ps. 48:13, 21.

131. Ibid., 50:7.

132. Office of Holy Saturday.

133. St. Thomas, *Summa theologiae,* Ia, q. 64, a. 2, c; *Q.D. de Malo,* q. 16, a. 5.

134. St. Thomas, *Summa theologiae,* IIaIIae, q. 30, a. 2, c.: ". . . cum misericordia sit compassio super miseria aliena, ut dictum est, ex hoc contingit quod aliquis misereatur ex quo contingit quod de miseria aliena doleat. Quia autem tristitia seu dolor est de proprio malo, intantum aliquis de miseria aliena tristatur aut dolet inquantum

miseriam alienam apprehendit ut suam. Hoc autem contingit dupliciter. Uno modo, secundum unionem affectus, quod fit per amorem. Quia enim amans reputat amicum tanquam seipsum, malum ipsius reputat tanquam suum malum, et ideo dolet de malo amici sicut de suo. Et inde est quod Philosophus, in IX Ethic., inter alia amicabilia ponit hoc quod est *condolere amico.* Et Apostolus dicit, ad Rom. XII, *Gaudere cum gaudentibus, flere cum flentibus.*—Alio modo contingit secundum unionem realem: utpote cum malum aliquorum propinquum est ut ab eis ad nos transeat. Et ideo Philosophus dicit, in II Rhet, Homines miserentur super illos, qui sunt eis conjuncti et similes, quia per hoc fit eis aestimatio quod ipsi etiam possint similia pati. Et inde est etiam quod senes et sapientes, qui considerant se posse in mala incidere, et debiles et formidolosi magis sunt misericordes. E contrario autem alii, qui reputant se esse felices et intantum potentes quod nihil mali putant se posse pati, non ita miserentur.—Sic igitur semper defectus est ratio miserendi; vel inquantum aliquis defectum alicujus reputat suum, propter unionem amoris; vel propter possibilitatem similia patiendi."

135. "What the Apostle says is beautiful, *exinanivit,* he emptied himself. The empty is opposed to the full, but the divine nature is fullness itself since in it is found the full perfection of the good. *I will show you every good.* But human nature, the soul, is not full; it only has a potentiality for fullness, since it was created as a *tabula rasa.* Thus human nature is empty. That is why the Apostle says, He has emptied himself: because he has assumed human nature. Thus, of the assumption of human nature he said first off, *Taking the form of a slave.* Man is indeed by his very nature the slave of God, and human nature is the form of a slave" (St. Thomas, *Super Epist. Ad Philippenses Expositio,* c. 2, lectio 2).

136. John 2:4.

137. St. Augustine, *In Ioannem,* tract. VIII, c.9, t. 4, col. 1455: "*Quid mihi et tibi est, mulier? Nondum venit hora mea*... tamquam dicens: Quod de me facit miraculum, non tu genuisti, divinitatem meam non tu genuisti: sed quia genuisti infirmitatem meam, tunc te cognoscam, cum ipsa infirmitas pendebit in cruce."

138. Canticle 2:2.

139. St. Albert, *Mariale,* q. 24, p. 53a: "Ut sapientia Dei misericordior appareret. Est enim exitus quo de bono bonum, et exitus est quo de malo malum. De primo, *Vidit Deus cuncta quae fecerat: et erant valde bona.* De secundo: *Initium omnis peccati est superbia....* Tertius exitus est, ut de bono malum, ut de muliere initium peccati. Quartus exitus est, ut de malo bonum, et hoc solius Dei est proprium, cujus sapientia vincit malitiam, attingens a fine usque ad finem."

140. Canticle 4:7.

142. St. Albert, *Mariale,* q. 24, p. 53a: "... ut confidentiam peccatoribus augmentaretur, per quod mediatrix eorum eadem propinquitate utramque extremitatem conjungeret, ut sicut mater esset et filia Dei, sic esset mater et soror nostra, et sic a natura inclinaretur ad miserendum peccatori."

142. Cornelius a Lapide, *In Canticum Canticorum,* 1:4, t. 7, p. 495b: "Beata virgo nigra fuit non in se, sed in patre suo Adam, qui peccavit et peccato totam suam posterioritatem (excepta B. Virgine) infecit. Est ergo ipsa nigra per denominationem extrinsecam, quia filia peccatoris; in se tamen est formosa per gratiae plenitudinem."

143. Cornelius of Lapide, *In Proverbia Solomonis,* 8:13, t. 5, p. 210a.

144. Luke 2:35.

145. Denis the Carthusian, *Enarratio in Canticum Canticorum Salomonis,* 1:5 (Cologne edition), t. 7, p. 324 D: "*Nigra sum, sed formosa.* Fuisti enim, o suavissima Virgo, nubilo dolorossimae compassionis, tristitiae ac moeroris, in omni dilectissimi Filii tui passione salvifica, cooperta, impleta ac penetrata, imo doloris gladio perforata; eratque tunc in te duplicatus parturientem dolor. Et nisi te omnipotentia Filii tui conservasset, prae doloris vehementia, rupto (ut creditur) corde, mox expirasses; sed reservavit te Filius tuus Ecclesiae suae in magnum profectum credentium. Verumtamen in tot tribulationibus atque doloribus, in tantis angustiis ac pressuris, mansisti formosa; quae et haec tanta compassio et afflictio tua, fuit ipsi ineffabiliter meritoria, et per eam promeruisti facultatem efficaciamque praecipuam succurrendi omnibus nobis. Nempe si Paulo Apostolo dicere licuit, Adimpleo quae desunt passionum Christi, in carne mea, pro corpore ejus quod est Ecclesia: quanto plus sacratissimae Matri Christi hoc dicere fas est? In hoc quoque mansuetudinis tuae, claritas patientiae tuae, profundissima tua humilitas, ac tuae sanctitas caritatis quod in tota tui unici praedelictissimi Filii amarissima et ignominiossima passione, nullatenus mota est aliqua indignatione, aversio, impatientia adversus crudelissimos scleratossimosque tui pretiossimi Filii perecutores et crucifixores: a quibus et tu reputaberis vilis, iniqua, deformis, sicut tabernacula Cedar, videlicet ut impiissimi seductoris mater infelix, quum esses pulchra in anima instar pellium Salomonis, eo coeliformi pulchritudine perornata, assimilat decori veri Pacifici, qui extendit coelum sicut pellem."

146. Cornelius a Lapide, *In Canticum Canticorum,* 1:4, t. 7, p. 496a: "Sicut filius non misere, sed misericorditer moriens, hunc indignum et indecorum sibi luctum dedignabatur impendi: sic et mater felicissima pro affectu commoriens filio, et quodam modo in ipso moriens, quia os ex ossibus ejus, et caro ex carne ejus ipse: Quid, ait, fletis super me, quasi miseram mulierem, et matrem miseri hominis? Nigra nunc sum, quia oportet me cum filio despecto despici, et cum reputato leproso leprosam reputari. Ille juxta Prophetam sol meus, nunc factus quasi saccus cillicinus in oculis vestris, et cui non est species neque decor; me quoque decet illi conformari, et pullo atroque habitu reorum assimilari: sicut tabernacula Cedar, quasi aliqua de peccatricibus, ait Honorius."

147. John of St. Thomas, *Cursus theologicus,* t. 8, d. 10, a. 3, p. 272.

148. St. Albert, *Mariale,* q. 75, p. 131a: "Beata Virgo, secundum universalem usum Ecclesiae appellatur est *mater misericordiae,* quod non convenit proprie alicui alteri creaturae. Dicuntur autem et aliquando viri misericordiae, id est viri humiles misericordia, et sic habent omnes alii comparationem ad misericordiam per modum principalem et per modum accidentalem, ipsa vero habet comparationem ad misericordiam per modum originis essentialis, quia per modum matris: convenientia autem essentialis improportionabiliter excellit modum inhaerentia et modum accidentalem: ergo beatissima Virgo improportionabiliter excedit in misericordia universos."

149. Ibid., q. 162, p. 236a: "Omnis qui est in regno Dei est in misericordia: sed non omnis qui est in regno, est in gloria, vel in gratiam vel in justitia: ergo sola misericordia comprehendit totum regnum: ergo regina super totum regnum potissime dicetur regina misericordiae."

150. Ibid., p. 236b: "Si construitur intransitive, sensus erit: ipsa est regina misericordiae, idest, ipsa misericordia: sed adhuc dicitur vere regina misericordiae: unde et Esther quae est in figura beatae Virginis, alio nomine Edissa vocatur, quod est interpretatum *misericordia.* Item, Isa. xvi, 5: *Et praeparabitur in misericordia solium*" sed solium proprie est in proprio loco regni: ergo misericordia est proprius locus regni: sed in utero beatae Virginis vere et proprie requievit et collocata fuit tota divinitas et humanitas ipsi per Spiritum Sanctum praeparato: ergo ipsa proprius locus fuit regni: ergo ipsa fuit misericordia, et cum hoc non immerito regina misericordiae: quia ipsa seipsam perfecte possidendo semper bene rexit, nam ab ipsa numquam aliquid indirectum processit."

151. Ibid., p. 237a: ". . . propriissmum nomen quod beatissimae Virgini secundum suam dignitatem summa debetur est regina misericordiae. . . . Nec dicitur proprie regina pacis et dilectionis: quia hoc est omnibus sui regni universale. Melius dicitur *regina misericordiae* quam regina potentiae vel sapientia. In sapientia enim intelligitur potentia, et non e converso. In misericordia sapientiae et potentia, et non e converso: unde misericordia illa omnia tria claudit in se: unde regina misericordiae et regina potentiae et sapientiae, et non e converso."

152. John 18:37.

153. Ps. 21:7 and Apoc. 19:16.

154. Ecclus. 43:2.

155. Luke 1:29.

156. "A great check is put on our appetite when it tends toward an object surpassing our dignity, and which cannot be attained or retained without another's gratuitous help. This is why we sing on the subject of the holy angels: *Tremunt Potestates*: the Powers tremble. For, when this gift that they have been given by God, and that they are certain to retain for all eternity, is envisaged with respect to what they are in themselves, as created from nothing, and so on, then one can say that a trembling arises in them for they see nothing in themselves that makes them worthy of this gift; rather one sees that one could be deprived of it, and that the gift was given them in a wholly gratuitous way and without merit on their part" (Cajetan, *In IIamIIae,* q. 161, a. 1, n. IV).

157. Ws. 8:11.

158. Ecclus. 24:3.

159. 1 Cor. 1:20 and 3:19.

160. Ibid., 1:25.

161. Mass *Salve sancta parens.*

162. St. Grignion de Montfort, *La vraie dévotion,* n. 30.

163. Ecclus. 24:24.

164. Ps. 110:10.

165. Canticle 6:3.

166. St. Grignion de Montfort, *La vraie dévotion,* n. 52.

167. Job 41:25.

168. Ibid., 14.

169. Ezek. 28:12, 15.

170. Isa. 14:12.

171. Job 41:24.

172. Ezek. 31:3–8.

173. Isa. 43:19–21.

174. The angel receives grace proportional to the perfection of his nature: St. Thomas, Ia, q. 62, a. 6. It is otherwise with man, IIaIIae, q. 24, a. 3.

175. John of St. Thomas, *Cursus theologicus,* t. 4, d. 23, a. 3, nn., xvii–xli, pp. 940 ff.

176. Ezek. 28:2.

177. Ibid., 17.

178. Isa. 14:13–14.

179. John of St. Thomas, *Cursus theologicus,* t. 4, d. 23, a. 3, nn. xxvii–xxix.

180. Ps. 32:3; Apoc. 5:9.

181. Ps. 144:9.

182. Ezek. 28, 17.

183. Luke 10:18.

184. Ezek. 31:10–11.

185. Phil. 2:8–9.

186. 1 Cor. 1:28.

187. Job 40:14.

188. 1 John 2:25.

189. Ws. 2:25.

190. Luke 13:30.

191. St. Bernard, *Sermo in Dominica infra Octavam Assumptionis B.V.M.,* n. 11, t. 3, p. 400a: "Merito facta est novissima prima, quae cum prima essent omnium, sese novissima faciebat. Merito facta est omnium domina, quae se omnium exhibebat ancillam. Merito denique super angelos exaltata est quae et infra viduas et penitentes, infra eam de qua ejecta fuerant septem daemonia, ineffabili sese mansuetudine inclinabat."

192. Ps. 137:6.

193. 1 Cor. 1:27.

194. Isa. 14:13.

195. 2 Thess. 2:4.

196. Matt. 24:24.

197. 2 Tim. 3:5.

198. Luke 21:8.

199. St. Bernard, *In Nativitate Domini,* sermo 1, t. 3, p. 33b: "Intolerabilis impudentia est, ut ubi sese exinanivit Majestas, vermiculus infletur et intumescat."

200. Ps. 49:21.

201. Apoc. 17:1–18.

202. Additamentum IV ad Opusculum S. Bonaventurae, *Vita mystica,* cc. xxv–xxvi, *Opera Omnia* (Quarrachi), t. 8, p. 203: "Carissimi, vetera cantica sunt haec, canticum superbiae, canticum detractionis, canticum dubitationis, canticum mendacii, et canticum excusationis. . . . Non ita cantavit novi cantici imitatrix mater Christi, verum lilium vere viridentibus foliis redimitum. Vis videre folia ejus? Vis audire canticum ejus? Exaltata, et supra quam dici aut cogitari potest, toti mundo coelisque

praelata, non ascendit supra se, sed cantavit canticum humilitatis, quod et charitatis est, quia charitas non inflatur. Eructavit de corde suo inebriato verbum bonum et suave, canticum novum virginibus cantandum. Et quid ait? *Magnificat anima mea Dominum.* Vide, quam contrarium est canticum hoc cantico angeli ruituri. Incoepit ille in alto, ac proinde in ima non descendit, sed corruit. Incoepit Maria ab imo, et in altum sublevaretur. Magnificat haec Dominum, non se; licet inaestimabiliter exaltata, servans quod scriptum est: *Quanto major es, humiliate in omnibus.* Unde et super omnes choros angelorum meruit elevari. Magnificavit se ille supra Dominum, unde infra omne quod est, meruit praecipitari. Exultavit spiritus humilis Mariae in Domino Jesu suo; unde oleo laetitiae prae consortibus suis iniungi promeruit: exultavit stultus ille angelus in se, unde luctu perpetuo meruit condemnari. Gloriatur ista humilitatem suam esse respectam, unde habebit fructum in respectione animarum sanctarum: gloriabatur ille in multitudine virtutis suae, unde perpetuam despectionem incurrit."

203. Luke 1:29.

204. How could the Blessed Virgin possibly humble herself to this degree? Is it not contrary to the truth? We find in St. Thomas (IIaIIae, q. 161, a. 6, ad 1) a response to this difficulty. "Someone can without falsehood believe and affirm that he is the most vile of men by reason of the hidden faults that he recognizes in himself and the gifts of God which exist in a hidden way in others. That is why Augustine says, in his book on virginity: Be convinced that some who are less than you on the outside are superior to you on the inside." See Cajetan's commentary on this text.

205. Cornelius of Lapide, *In Lucam,* 1:29, t. 16, p. 19a: "Cogitabat enim intra se: Ego mihi videor indigna omnis gratiae, quomodo ergo angelus me vocat gratia plenam? Ego paupercula cum pauperibus virginibus dego et conversor, quomodo ergo angelus mihi insonat Dominus tecum? Ego aestimo me feminarum omnium minimam et vilissimam, quomodo ergo angelus mihi ait: Benedicta tu inter mulieres?"

206. Apoc. 3:17.

207. 2 Tim. 3:7.

208. Mark 8:2.

209. Ps. 21:25.

210. St. Thomas, *Summa theologiae,* IIaIIae, q. 30, a. 2, ad 2: "Illi qui jam sunt in infimis malis non timent se ulterius pati aliquid: et ideo non miserentur."

211. St. Grignion de Montfort, *La vraie dévotion,* n. 48.

212. St. Bonaventure, "The blessed Virgin intercedes for us, not only in timely but in untimely ways." *De nocte surrexit deditque praedam domesticiis suis, etc.* (Proverbs 31:15). Hoc recte competit Virgini, quae non tantum opportune, sed etiam importune interpellat pro nobis, quod designatum fuit Joannis secundo: Deficiente vino, dicit Mater Jesu ad eum, vinum non habent, Et dicit ei Jesus, quid mihi et tibi est, mulier? (John 3–4) in quo insinuatur anima officiositas et sollicitudo, quam beatissima Virgo habet pro genere humano" (*De nativitate B.V.M.,* sermo 3, t. 9, p. 713a).

The Primacy of the Common Good against the Personalists

and

The Principle of the New Order

Charles De Koninck

Preface by Cardinal Villeneuve of Quebec

1943

To King Leopold III

Preface

This is no ordinary book. It is pure wisdom. We have lost the habit of regarding the practical world in the light of the most profound principles, with the exception perhaps of those inverted thinkers who shake up the order of thought the better to reverse the real order, the political and moral order, and organize with an innocent air the most radical revolutions which end by being both the bloodiest and most cynical. At the same time, good souls throw up their hands in terror and scandal but soon, little by little, begin to think like the revolutionaries, not taking into account the equivocations hidden in apparently acceptable formulas nor the fact that such a concession is a way of cooperating even in the shedding of blood.

The author of this book surely sees better than most the frightening perils and social disorder that arise from Nazism and communism. He sees them better because he has penetrated the false wisdom, the principles which lie latent in movements of attack and retreat by the organizers of disorder. He sees the principles in all their perfidy and in their twisted truth, a truth poisoned by the microbe of pride which uses the terms of truth in order to lead to error and the language of virtue to conceal sin and evil. What frightens and grips the soul is that good people, sometimes the best, become used to this, however much they dread the revolutions which unroll before their eyes, until they too misconceive the essential and become drunk on the most harmful formulas. The world, in a word, grows accustomed to thinking in a communist, a Marxist manner, radically negative, first unconsciously, thoughtlessly, then with tipsy enthusiasm, denying all that is true because it is being, all that is just because it is ordered, all that perfects man because he is subordinate to God and rectified only by his ordination to the true sovereign end.

The author has already shown in earlier studies the historical origins and evolution of this essentially deviant and corrosive philosophy. One must go back to Averroism which sought emancipation from the natural order, to voluntarism which sought the emancipation of appetite, to nominalism which led to the emancipation of human language, to the moralism of good will which sought the emancipation of feeling, to a posturing and methodical skepticism which sought the emancipation of purely human thought, to Kantian subjectivism which sought the emancipation of reason against intelligence and of right against the common good, and continued its avatars in the emancipated dialectic of Hegel which turned against all nature, in Marxism, which acquired the power of destruction in Bolshevism, and Nazism. It is by seeing how, little by little, even among traditional minds, revolutionary thought gains more or less conscious adherents, that the author is both moved and fired by zeal for the truth.

Nowadays it is personalism that is in vogue. The most sincere minds recommend it. The dignity of the human person is exalted, one demands respect for the person, one writes on behalf of a personalist order and works to create a civilization which will be for man. . . . All that is very well, but too brief, for the person, man, is not an end unto himself nor the end of all else. He has God for his end, and to want to borrow the language of others, even when one seems to correct it by the employment of the best of adjectives—haven't we heard of the 'dialectical materialism of Aristotle and St. Thomas' to designate their natural doctrine?—even if one does not exclude the presuppositions that orthodoxy presupposes, one admits the presuppositions of others, of a naturalist thought, atheist if only out of indifference, and radically humanist, which favors turning civilization upside down because this has already been done to language, and along with language, philosophy and theology. This is what the author opposes. He is not wrong. More than ever, it is time to cry Thief! Time for societies to reorganize themselves, not in function of the individual person, but in function of the common good in its various degrees, that is, the sovereign end—that is, in function of God.

The author openly attacks the personalists, but only in order truly to defend the dignity of the human person. His study insists on the grandeur of the person without flattering persons. It is opposed to any doctrine which, under the pretext of glorifying, diminishes and atrophies the human person, depriving him of his most divine goods.

Among Christian thinkers today there is agreement about the social facts of the contemporary era, but at the same time one discerns among them clearly contrary tendencies once they begin to interpret these facts.

All seem to see that political society increasingly fails in its duties, that it is dissolving and becoming less and less worthy of the most essential tasks. It no longer cares for God, for the human soul, for eternal goods, but wearies itself in purely economic preoccupations and bodily well-being. This is what increases the burden of family responsibility which must more and more, under the constraint of circumstances, supply the goods it should receive from public society. But, alas, the same authors also observe the ever-increasing dissolution of the family. The individual person is more and more isolated and abandoned to himself in the home as the family is with respect to society. What to do?

When it is a question of interpreting these facts, in order to correct evils, some, imbued with the idea of Progress, see in the increased dissolution a recognition of the value of true hierarchy. In the complete collapse of civil society, they find a good which inclines them to justify the collapse, namely, the occasion for the individual person, whom they take to be the aim of the whole social order, to mount his pedestal, the better to shine, forgetting that of his very nature man enters into this order, that he cannot fully and profoundly shine except by reason of the different common goods to which he is ordered as to his greatest goods and which have at their summit the sovereign common good that is God Himself. In short, they make an essential concession to Marxism. They twist the rule of Christian optimism, "God only permits evil in order that he can draw from it a greater good," and confuse evolving progress with perfection which is necessarily based on order and on what is essential and immoveable.

Others, on the contrary, and our author puts himself resolutely among them, see in social and familial dissolution, foreseen and deplored by the most authoritative voice, that of the Church, a pure and simple increase of human misery and a gradual impoverishment. And that these developments are the normal consequence of the exploitation of civil society and the family for the individualist profit of the person.

It is true that the role of the family should increase, it always should, but all the more so now when its existence is menaced. It is true that it is more than ever important to insist on the dignity of the individual person, to cry it aloud, both to public powers and to individual persons. The person must be saved despite the corruption of the family and social milieus. But that is not to say that the corruption of the milieu is a good, an occasion for the person to value his richest qualities. You've lost an eye? Then the one you still have requires greater care and particular prudence. Granted. But is the one-eyed man to console himself by saying it is better to have one eye than two because of the value accruing to the remaining one? Should we speculatively denounce better things because of the expedients that evil suggests in practice? Should the speculative order be subjected to the practical? Must one falsify the dignity of the person and promote personalism because a corrupt society no longer fulfills its role toward the common good, thus depriving the person of the supports natural to him when society and the family are centered on the notion of the common good?

That is the thesis of this book, the primacy of the common good, in society, in the family, for the soul itself, on condition that the notion of the common good be properly understood: it is the best good of the singular, not insofar as it is a collection of singular goods: it is the best for each of the particulars who participate in it, by reason of its very community. Those who defend the primacy of the singular good of the person presuppose a false notion of the common good, one that would be alien to the good of singulars. It is rather by nature and as his proper good that the individual desires more the good of the species than his own individual good. The person, an intellectual substance, being a part of the universe, in whom can exist, by way of knowledge, the perfection of the entire universe, will have as his most proper good insofar as he is an intellectual creature the good of the universe, an essentially common good. "Rational creatures, persons, differ from irrational beings in this that they are more ordered to the common good and can act explicitly for it. It is also true that they can perversely prefer the singular good of their person to the common good, or, as it is put nowadays, prefer their personality erected into a common measure of all good. Moreover, if the rational creature cannot be ordered entirely to a subordinate common good, to the good of the family, for example, or to the good of political society, this is not because his singular good, taken as such, is greater, but because of his ordination to a higher common good to which he is principally ordered. In this case, the common good is not sacrificed to the good of the individual, but to the good of the individual insofar as he is ordered to a more universal common good," that is, to God. "A society made up of persons who love their private good above the common good, or who identify the common good with the private good, is a society, not of free men, but of tyrants—'and thus the whole people become as it were one tyrant'—who would use force on one another and whose eventual chief would be the shrewdest and strongest of tyrants, his subjects being only frustrated tyrants."

That is the substance of the book. It makes its point with the hammer of reason, striking with multiple strokes on the anvil of evident and fundamental truths the hot iron of the truth, and on the other hand makes clear the inconsistency and absurdity of equivocation and error.

The dissolution of human societies would not be so great an evil if it were not the corruption of the best of human goods, the common good, and if that did not at the same time lead to forgetfulness of the very notion of the common good. One will not build a better society with personalism if one destroys the very principle of any society, the very first principle, which is the common good.

It is not in a "personalist conception of marriage" nor in a so-called "Christian and socialist personalism," both of which are the result of speculative and ethical concessions to error, that one will find the solution to the problems which are more and more tragically raised by deviations from the truth. It is always the truth that will set us free. But these notions only press to the point of exasperation the perilous solitude into which the person is plunged when he is detached and isolated, under the pretext of exalting him, from his natural basis, the common good.

Some have dared to see in the insistence of encyclicals on the dignity of the person a belated approbation of the emancipation of the person. It is even said that communism would have been a salutary experience to put in practice a new conception of society; there are those who think that the danger of its evil doctrines has been exaggerated and that it is in the logic of things that human nature will always emerge victorious.

Thus one now submits the most evident truths and the best established principles to a historical dialectic. The errors that the popes have never ceased to condemn will receive, after mature reflection and thanks to the new perspectives furnished by the advantage of an acquired experience, a very just vindication. Certain Catholics, forgetting that Pius XI denounced communism as a false and intrinsically perverse redemption (in the encyclical *Divini Redemptoris*) would have it understood that the Church has not made the concessions it should have, concessions which would have been able to save many generous souls vexed by these rebuffs. Blasphemy and ingratitude!

It is the holy Church that showed itself heroic that is found at fault. For the Church defends the person against the very consequences of doctrines which, because of a false conception of the state, because of an exclusive preoccupation of states with the purely personal good, because of the disinterest of persons in the common good, have given to the state a growing, blind, and crushing power.

There would be no greater manifestation of the divine mercy than to save the person from the solitude into which men have been thrown. Karl Marx saw deprivation increasing in the world, but according to him this deprivation would be only the occasion for man to show his proper power; he called for a purely human man to correct such deprivation, and not at all for a man ordered to the common good, ordered to God.

Faced with the greatest menace, we still sustain the truth, namely, that the person ought to appeal to the family and to society, and that the whole created order ought to appeal to God. To the scandal of a world which disregards those who hunger and thirst for justice, we appeal to God, to his mercy whose ways are hidden. And faced with a world that thinks evil a better way to adapt to facts, that seeks the good in evil, we have no easy solutions, we only have those which correct the facts according to the principles of the good.

Evil would not exist if God could not draw good from it; evil could not be so great if God could not draw from it a greater good. But woe to those who, by teaching or by action, push man into this extreme need, into this infernal solitude where the person himself will perish if the pure divine liberality does not come to save him. Woe to those who encourage evil *ut eveniat bonum.*

When the Tempter addresses men he knows that he must speak of divinity: "You shall be as gods." Since then, all attacks on religion and truth against the rights of God and the true dignity of the person are made in analogous terms. Even Karl Marx could only make himself heard by proclaiming "human conscience is the highest divinity." But just as the divine mercies increase in the course of time, so too does the shrewdness of the demon sharpen. Let us heed attentively the warning of the Apostle "I indeed fear that, as Eve was seduced by the astuteness of the serpent, so your thoughts corrupt and lose that simplicity with regard to Christ" (2 Cor. 11:3).

In circuitu impii ambulant, according to the book of Psalms (11:9), the impious go around in circles. And they always return to the charge. When they have been chased out one door, they try to come back by another, above all that one where they are least expected.

We must expect an ever more veiled return of the most nefarious doctrines of the past. Perhaps there is none that has had more rebirths than the polycephalous monster of Pelagianism. All the more reason for Christians to proclaim the necessity of grace to save man from sin and heal his wounds, to proclaim that the person is nothing save by imitating God, by participating in uncreated Being, by his ordination to the divine common good, by his supernatural vocation to participate in the life and splendor of the Savior.

May the sons of St. Thomas who react even to the appearance of the shadow of this peril obtain from God untiring vigilance.

That is the authorized warning of the author of *The Primacy of the Common Good against the Personalists.*

Some will contest, under the pretext of prudence, the opportuneness of questioning without respite the irreducible doctrinal divergences which are the object of this study. To them we recall again the words of the Apostle that one finds in the Mass for a Doctor: "I charge thee, in the sight of God and Christ Jesus, who will judge the living and the dead by his coming and by his kingdom, preach the word, be urgent in season, out of season; reprove, entreat, rebuke with all patience and teaching. For there will come a time when they will not endure sound doctrine, but having itching ears, will heap up to themselves teachers according to their own lusts, and they will turn away their hearing from the truth and turn aside rather to fables" (2 Timothy 4:1–4).

+ J.-M.-Rodrigue Cardinal Villeneuve, O.M.I.
Archbishop of Quebec

Foreword

Human society is made for man. Any political doctrine that ignores man's rational nature, and consequently denies his dignity and liberty, is vitiated at its root and subjects man to inhuman conditions. That is why one rightly opposes totalitarian doctrines in the name of the dignity of man.

Does this mean that we are in agreement with all those who invoke the dignity of man? It should not be forgotten that, far from having denied the dignity of the human person, the philosophies which engendered modern totalitarianism exalted that dignity more than has ever been done before. It is therefore important to determine carefully in what the dignity of man consists.

Marxists push the dignity of the human person to the negation of God. "Philosophy has not concealed it," says Marx. "The profession of Prometheus: 'in a word, I hate all the gods . . .' is its proper profession, the case that it makes and will always make against all the gods of heaven and earth who do not recognize the human conscience as the highest divinity. This divinity will brook no rival."[1]

Let us not forget that the sin of *him who sinned from the beginning* consisted in the exaltation of his personal dignity and of the good proper to his nature: he preferred his proper good to the common good, to a beatitude participated in and common to many; he refused this because it was participated in and common. Even though he possessed the natural good and excellence of his person, not by a special favor, but by a right founded in his very creation—he owed his creation to God but all the rest was his due—he felt wounded in his proper dignity by this invitation to participate. "Seizing on their proper dignity [the fallen angels] desired their 'singularity,' that which is most proper to the proud."[2]

The dignity of the created person is not without limits, and our liberty has for its end, not to break those limits, but to liberate ourselves by strengthening them. The limits are the principal cause of our dignity. Liberty itself is no guarantee of dignity and of practical truth. "Aversion from God has the mark of an end insofar as it is desired under the note of liberty, according to that verse of Jeremiah (2:20): 'For too long you have broken the yoke and burst the bonds, and have said, 'I will not serve!' "[3]

One can affirm the dignity of the person and find himself in pretty bad company. Would it suffice to exalt the primacy of the common good? Not at all. Totalitarian regimes seize on the common good as a pretext for enslaving persons in the most ignoble fashion. Compared to the slavery to which they threaten to reduce us, the servitude of the beasts is liberty. Shall we commit the cowardice of conceding to totalitarianism this perversion of the common good and its primacy?

Might there not be between the exaltation of the quite personal good above every truly common good and the negation of the dignity of persons some link of a quite logical consequence and realization in the course of history? The sin of the angels was a practically personalist error: they preferred the dignity of their proper person to the dignity that would have come to them in the subordination to a superior good, but one common in its very superiority. The Pelagian heresy, John of St. Thomas says, can be seen as a reflection of the sin of the angels. It is only a reflection since, while the error of the angels was purely practical, the error of the Pelagians was at the same time speculative.[4] We believe that modern personalism is only a reflection of that reflection, even weaker speculatively. It elevates into a speculative doctrine an error which at the outset was only practical. The enslavement of the person in the name of the common good is like a diabolical vengeance at once remarkable and cruel, a sly attack on the community of the good to which the demon refused to submit. The denial of the superior dignity that man receives by the subordination of his quite personal good to the common good assures the negation of all human dignity.

We do not intend to maintain here that the error of all those who call themselves personalists nowadays is more than speculative. Let there be no ambiguity about that. No doubt our insistence will wound those personalists who have identified this doctrine with their person. That is their quite personal responsibility. But there is also ours—we think this doctrine pernicious in the extreme.

PART ONE

The Primacy of the Common Good against the Personalists

*Although the [fallen] Angel in fact abased himself by this abandonment of superior goods, and has, as St. Augustine says, fallen to the level of his proper good, nonetheless he raised himself in his own eyes, forced himself to prove exhaustively to others with a huge budget of arguments (*magna negotione*) that he came thereby only to a greater resemblance to God, because he thus acted less in dependence on his grace and favors, and in a more personal manner (*magis singulariter*), having nothing in common with inferiors.*

—John of St. Thomas

De Angelorum malitia

Be sure of this, I would exchange nothing of my miserable lot for your servitude. I prefer to be bound to this rock rather than be the faithful valet and messenger of Zeus the Father. . . .

—Prometheus to Hermes

cited by Karl Marx

The Common Good and Its Primacy

The good is that which all things desire insofar as they desire their perfection. Thus, the good has the note of final cause. Thus, it is the first of causes and consequently diffusive of itself. "The more elevated a cause is, the more its causality extends to many beings. Indeed, a more elevated cause has a more elevated proper effect, which is more common and is found in more things."[5]

Thence it follows that the good, which has the note of final cause, is all the more efficacious insofar as it communicates itself to a greater number

> of beings. That is why, if the same thing is a good for each individual and for the city, it is clear that it is much greater and more perfect to have it in mind, that is, to procure and defend that which is the good of the whole city, than that which is the good of only one man. Certainly, the love which ought to exist among men has as end to conserve the good even of the individual. But it is much better and more divine to show that love for the sake of a nation and cities. Or, if it is certainly desirable sometimes to show his love for only one city, it is much more divine to show it to the whole nation, which contains many cities. We say this is more 'divine' because it is more like God, who is the ultimate cause of all goods.[6]

The common good differs from the singular good by that very universality. It has the note of superabundance and is eminently diffusive of itself insofar as it is more communicable; it extends to the singular more than the singular good does: it is the better good of the singular.

The common good is better, not insofar as it comprises the singular good of all the singulars: it would not then have the unity of the common good insofar as it is universal but would be a pure collection. It would be only materially better. The common good is better for each of the particulars that participate in it insofar as it is communicable to the other particulars: communicability is the very reason of its perfection. The particular does not attain the common good under the note itself of common good if it does not attain it as communicable to others. The good of the family is better than the singular good, not because all the members of the family find in it their singular good: the good of the family is better because, for each of its individual members, it is also the good of the others. That does not mean that the others are the reason for the proper loveableness of the common good. On the contrary, under this formal aspect, the others are loveable insofar as they can participate in this good.

Hence, the common good is a good which is the good of particulars, not the good of the collectivity envisaged as a sort of singular. In that case, it would be common only accidentally and would be properly singular, or, if one wishes, it would differ from the singular good of the particulars in being the good of none of them (*nullius*). But, when we distinguish the common good from the particular good, we do not mean by this that it is not the good of particulars. If it were not the good of particulars, it would not be truly common.

The good is that which all things desire insofar as they desire *their perfection.* That perfection is for each of them its good—*bonum suum*—and, in this sense, its good is a proper good. But then the proper good is not opposed to the common good. In fact, the proper good to which a being naturally tends, the *bonum suum,* can be understood in different ways, according to the diverse goods in which it finds its perfection.[7]

It can be understood first of all as the proper good of a particular insofar as it is an individual. This is the good that the animal pursues when he desires nourishment for the conservation of his being.

In a second way, the proper good can be understood as the good of a particular by reason of that particular's species. This is the good the animal desires in generation, nutrition, and the defense of individuals of its species. The animal prefers 'naturally', that is to say in virtue of the inclination which is in it by nature (*ratio indita rebus ab arte divina*), the good of the species to its singular good. "Every singular naturally loves the good of its species more than its singular good."[8] That is because the good of the species is a greater good for the singular than its singular good. Therefore, this is not a species which abstracts from individuals and desires its proper good against the natural desire of the individual; it is the singular itself which, by nature, desires the good of the species rather than its singular good. This desire for the common good is in the singular itself. Hence, the common good does not have the note of an alien good—*bonum alienum*—as in the good of the other taken as such.[9] Isn't it this which, on the social plane, separates us profoundly from collectivism which sins by abstraction, which demands an alienation from the proper good as such, and, consequently, from the common good since it is better than proper goods? Those who defend the primacy of the singular good of the singular person themselves presuppose this false notion of the common good.

In a third way, the good of a particular can be understood as the good which belongs to it because of its genus. This is the good of equivocal agents and intellectual substances, whose action can of itself attain, not only the good of the species, but also a greater good communicable to many species.

In a fourth way, the good of the particular can be understood of the good which belongs to a cause by that similarity of analogy which 'principled' things (that is, which proceed from a principle) have to their principle. It is thus that God, a purely and simply universal good, is the proper good that all things desire naturally as the highest and best good, and which confers on all things their entire being. In short, "nature returns on itself not only in that which is singular to it, but even more in that which is common; in fact, every being tends to conserve not only itself as an individual, but its species. And every being is even more naturally borne toward that which is the absolutely universal good."[10]

This enables one to see how profoundly nature is a participation in intelligence. It is thanks to this participation in intelligence that every nature tends principally to a universal end.

In the appetite which follows on knowledge, we will find a similar order. Beings will be perfect insofar as their appetite extends to a good more remote from their singular good alone. Since the knowledge of brutes is linked to the sensible singular, their appetite can extend only to the sensible singular and their private good. Explicit action for a common good presupposes universal knowledge. The intellectual substance, being *comprehensiva totius entis*, comprehensive of all being,[11] being a part of the universe in which, by way of knowledge, the perfection of the universe in its entirety can exist,[12] its most proper good insofar as it is an intellectual substance will be the good of the universe, an essentially common good. In fact, it is well to mark here the radical difference between knowledge and appetite: "the known is in the knower, the good is in things." If, as known, the good were in the one loving, we would ourselves be the good of the universe.

Consequently, inferior beings differ from superior in that the most perfect good they know is identical to their singular good, and also in that the good they can diffuse is restricted to the good of the individual.

> The more perfect the power of a being and its degree of eminent goodness, the more its appetite for the good is universal and the more it seeks and does good to beings which are far from it. For imperfect beings tend to the good of the individual alone properly speaking; perfect beings tend to the good of the species; more perfect beings to the good of the genus. But God, who is of a goodness absolutely perfect, tends to the good of being in its entirety. Thus, it is not without reason that the good as such is said to be diffusive, because, the more a being is good, the more it diffuses its goodness to beings which are more distant from it. And because that which is most perfect in any genus is the exemplar and measure

> of all that are comprised in that genus, it follows that God who is the most perfect goodness and who shares it in the most universal fashion, should be, in the diffusion of his goodness, the exemplar of all beings which share their goodness.[13]

It is the created common good, of whatever order it be, that most properly imitates the absolute common good.

From this it is clear that the more perfect a being is, the more it is related to the common good, and the more it acts principally for this good which is, not only in itself, but for him, better. Rational creatures, persons, differ from irrational beings in this that they are more ordered to the common good and can act explicitly for it. It is also true that they can perversely prefer the singular good of their person to the common good, or, as it is put nowadays, prefer their personality erected into a common measure of all good. Moreover, if the rational creature cannot be ordered entirely to a subordinate common good, to the good of the family, for example, or to the good of political society, this is not because his singular good, taken as such, is greater, but because of his ordination to a higher common good to which he is principally ordered. In this case, the common good is not sacrificed to the good of the individual, but to the good of the individual insofar as he is ordered to a more universal common good. Singularity alone cannot be the per se reason. In every genus the common good is higher. Comparison by transgression of genera, far from weakening this principle, presupposes and confirms it.

It is in the most perfect created persons, the pure spirits, that one best sees this profound ordination to the common good. Indeed, insofar as they are more intelligent, the common good is more their good. "Since desire follows knowledge, the more universal the knowledge, the more the desire that follows on it bears on the common good, and the more particular the knowledge, the more the desire that follows on it bears on the private good. Thus it is that in us love of the private good follows sense knowledge, but the love of the common and absolute good follows intellectual knowledge. Therefore, because angels have more universal knowledge to the degree that they are more perfect, their love tends especially to the common good."[14] This love of the common good is so perfect and so great that the angels love their inequality and the very subordination of their singular good, which is always more distant from their common good, and are more subject and more conformed to it in proportion as they are more elevated in perfection. "Therefore they love the others more because of their specific difference, which belongs more to the perfection of the universe than they would if they were all of the same species, which belongs to the private good of a single species."[15] And this, because "their love is rather for the common good."

In sum, according to the authors who put the common good of persons second, the more perfect angels would also be the most subject and least free. Because of his attachment to the common good, the citizen would be in truth a slave, and the slave would be a free man. The slave too lives principally on the margin of society, as the stone in a pile is free from the order of the building. "It is with the world," Aristotle says, "as with a household, where free men are least at liberty to do this or that as occasion arises but all their functions, or most of them, are regulated; for slaves and beasts, on the other hand, there are few things which relate to the common good and most things are left to their choice."[16] In Marxist personalism, which is achieved in the final phase of communism, the citizen is nothing but a slave to whom has been given, in his very condition as slave, the title of apparent liberty by which one can take from him participation in true liberty.[17]

The common good is in itself and for us more loveable than the private good, but an equivocation could remain, since one can love the common good in two ways. One can love it in order to possess it or one can love it for its conservation and diffusion. One could in effect say: I prefer the common good because possession of it by me is for me a greater good. But that is not a love of the common good insofar as it is a common good. It is a love which looks on the common good as a private good, which identifies the common good with the good of the singular person taken as such. "To love the good of a city in order to appropriate and possess it oneself does not make a good politician; for it is thus that the tyrant too loves the good of the city, in order to dominate it, which is to love himself more than the city: in fact it is for himself that he desires the good, not for the city. But to love the good of the city in order that it might be preserved and defended is truly to love the city, and it is that which makes a good politician, so much so that, with an eye to conserving and augmenting the good of the city, he exposes himself to danger and neglects his private good."[18] Thus one cannot love the common good without loving its shareability with others. The fallen angels did not refuse the perfection of the good that was offered them, they refused its community and they scorned that community. If the singular good of the person truly came first, how would they have sinned against the common good? Above all, how would the most naturally worthy rational creature have been able to turn away from the most divine good of all?

A society made up of persons who love their private good above the common good, or who identify the common good with the private good, is a society, not of free men, but of tyrants—"and thus the whole people become as it were one tyrant"[19]—who would use force on one another and whose eventual chief would be the shrewdest and strongest of tyrants, his subjects being only frustrated tyrants. Refusal of the primacy of the common good proceeds, at bottom, from the distrust and scorn of persons.

Appeal has been made to the absolute transcendence of supernatural beatitude to sustain the thesis that the good of the singular person is purely and simply superior to the common good, as if this good were not, in its transcendence, and indeed precisely because of that, the most universal common good which ought to be loved for itself and for its diffusion. The ultimate end is not distinguished from lesser common goods because it is the singular good of the individual person. One can in effect play on the ambiguity of "particular" and "proper" and "singular." "The proper good of man can be understood in different ways. For the proper good of man as man is the good of reason, since to be a man is to be a reasonable being. But the good of man as artisan is the good of art, and, insofar as he is political, his good is the common good of the city."[20] Just as the good of man as citizen is not the same as the good of man simply, so the good of man as a citizen of civil society is not the same as his good as a member of the celestial city. "To be a good politician one must love the good of the city. But if a man, insofar as he is admitted to participate in the good of some city and is made a citizen, has need of certain virtues to accomplish the things that fall to citizens and to love the good of the city, it is the same with a man who has been admitted, by grace, to participation in celestial beatitude, which consists in the vision and enjoyment of God; he becomes in a way a citizen and member of this blessed society which is called the heavenly Jerusalem, according to Saint Paul to the Ephesians 2:19, *You are citizens of the city of the saints, and members of the family of God.*"[21] And just as the virtues of man purely as man do not suffice to rectify us with respect to the common good of civil society, so too quite special virtues, higher and more noble, are necessary to order us to beatitude, and that under the quite formal aspect of common good. "Therefore, certain gratuitous virtues are necessary to man when admitted to the celestial society: these are the infused virtues whose proper exercise presupposes love of the good common to the whole society, namely, the divine good insofar as it is the object of beatitude."[22] And it is here that St. Thomas makes the distinction cited earlier between the love of possession and the love of diffusion. *You are citizens,* above all in a beatitude where the common good most fully has the note of the common good.

Elevation to the supernatural order only augments the dependence of the good of the person taken as such on a more distant good. If a monastic virtue cannot accomplish an act ordered to the common good of civil society save insofar as it is elevated by a higher virtue which properly looks to the common good, it will be even less able to do it when the good is properly divine, "since no merit can exist without charity, the act of the acquired virtue cannot be meritorious without charity. . . . For the virtue ordered to a lower end cannot make its act to be ordered to a higher end, save by means of a superior virtue; for example, fortitude which is a virtue of man as man cannot order human action to the political good, save by the mediation of the fortitude which is a virtue of man insofar as he is a citizen."[23] The fortitude of man as man by which he defends the good of his person does not suffice to defend the common good reasonably. That society is very corrupt which cannot appeal to the love of the arduous common good and to the higher fortitude of the citizen as citizen for the defense of this good, but which must present its good under the color of the good of the person.

Let us not treat the virtues of the political as accessory complements of the virtues of man as purely man. On the one hand, one pretends that these are more profound and at the same time, on the other hand, would have it that a bad man in his monastic and domestic life can be a good political man. This is a sign of the contempt in which one holds everything that formally regards the common good. And yet

> those who fulfill in a worthy and laudable manner the ministry of kings will attain celestial beatitude to an eminent degree. Indeed, if the happiness that virtue procures is a reward, it follows that the greater the virtue, the greater is the degree of happiness due it. But the virtue by which a man directs not only himself but also others is a higher virtue;

> and it is the higher insofar as it can direct a greater number of men, just as someone is reputed virtuous in the sense of bodily virtue when he can vanquish a greater number of adversaries or lift heavier weights. Thus, a greater virtue is required to direct the family than to direct oneself, and a still greater virtue to govern a city and a kingdom. . . . But someone is that much more pleasing to God who most imitates God, whence that admonition of the Apostle to the Ephesians, 5, 1, "Be ye imitators of God, as well-beloved sons." But, in the word of the Sage: "Every animal loves its like, insofar as effects have a certain resemblance to their cause; it follows then that good kings are most pleasing to God and they will receive from Him a very great reward."[24]

The position according to which the good of the singular person would be, as such, superior to the good of the community becomes abominable when one considers that the person is himself the principal object of the love of the singular good.

> As love has good for its object, love is diversified according to the diversity of the good. But there is a good proper to man insofar as he is a singular person; and as regards the love that has this good for its object, each is for himself the principal object of such love. But there is a common good which belongs to this one and that one insofar as they are parts of some whole, for example to the soldier insofar as he is a part of the army, and to the citizen insofar as he is part of the city: as regards the love that has this good for its object, its principal object is that in which this good principally consists, as the good of the army in the general, and the good of the city in the king; that is why it is the duty of the good soldier to neglect even his proper safety in order to preserve the good of his general, as a man naturally exposes his arms in order to protect his head. . . .[25]

In other words, the most elevated end of man belongs to him, not insofar as he is in himself a certain whole where the self is the principal object of his love, but "insofar as he is a part of a whole," a whole accessible to him because of the very universality of knowledge. You say that the notion of a part does not belong to man envisaged in his relation to the ultimate end?

Here is the immediate sequel of the text we have just quoted: ". . . and it is in this way that charity has for its principal object the divine good, which is the good of each, insofar as each participates in beatitude."[26] It is then indeed as part of a whole that we are ordered to the greatest of all goods which cannot be ours except in its communicability to others. If the divine good were formally "a good proper to man insofar as he is a singular person," we would be ourselves the measure of this good, which is very properly an abomination.

Even the love of the proper good of the singular person cannot be removed from the love of the common good. We have, in fact, so perfectly the note of a part that the rectification with respect to the proper good cannot be true unless it is conformed and subordinated to the common good. "The goodness of any part resides in its relation to its whole: that is why Augustine says . . . that 'every part is bad which does not conform to its whole.' Therefore, since every man is part of the city, it is impossible for any man to be good if he is not perfectly proportioned to the common good; and the whole itself cannot exist as it should (*bene*), save by means of the parts which are proportioned to it."[27] This ordination is so integral that those who pursue the common good pursue their proper good *ex consequenti:* "because, first of all, the proper good cannot exist without the common good of the family, the city, or the realm. That is why Valerius Maximus says of the ancient Romans that 'they preferred to be poor in a rich empire than to be rich in a poor empire.' Because, in the second place, as man is part of the household and of the city, it is important that he judge what is good for him in the light of the prudence which has for its object the good of the multitude: for the good disposition of the part is drawn from its relation to the whole."[28] And this is even more evident in the common good which is beatitude, where the very universality of the good is the principle of beatitude for the singular person. In fact, it is by reason of its universality that it can beatify the singular person. And this communication with the common good grounds the communication of singular persons among themselves *extra verbum:* the common good insofar as it is common is the root of the communication which is only possible if the divine good was already loved in its communicability to others: "*praexigitur amor boni communis toti societati, quod est bonum divinum, prout est beatitudinis objectum:* love of the common good is prerequisite in the whole society, and it is the divine good insofar as it is the object of beatitude."[29]

If one concedes that single persons are ordered to the separated common good insofar as this has the note of common good, it is not so readily conceded that, in the universe itself, persons are willed only for the good of the order of the universe, the intrinsic common good that is better than the single persons who materially constitute it. One would rather have it that the order of the universe is only a superstructure of persons that God wills, not as parts, but as radically independent wholes; and these wholes would be parts only secondarily. In fact, do not rational creatures differ from irrational creatures in this that they are willed and governed for themselves, not only with respect to the species, but also with respect to the individual? "The acts of the rational creature are directed by divine providence, not only by reason of their belonging to a species, but also insofar as they are personal acts."[30] Therefore, it is concluded, individual persons are themselves goods willed first of all for themselves and are in themselves superior to the good of the accidental whole they constitute by way of consequence and complement.

But what is the end God proposes to Himself in the production of things?

> God has produced the being of all things, not by necessity of nature, but by His intelligence and will. His intelligence and will cannot have for ultimate end anything other than His proper goodness, which He effects by communicating this goodness to things. Things participate in the divine goodness by way of similitude, insofar as they are in themselves good. But, that which is better in created things is the good of the order of the universe, which is the most perfect, as the Philosopher says; and this is in agreement as well with Sacred Scripture where it is said, 'And God saw all that He had made, and behold it was very good,' whereas of works taken separately he said simply that 'they were good.' Consequently, the good of the order of the things created by God is also the principal object of God's will and intention (*praecipue volitum et intentum*). To govern a being is nothing else than to impose an order on it. . . . Moreover, whatever tends to an end is chiefly concerned (*magis curat*) with that which is more proximate to the ultimate end, for that is as well the end of all the others. But the ultimate end of the divine will is His proper goodness, and, in created things, it is the good of the order of the universe which is closest to that goodness (*cui propinquissimum*), for every particular good of this thing or that is ordered to the good of the order of the universe as

> to its end, as the less perfect is ordered to the more perfect. Hence, each part is found to exist for its whole. Consequently, in created things, God's chief concern is the order of the universe. . . .[31]

Why does God will the distinction of things, their order and their inequality? "The distinction of things, and their number, is the intention of the first agent who is God. In effect, God has given being to things in order to communicate His goodness to creatures, and manifest it through them; and because this goodness cannot be sufficiently manifested by one creature alone, He has produced many and diverse creatures, in order that what one fails to manifest of the goodness of God may be supplied by others. For the goodness which exists in God in a simple and uniform manner exists in a multiple and divided manner in creatures; that is why the entire universe participates more in the divine goodness and manifests it more perfectly than any other creature."[32] "In every effect it is the ultimate end which is properly willed by the principal agent, as the order of the army is willed by the general. But, that which is most perfect in things is the good of the universal order. . . . Therefore, the order of the universe is properly willed by God, and is not an accidental product of a succession of agents. . . . But . . . this same universal order is, in itself, created and willed by God. . . ."[33]

> The end for the sake of which an effect is produced is that which in it is good and better. But, that which is good and the best in the universe consists in the order of its parts among themselves, which order could not exist without distinction; it is in effect this very order which constitutes the universe in the manner of a whole, and which is the best in it. Therefore, the very order of the parts of the universe, and their distinction, is the end for the sake of which it has been created.[34]

To be sure, one will rebel against this conception if one considers the single person and his singular good to be the first root, as the intrinsic ultimate end, and, consequently, as the measure of every good intrinsic to the universe. This rebellion arises either from speculative ignorance or from practical ignorance.

To consider the common good as a foreign good is to misunderstand it speculatively, to consider it a *bonum alienum* opposed to the *bonum suum:*

one thereby limits the *bonum suum* to the singular good of the single person. On this view, the subordination of the private good to the common good would mean subordination of the most perfect good of the person to an alien good; the whole and the part would be strangers one to the other; the whole of the part would not be 'its whole.' This error disparages the person in his most fundamental capacity, that of participating in a good greater than the singular good; it denies the most splendid perfection of the universe: that which God Himself principally wills and in which persons can find *their* greatest created good. This error rejects the created common good, not because it is only created, but because it is common. And it is in this that the gravity of this error lies: it ought as well reject the most divine common good which is essentially common.

A pernicious practical ignorance can nonetheless coexist with a correct speculative understanding of the common good. One could refuse the primacy of the common good because it is not first of all the singular good of the single person and because it demands a subordination of it to a good which is not ours by reason of our singular personality. By a disordered love of singularity, one rejects the common good practically as an alien good and one judges it incompatible with the excellence of our singular condition. One thus withdraws himself from order and takes refuge in himself as in a universe for his own sake, a universe rooted in a very personal free act: one freely abdicates his dignity as rational creature in order to establish it as radically independent. "Turning away from God has the note of an end insofar as it is desired under the aegis of liberty, according to the text of Jeremiah (2:20): *Long have you kicked against the goad, you have broken out of your place, and you have said: I will not serve.*"[35] One would not refuse the common good if one were oneself the principle of it or it derived its excellence from that which we have freely chosen: one grants primacy to liberty itself. Above all, one wants to be a whole so radically independent that one needs God only in order to exist, after which one would enjoy the right to submit to order or not. And when one is pleased to submit, this submission would be an act emanating from the abundance of a pure 'for itself', and a burst of its own generosity so great that it is not repugnant to it to expand and diffuse itself; on the contrary, the personality would blossom and pour forth beyond the good that it already possesses in itself.[36] It would blossom, that is to say its good itself would come from within; it would owe to the outside only the generosity of space. It would recognize willingly its dependence on unformed matter, as a sculptor recognizes his dependence on stone. One would even allow himself to be directed by another and would recognize a superior, provided that he was the fruit of his choice and the vicar, not of the community, but first of all and principally of me. Every good other than that due to our singular nature, every good anterior to it and to which we ought freely to submit ourselves lest we do wrong, is abhorred as an insult to our personality.

It is the very idea of order against which one rebels, even though a creature is more perfect in the measure that it participates more in order. The separated substances are more perfect than us because they are more ordered and by nature participate more profoundly in the perfection of the universe which clothes them with its splendor, thanks to that ordination. "'The things which are of God are ordered.' But it is necessary that the superior parts of the universe participate more in the good of the universe, which is its order. Things among which order exists participate as such more perfectly in order than those in which one finds only an accidental order."[37] Why this disdain for the order which is the work of divine wisdom? How could the angels love their inequality if it were not rooted in the common good, and if this common good were not their greatest good? "If, on the contrary, the very being of their person were for them the greatest intrinsic good of the universe, inequality would be a principle of discord, both among the angels and for every individual person, and the common good would be an alien good; this inequality would proceed, not from divine wisdom, but either from free will and the contrariety of good and evil, or from the primacy accorded to material distinction."[38] The fact that the principal parts materially constituting the universe are ordered and governed for themselves can only make shine forth all the more the pre-eminent perfection of the ensemble which is the first intrinsic reason of the perfection of the parts. And "when we say that divine providence orders intellectual substances for themselves, we do not mean that these substances have no further relation with God and with the perfection of the universe. We say then that they are ruled for themselves and that other creatures are for their sake because the goods they receive thanks to divine providence are not given them for the utility of other creatures; on the contrary, the goods conferred on them are ordered by divine providence for the use of the intellectual substances."[39] It is then one thing to say that rational creatures are governed and ordered for themselves, and quite another to say they are for themselves and for their singular good: they are ordered for their own sakes to the common good. The common good is for them, but it is for them as a common good. Rational creatures can attain by themselves in an explicit manner the good to which all things are ordered; they differ in this from irrational creatures, who are pure instruments, who are useful only and who do not attain by themselves in an explicit manner the universal good to which they are ordered. It is in this that the dignity of the rational creature consists.

Objections and Replies

The Dignity of the Person and Liberty

1. It seems that the dignity of the person is opposed to the notion of part and to this ordination to the common good. Indeed, "dignity means goodness for one's own sake; utility, goodness for the sake of something other than oneself: *dignitas significat bonitatem alicujus propter seipsum, utilitas vero propter aliud.*"[40] Besides, "dignity pertains to that which is said absolutely: *dignitas est de absolute dictis.*"[41] Is it not for this reason that persons are ordered and governed for their own sakes?[42]

Ad 1. To this we reply that the rational creature takes his dignity from this that, by his own proper activity, by his intelligence and love, he can attain the ultimate end of the universe. "Intellectual and rational creatures take precedence over others both by the perfection of their nature and by the dignity of their end. By perfection of nature, since the rational creature alone is master of his acts and determines himself freely in his operations, whereas other creatures are rather acted upon than act. By the dignity of their end, for only the intelligent creature attains to the very last end of the universe, by knowing God and loving Him, whereas other creatures attain the ultimate end only by a certain participation in its likeness."[43]

Moreover, the rational creature, insofar as it can itself attain the end of God's external manifestation, exists for its own sake. Irrational creatures exist only for the sake of that being which can itself attain to the end which is only implicitly theirs. Man is the dignity which is their end. But that does not mean that the reasonable creature exists for the dignity of his proper being and that he is himself the dignity for the sake of which he exists. He draws his dignity from the end which he can and ought attain; his dignity consists in this that *he* can attain the end of the universe, the end of the universe being, in this respect, for rational creatures, that is, for each of them. However, the good of the universe is not for them as if they were the end for the sake of which it is. It is the good of each of them insofar as it is *their* good as a common good.

But the dignity in which he is clothed because of his end is so dependent on that end that the rational creature can fall from this dignity as he can fall from his end. "In sinning, man departs from the order of reason and consequently falls from human dignity, which consists in this that a man is naturally free and that he exists for his own sake, and puts himself in a certain manner in the slavery of the beasts. . . . For a bad man is worse than a beast."[44]

Far from excluding or rendering indifferent the ordination of his private good to the common good (or of his proper good when this does not already embrace the common good), as if this ordination were an affair of pure liberty of contradiction, the dignity of the intelligent creature rather entails the necessity of this ordination. Man falls from human dignity when he refuses the very principle of this dignity: the good of intelligence realized in the common good. He subjects himself to the servitude of the beasts when he judges the common good to be an alien good. The perfection of human nature so little assures dignity that for a man to close himself on his proper dignity is a sufficient reason and first root of his falling from his being-for-himself.

Because *dignitas est de absolute dictis*, dignity cannot be a proper attribute of the person envisaged as such, but of the nature of persons. Indeed, to be absolute does not pertain as such to the person. The divine persons are subsistent relations. "Paternity is the dignity of the Father, as it is the essence of the Father: for dignity is an absolute and belongs to the essence. Thus just as the same essence which, in the Father, is paternity, is, in the Son, filiation, so the same dignity which, in the Father, is paternity is, in the Son, filiation."[45] Likewise, in man, dignity is not an attribute of the person envisaged as such, but of rational nature. It happens that the created person is an absolute because of his imperfection in the very notion of person. For the rest, in the reasonable being, purely created personality is subordinated to nature.[46] It is also important to remark that the person himself is defined universally by communicability: *rationalis naturae individua substantia:* an individual substance of a rational nature, where *nature* must be understood in the sense of principle of operation. The incommunicability of the person itself does not have the note of term, as if the person existed for his incommunicability; on the contrary, far from being 'for itself' in this incommunicability, this opens nature to communication—*actiones sunt suppositorum.* The divine persons are essentially expressive of the fecundity of the divine nature. In the case of the created person, communication is accomplished in the vital participation in the common goods.

The being-for-itself of every created person is for the sake of its end which is God. Nothing is anterior to this indissoluble being-for-itself-for-God. The only thing that can dissolve it is evil. Since he has from God all that he is—*secundum hoc ipsum quod est, alterius est*—the created person ought to advance toward the end by a direct movement. In this fundamental perspective, and there is no other as fundamental, any deliberately reflexive regard for himself is a nocturnal regard and aversion from God. If the human person were truly what the personalists say, man would have to be able to find in himself a loveableness which would be his as opposed to his end: the self would be itself the principle of its destiny and would be as well the term: he would not subordinate himself to an end other than himself except to subordinate it to himself; he would be borne toward things other than himself only in order to make them his own as the end. In truth, the end of persons would consist in the expansion of their personality.

Let us now consider the intelligent creature in his perfection as free agent. The perfection of nature which is the root of liberty has no other end than God. Moreover, God is not called free save with reference to things which are below Him. Liberty does not bear on the end as such, but on means; when it bears on an end, this is because that end is a subordinate end and can take on the character of means. God is necessarily the end of all that He freely makes, His liberty only looks to what He has made for that end, which is supreme goodness: only the dignity of God is identical with His being and inalienable. Because no other being is itself its supreme end and because the proper end of its nature can be ordered to a higher end, the rational creature is defectible and can lose his dignity: his dignity is only assured insofar as it is kept within order and acts in conformity with that order. Unlike irrational creatures, the rational creature must keep himself in the order established independently of him; but, to keep within that order is to submit oneself to and to be measured by it. Dignity is linked to order; to turn away from it is to lose one's dignity. If dignity belonged absolutely to the rational creatures, if it was assured by his liberty of contrariety, in his capacity to submit to order or not to submit, his dignity would be inalienable. The excellence of the rational creature does not consist in the ability to escape order, but in his ability to will that order in which he ought to be; he does not have the right to leave it.

> Just as there is an order in active causes, so there is one in final causes, such that the secondary end depends on the principal end, as the secondary agent depends on the primary agent. But sin occurs in agent causes when the secondary agent leaves the order fixed by the principal agent; thus, when the leg, because of a handicap, does not execute the movement commanded by the appetitive power, the defect produces a faulty walk. It is the same therefore with final causes, every time the secondary end departs from the order of the principal end, the will sins, even though its object is good and constitutes an end. But every will naturally wills the proper good of the one who wills, that is, the perfection of his very being, and it cannot wish the contrary. Therefore, the sin of the will is impossible in an agent who wills and whose good is the supreme end; for this end is not subordinated to others but all the others are subordinate to it. Such is the will of God, whose being is the sovereign good, which is the last end. Therefore, the sin of the will is impossible to God. But if one considers them in their nature, every other being which possesses this faculty and whose proper good is necessarily subordinated to another good, a sin of the will is possible. For, although there is in each of these beings a natural inclination of the will to will and love its perfection, so much so that it cannot will the contrary, this inclination nonetheless is not naturally of such a kind that it subordinates its perfection to another end without being able to depart from this order: the higher end is not the proper end of its nature, but is of a more elevated nature. Therefore, it depends on its free will to subordinate its proper perfection to the higher end; for beings endowed with will differ from those that lack it in this that the former subordinate themselves and all they have to the end; which comes down to saying that they have

> free will; whereas the latter do not subordinate themselves to the end, but are subordinated by a higher agent, and are directed and do not direct themselves to that end.[47]

The angel cannot by himself fail with respect to the end of his person or the common good proper to his nature. But the good of the angelic nature is not the supreme good that is God as He is in Himself. But God has commanded him to order himself to this sovereign good. As the proper end of the angelic nature takes on under this relation the character of an end to be ordered to a higher end, something which is not assured by the nature of the agent, his will can fail with respect to this end, and, by way of consequence, can fail as well with respect to his proper end.

If the angel is not of himself defective save by relation to the supernatural end, man of himself can fail with respect to his natural end.

> There is this difference between man and the separated substances, that the same individual has several appetitive powers, of which one is subject to the other; something which is in no wise the case with separated substances, which are however subject one to another. Sin enters the will when the inferior appetite diverges in whatever manner. Just as there would be sin in the separated substances either when they are deflected from the divine order or when one of lower rank departs from the order established by a higher substance which remains subject to the divine order, so man sins in two ways: first of all, when the human will does not subordinate its proper good to God, and this sin is common to the separated substances; second, when the good of the lower appetite is not ruled by the higher appetite; that is what happens when without observing the order of reason one seeks pleasures of the flesh, which are the object of the concupiscible part. This second kind of sin has no place with separated substances.[48]

In the very interior of man, the good of intelligence is higher than the good of sense. The union of intellectual nature with sensible nature subjects man to a certain contrariety. Sensible nature carries us toward the sensible and private good, intellectual nature has for object the universal good under the very notion of the good, which is found principally in the common good. The good of intelligence whence man derives his dignity as man is not assured by the very nature of man. Sensitive life is basic in us: we cannot attain to the acts of reason save by passing through sense which, in this respect, has the note of a principle. So long as a man is not rectified by the cardinal virtues, which he ought to acquire, he is chiefly drawn toward the private good against the good of intelligence. There exists for man, envisaged even in the purely natural order, a liberty of contrariety which makes him as such defectible with regard to his purely natural end. In order to use to advantage his dignity, he must submit his private good to the common good.

One could still object that if the dignity of the rational creature is linked to his subordination to God from whom the person has all that he is, his dignity is not linked to his subordination to other ends no matter how superior. Moreover, this dignity is anterior to any subordination other than that to God and independent of the order among created things. Indeed, "when the proper good of a being is subordinated to several higher goods, the agent endowed with will is free to depart from the order attached to one of these and remain in the order which terminates in the other, whether this be more or less elevated."[49]

To that we reply that when an agent endowed with will should subordinate his proper good to a created higher end, this can only be insofar as this is itself conformed to the divine order. Hence, when the lower ought to withdraw from that which is higher than himself, this is because the higher has himself departed from the order in which he ought to remain. But as long as the higher remains within that order, he is a higher good to which the lower ought to submit. "For example, the solder who is subject to the king and to a general of the army can subordinate his will to the good of the general and not to that of the king, and conversely: but in the case where the general transgresses the order given by the king, the will of the solder will be good if he detaches himself from the will of the general to submit himself to the king; he will be bad if he executes the will of the general contrary to the will of the king; for the order of a lower principle depends on the order of the higher principle."[50] However, "there would be sin in the separated substances if one of lower rank departed from the order of a higher substance who remains subject to the divine order." Hence, the revolt of the lower against a non-submissive superior is a revolt against disorder.

Order and Liberty

2. Looked on as such, free acts are outside and above the order of the universe, for only the cause of being in its entirety can act on our will. Therefore, persons are not, in the whole of themselves, included in the order of the universe. Besides, to be free is to be the cause of oneself. Hence, the person should derive his perfection from himself and not from the universe of which he is a part.

Ad 2. In reply to these difficulties, let us note first of all that free action does not itself have the note of term. The free agent differs from the purely natural agent in this that he moves himself to judge and then moves himself to an end in virtue of this very judgment. He is master of his action for the end, but he is not master of the end as such. His judgment should be correct: the truth of this judgment will depend on the conformity of appetite to the good that is the end. But the good for which the rational creature ought principally to act and by which his judgment ought to be ruled is the good which for him is naturally the best, the common good. But the common good is essentially shareable by many. Therefore, with respect to this good, every intelligent creature has the note of part. Free action ought to be ordered by the agent himself with a view to a participated good.

For the rest, the perfection of the universe requires that there be intelligent creatures and, by way of consequence, creatures who are masters of their acts, which carry them to their good. The perfection of the good that they ought to pursue is such that they ought to direct themselves toward it. If free action cannot have the note of part of the universe envisaged in itself, it must nevertheless in the last instance be ordered to an end before which the intelligent creature has the note of part. And the end is the first of all causes.

Moreover, the order of the universe can be understood in two ways: either as the order which is the form of the universe (this form is the intrinsic good of the universe) or as the order of the universe to its very first principle (the separated good that is God). The order of the universe is for the order to the separated principle. And, since that order is purely and simply universal, it includes even free acts: God governs free agents and their acts just as he governs fortuitous and chance events which have no determinate cause inherent in the order of the universe.[51]

Both the good inherent in the universe and the separated good bear the note of common good. Hence, in relation to the one or the other, the rational creature has the note of part: he can only have the note of whole in relation to the singular good of the single person. But to exist fully, the person must needs participate. Certainly, to attain fullness depends on my liberty, but the plenitude does not owe its plenitude to my liberty. My free act ought to be ordered to the plenitude that is common. My free act is my singular act; my end is not an end because it is mine.

To the second part of the objection we reply that the proposition *liberum est quod causa sui est* should not be understood in the sense that the free agent is the cause of himself or that he would be, as such, the perfection for the sake of which he acts, but in this sense that he is himself, by his intelligence and will, the cause of his action for the end to which he is ordered. One could also say that he is the cause of himself in the line of final causality insofar as he directs himself to the end to which he is called as an intelligent and free agent, that is to say, in conformity with the principles of his nature. But this end chiefly consists in the common good. The agent will be the more free and noble as he orders himself more perfectly to the common good. One sees thus how the latter is the first principle of our condition of liberty. The free agent puts himself into the condition of a slave if he could not or does not will to act for anything but the singular good of his person. Man does not have the condition of liberty less when, by his own reason and will, he submits to a higher reason and will. It is thus that citizen subjects can act as free men, for the common good.

One might even push farther the first part of the objection. Not only is the free act outside the universe, but every intelligent creature can keep to himself and conceal from every other created intelligence the term of his free thought: only God knows the secrets of the heart. Moreover, every created person can thus constitute a universe of radically independent objects for himself and freely withdraw himself from the order of the universe. Isn't this what constitutes the sovereign perfection of the person? Behold, here is one who regards only the person and in no wise the universe.

We reply that neither the faculty of keeping an object outside the order nor the object thus kept back and taken as such could have the character of an end. Even the secrets of the heart should be conformed and ordered to the common good; they are of the type of pure means; they should always be conformed to the order established by God. Even in our secret cogitation we do not have ourselves as the supreme rule; otherwise the secret thought would be good for the sole reason that it is our singular thought and engages only ourselves. If "the fool says in his heart, there is no God," or if he says, "my singular good is better than any common good," if he withdraws himself from every order, he is in no way protected in his singularity: he will be subjected to the disorder into which he withdrew.

For the rest, the object as object acquires no perfection from the simple fact that it is held secret. Even though one reveals it to another, it would not be for all that illuminating. Not all speech (*locutio*) is enlightening (*illuminatio*). "The revealing of things which depend on the will of the knower cannot be called illumination but only talk; for example, when one person tells another, 'I want to learn this, I want to do that, etc.' The reason for this is that the created will is not light, nor the measure of truth, but it participates in the light; that is why to communicate things which depend on the created will does not, as such, illumine. In fact, it does not pertain to the perfection of my intelligence to know what you want or what you understand, but only what is the truth of the thing."[52] Because the divine will alone is the measure of truth, it is only divine talk that is always enlightening.

Besides, to delight in secret thoughts insofar as they owe to us that secret character is perverse: one takes pleasure then in one's own originality for its own sake instead of ordering it to the greatest good, and thus one enjoys singularity in a disordered manner.

More, if the secrets of the heart escape the order inherent in the universe, they remain within the universal order envisaged with respect to the separate principle. Just as He orders chance and fortune, God can order them to the intrinsic good of the universe.

Common Good and Generic Community

3. The primacy of the common good entails that egalitarian leveling for which one reproaches the personalists: the commonness of this good entails a sort of confusion of persons before their ultimate end: to attain the end would thus be the deed of a body constituted of persons, and not the deed of persons taken as such.

Ad 3. We reply that the commonness of this good ought not be understood as one of predication, but rather as one of causality. The common good is not common as "animal" is with respect to "man" and "beast," but common in the manner of the universal means of knowing which in its unity attains the known in that which is most proper to them. It extends to many, but because of its very elevated determination which extends principally to that which is most elevated in the inferiors: "a higher cause has a more elevated proper effect." It extends to Peter, not first of all insofar as Peter is animal, nor even insofar as he is only a rational nature, but insofar as he is *this* rational nature: it is the good of Peter envisaged in his most proper personality. That is why the common good is also the most intimate link of persons with one another and the noblest.

Common Good and Beatitude

4. The beatitude of the single person does not depend on the communication of this beatitude to many. What is more, one ought to love God in the first place and one's neighbor *ex consequenti*. Therefore, the common character of beatitude is secondary; it is first of all and chiefly the good of the single person.

Ad 4. We reply that if as such the beatitude of the single person does not depend on the actual communication of this beatitude to many, it nonetheless does depend on its essential communicability to many. The reason for this is the superabundance of the good that is beatitude and its incommensurability with the singular good of the person. The sin of the angels consisted in wanting every good to be commensurable with their proper good. Man sins when he wishes the good of intelligence to be commensurable with his private good. Hence, even though only one person enjoyed beatitude he would always have the note of part before this superabundant good: even if in fact he was alone in enjoying it, the single person could never consider the common good as his singular good.

Society Is an Accidental Whole

5. It is said that the good of an accidental whole is less than the good of a substantial whole. But, society is an incidental being and exists only by accident. Therefore, the common good ought to be subordinate to the good of the person.

Ad 5. This difficulty presupposes a false notion of the common good. In fact, the common good does not concern society formally insofar as it is an accidental whole: it is the good of the substantial wholes who are members of society. But it is the good of these substantial wholes only insofar as they are members of society. And, if we consider the intrinsic common good of society as an accidental form, it would in no wise follow that it is inferior to that which is substantial. We are speaking of the good. But, the division of the good is not the same as that of being. "It is by reason of its substantial being that each thing is called being absolutely (*simpliciter*), whereas it is by reason of acts superadded to its substance that a thing is said to be in a certain respect (*secundum quid*). . . . But the good means the perfect, which is desirable and consequently has the note of end. That is why a being that possesses its full perfection is called good absolutely, but the being that does not have the full perfection that it ought to have, although it has a certain perfection insofar as it actually exists, is still not called perfect absolutely nor good absolutely, but only good in a certain respect."[53]

Morever, if to determine the superiority of a good one based oneself on its union with us according to our substance taken absolutely, one would have to conclude that each thing is loved above all things and that the singular good is the measure of the common good. That would presuppose, further, that created persons are above all wholes, absolutes, and that their being a part is secondary. But that is not the case. We are first of all and principally parts of the universe. That is why we love naturally and more the good of the whole.

> Among natural things, every being which is, according to nature and in its very being, for another (*quod secundum naturam hoc ipsum quod est, alterius est*) is principally inclined to that from which it receives its being (*in id cujus est*) rather than to itself. And this natural inclination is evident in things which are naturally made, because, as is said in *Physics* II, every being is born apt to act in the manner that it acts naturally. We see in fact that the part naturally exposes itself for the protection of the whole, as, for example, the hand wards off blows without deliberation for the protection of the whole body.

But here, one might say, is something true of natural acts insofar as they are natural, but it is quite otherwise with acts which are done freely and not by nature alone. But let us read the immediate sequel of the text. "And because reason imitates nature, we find a similar inclination in the political virtues: it is the deed of the virtuous citizen to expose himself to the peril of death in order to save the republic; and if man were a natural part of the city this tendency would be natural."[54] Because the human person is in his very being of another he is radically dependent, he is radically, *primo et per se*, a part. Consequently, he is principally and more inclined toward that from which he participates his very being.

It is this principle, observed first of all in nature and in political virtues imitating nature, which serves as basis for the conclusion that we love God according to natural love more than ourselves. "[T]he nature and substance of the part, because of what it is, is above all and essentially for the whole and the being of the whole. That this is so of every creature envisaged with respect to God is evident. For every creature is, according to its nature, a natural part of the universe, and because of that, naturally loves the universe more than itself. . . . Therefore, *a fortiori*, he will rather love the universal good itself, because the entire universe is more eminent both because it is all good and because the universal good itself, which is the glorious God, is the end and good of this universe, and, consequently, he who loves the universe more loves God more; as we see in the case of the army and its leader, according to the doctrine of Book 12 of the *Metaphysics*, c. 10."[55]

One could, basing himself on the Philosopher (IX Ethics, cc. 4 and 8), push the objection farther. "The testimonies of friendship that one gives to others are not those testimonies of friendship one gives to oneself." To this objection, St. Thomas replies "that the Philosopher speaks here of the testimonials of friendship shown another in whom is found the good that is the object of friendship in a certain manner; he does not speak of the testimonials of friendship shown to one in whom is found the good under the note of good of the whole."[56] That is why, in the political order, all civic friendship anterior to the common good is a principle of corruption: it is a conspiracy against the common good, as one sees in politicians who favor their friends precisely under the pretext of civic friendship.

Moreover, if, according to natural love, every being loved its proper good more, and the common good as its singular good, charity could not perfect natural love, but would be contrary to it and destroy it.[57]

The Speculative Life and Solitude

6. The practical order is in its entirety ordered to the speculative. But the perfect good consists in the speculative life. But the speculative life is solitary. Therefore, the practical happiness of society is ordered to the speculative happiness of the single person.

Ad 6. We reply that the practical happiness of the community is not, in itself, ordered to the speculative happiness of the single person, but to the speculative happiness of the person insofar as he is a member of the community.[58] It would in fact be contradictory if a common good were as such ordered to the single person as such. It is true enough that the speculative life is solitary, but it is also true that sovereign beatitude which consists in the vision of God is essentially a common good. The apparent opposition between the solitary life and the common good which is the object of this life is explained by the fact that happiness can be considered either on the side of those who enjoy it or on the side of the very object of this happiness. This object is, in itself, communicable to many. Under this aspect, it is the speculative good of the community. The practical common good should be ordered to this speculative good which extends as a common good to persons. The independence of persons one from another in the vision does not exclude this universality from the object, which means, for every created intelligence, essential communicability to many. Far from excluding it, or abstracting from it, the independence presupposes this communicability.

The Good of Grace and the Good of the Universe

7. One might also object that "the good of grace of one person is greater than the natural good of the whole universe,"[59] in order to conclude that the intrinsic common good of the universe envisaged in its nature is subordinate to the good of the single person.

Ad 7. This objection is based on a transgression of genera which permits only an accidental comparison. But it should be noted that St. Thomas does not oppose the good of grace of one person to the good of grace of the community, but to the good of the nature of the universe. And, if the spiritual good of the person is higher than any created common good, and if, according to this spiritual good, the person ought to love himself more, it in no wise follows that the created common good, as such, is subordinate to the single person. Once again, the spiritual good of man involves an essential relation to the separated common good and, in this order, man has more reason of a part than anywhere else. The supernatural good of the single person is essentially ordered to the supernatural common good, so much so that one could not distinguish between the supernatural virtue of a man and the supernatural virtue of the man as part of the celestial city.

Society and the Image of God

8. The single person is in the image of God. But no society is properly in the image of God.[60] Therefore, the single person is purely and simply superior to any society.

Ad 8. Like the preceding objections, this one assumes as given the collectivists' interpretation of our conception of society. But society is not an entity separable from its members: it is constituted of persons who are the image of God. It is this society constituted of persons, and not some quasi-abstract entity, that is the principal intention of God. That its members are in the image of God is a sign of the perfection of their ensemble. Why would He have made many and ordered persons? Does not the divine good shine forth more brilliantly in a multitude and an order of rational creatures than in a single person as such? Is not truth better communicated in the contemplative life of the community than in the contemplative life of a single person? Does not beatitude have the note of a common principle? And does the love that bears on that object bear on the universal good as such or on that good as appropriated by a single person? Is this good like an inferior common good whose distribution entails, by way of consequence, a division from itself and a particularization where it is of the part as such and loses its note of community?

Let us recall again that the common good is called common because of its superabundance and its incommensurability with the singular good. The properly divine good is so great that it could not be the proper good even of the entirety of creation: the latter always retains in some fashion the note of part. It is true enough that before the common good the single person can call it *mine,* but it is not for all that appropriated to the single person. The good that he calls mine is not in itself taken as end. If that were the case, the good that is the person himself would be the end for which he wills. "When it is said that the angel loves God insofar as God is a good for himself, if 'insofar as' signifies the end, this proposition is false; in fact the angel does not love God naturally for his own good but for God Himself. But if 'insofar as' signifies the reason for loving on the side of the one loving, then the proposition is true. Indeed, it could not be the nature of anything to love God except because he depends on the good God is."[61]

Man as Whole and Society

9. "[M]an is not ordered to political society according to all of himself and with regard to everything that he is."[62]

Ad 9. Some have wished to conclude from this isolated text that political society is in the last instance subordinated to the single person as such. And whoever dares contradict this crude inference turned in favor of personalism is considered totalitarian. But, as we have seen, it is contrary to the very nature of the common good to be, as such, subordinated to a singular unless that singular has itself the note of common good. Saint Thomas means to say only that man is not ordered only to political society. He is not according to all that he is part of political society since the common good of that society is only a subordinate common good. Man is ordered to this society only insofar as he is a citizen. Although man, the individual, member of a family, civil citizen, celestial citizen, etc. are the same subject, they are formally different. Totalitarianism identifies the formality 'man' with the formality 'citizen'. For us, on the contrary, not only are these formalities distinct, they are subordinated one to another according to the order of goods. But it is this order of goods, first and final causes, and not man purely as man that is the principle of the order of these formalities in the same subject. Personalism reverses the order of goods and accords the greatest good to the lowest formality of man. What the personalists understand by person is, in truth, what we understand by pure individual, quite material and substantial and enclosed in itself, and they reduce rational nature to sensible nature which has the private good as its object.

Man cannot be subordinated to the good of political society alone; he should order himself to the good of the perfectly universal whole to which every lower common good should be expressly ordered. The common good of political society should be expressly ordered to God, both by the chief citizen and by the citizen who is a part, each in his own way. This common good requires, of itself, this ordination. Without this express and public ordination, society degenerates into the State, rooted and enclosed in itself.

The City Is for the Sake of Man

10. "The city exists for man, man does not exist for the city."[63]

Ad 10. In order to turn this text into an objection to our position, it would have to be translated as "The common good of the city exists for the private good of man." We would then be able to quote the immediately following passage in this same encyclical: "This does not mean to say, as liberal individualism would have it, that society is subordinate to the egoist utility of the individual."

The city exists for man. This can be understood in two ways. First, the city, when we consider it as an organization with a view to the common good, should be entirely subject to that good insofar as it is common. Considered in this way, it has no other reason for being than the common good. But the common good is itself for the members of society, not for their private good as such: it is for the members insofar as it is a common good. And, since it is a question of a common good of rational creatures, it should be conformed to reason, it should treat rational creatures as rational. The city is not, and cannot be for itself [*pour soi*], rooted and enclosed in itself, opposed as a singular to other singulars: its good should be identical to the good of its members. If the common good were the good of the city insofar as the city is a sort of individual, it would then be a particular good properly alien to the members of the society. It would even accord to the organization the intelligence and will stolen from its members. The city would then be like an anonymous tyrant which subjects man to itself. Man would be for the city. This good would be neither common nor the good of rational natures.

Second, the city, as a common good, is for man insofar as he comprises formalities which order him to higher common goods, formalities which are in man higher than that which orders him to the common good of the city. But the identity of the subject of these diverse formalities can invite confusion. The private good and the common good are both goods of man. And yet, not every good of man is the good of man purely as man. The good of man purely as man, according to the meaning Saint Thomas gives in the texts already cited, is nothing other than the good which belongs to him insofar as he is an individual. The common good can never be subordinate to this man purely as man. The formality 'man purely as man' cannot be identified with the formality 'citizen,' any more than it can be with the subject 'man.' Hence, when we say a common good is subordinate to man, this can only be by reason of a formality which looks to a higher common good. It is only the most perfect common good that cannot be subordinated to man.

Moreover, when we say that the common good can never be considered as a pure extension of the good of man in the line of his singular good such that the common good would be only a detour to rejoin the singular good, we do not thereby mean that the singular good is contemptible, that it is nothing, that it ought not be respected or that it is not in itself respectable. However, a higher respect is due the person when we consider him in his ordination to the common good. Even the singular good of the person is better when we regard it as ordered to the common good of the person. For the rest, a city which does not respect the private good or the good of families acts against the common good. Just as intelligence depends on well-disposed senses, so the good of the city depends on the integrity of the family and its members. And just as a sensible nature well subject to reason is more perfect in the very line of sensible nature, so in a well-ordered city, the singular good of the individual and the common good of the family ought to be most perfectly realized and protected. Moreover, if the common good of the city were subordinate to these, it would no longer be their common good and man would be deprived of his greatest temporal good: the city would not be a city. It would be like an intelligence subordinated to sense and reduced to the condition of instrument of the private good.[64]

Most of these objections involve a transgression of genera, they exploit the *per accidens*. From the fact that some private good is better than some common good, as is the case with virginity which is better than marriage, one concludes that any private good taken as such is better than any common good taken as common good, that the private good as such can have an eminence which escapes the common good as such, that one can then prefer a private good to a common good because it is private. What could be easier then than to deny all first principles?

Thus one wants to destroy a *per se nota* proposition which follows on what common good means.

Personalism and Totalitarianism

As Jacques de Monleon once said to me, "Notice how the self-acclaimed personalists who put the person above the common good can no longer see in this the link between persons. Thus they replace this link with another, by a sordid fraternity that would immediately unite persons among themselves, as if each person were a common good for all the others. This results in each citizen being a tyrant, *amans seipsum magis quam civitatem.*" That was indeed the ideal of Karl Marx. In the final phase of communism, each individual person will be substituted for the common good, he will have appropriated to himself "his essence of multiple aspects in a multiple fashion, that is, as a complete man"; the individual man "will become a generic being"; each individual himself will have become the good of his species: "this will be the true end of the quarrel . . . between the individual and the species." The common good will no longer be distinct from the singular good, the individual will have become, himself, the first principle of the social order and of all political power; as generic "he will have recognized his own social powers and have organized them himself as such . . . he will no longer separate from himself social power under the form of political power."[65] But this "integral development of the individual" cannot be accomplished without the complicity of the confused mass: the self cannot alone establish the totalitarianism of the self. There must be a fraternity of men, born of self-love and the need for anonymous power, blind and violent, for the realization of the self who is himself his proper end; a fraternity very logical in its cynicism. The obstacle that would be the person of the other who also acts only for himself is vanquished by the confusion of him with the indistinct mass. By this ruse, each person can enslave all the others without anyone enslaving himself.

By their false notion of the common good, the personalists are, at bottom, in agreement with those whose errors they pretend to combat. To individualism, they oppose and recommend the generosity of the person and a fraternity outside any common good, as if the common good had its principle in the generosity of persons, as if it were not above all that for the sake of which persons ought to act. To totalitarianism, they oppose the superiority of the whole-person and a common good reduced to the status of the particular good of persons. Their protest is not made in the name of the person as citizen, but of the citizen as person, as if the person were not greater as ordered to the common good than as ordered to his personal good.

In fact, personalism makes its own the totalitarian notion of the State. Under totalitarian regimes, the common good is singularized and is opposed as a more powerful singular to singulars purely and simply subject. The common good has lost its distinctive note, it becomes an alien good. It has been subordinated to that monster of modern invention that is called the State, not the state taken as synonymous with civil society or the city, but the State which means a city elevated into a sort of physical person. Let us note, indeed, that the person, "an individual substance of a rational nature," can be applied to civil society only by metaphor and not by analogy.[66] In this reduction of the moral person to a physical personality, the city loses its note of community. What is due to the common good is converted into what is due to the singular good, a singular which orders everything to itself. Legal justice is destroyed. Being turned away from the common good, the State acquires the status of the personalist person. It loses all ordination to a higher common good "such that one takes as end the notion of the common state, which is the ruin of a well-ordered republic."[67]

This kind of State arises either when the chief, in his quality of person-member of the society, appropriates the common good as his own or when the moral personality of the society is elevated into a physical person. In the two cases, the State is a power alien to individuals, a power of alienation against which they must constantly defend themselves. This totalitarian conception establishes a tension between the person and society, an inevitable conflict, a contest in which certain sociologists pretend to find a principle of fecundity. Society is openly totalitarian when the State thus acquires liberty by victory over individuals; it is openly individualist when the individuals dominate the State. But in the one case as in the other, the city is personalist and totalitarian.

In short, the State, taken in this sense, that is to say a city rooted and closed on itself, is by nature tyrannical. In the condition of liberty of this State, obedience is substituted for the legal justice of the citizen-subjects.[68]

The State absorbs the citizen and substitutes for him an abstract citizen and an abstract liberty.[69]

The totalitarian State, founded on the negation of the common good and elevated into a person for itself, cannot be ordered to a higher common good, cannot be referred to God. The negation of the very notion of the common good and of its primacy is a negation of God. By denying the universality of the end to which man is ordered, one denies the dignity that man derives from this ordination, one leaves him only the inalienable personality that a man can also take with him to hell, *ubi nullus ordo.* Even Marxists can applaud this invincible soul.

Is not society corrupted in its very root when those who have charge of the common good do not order it explicitly to God? Why indeed do we require that, in principle and as an essential condition, the directors of society should be purely and simply good men? How could one admit that a bad man can be a good politician? Certainly it was not only in the past that subjects have been governed by bad men, to whom one owes nonetheless obedience in matters which are relevant to their authority.[70] What is new is the manner of accepting and defending them. If a politician ought truly to possess all the moral virtues as well as prudence, is it not because he is at the head and that he ought to judge and order all things to the common good of political society and that to God? Isn't it for that reason that, according to Cajetan and John of St. Thomas, the legal justice of the prince is more perfect than the virtue of religion?[71] Without doubt, the reasons for which we ignore the common good are the same that make us forget political prudence. "We have been mistaken for too long about the role of intelligence. We have neglected the substance of man. We have believed that the virtuosity of base souls could bring about the triumph of noble causes, that clever egoism could exalt the spirit of sacrifice, that dryness of heart could, by gusty discourse, bring about the fraternity of love."[72]

Intelligence has succumbed to sense, to sense riveted on the singular good. The conflict which exists between man and society does not come from the perfection of the person nor from the common good to which the person is thought to have a relation of contrariety; it comes properly from the sensible part of man, from the revolt of this lower part of man against the good of intelligence. As for what concerns the intelligence as such, the ordination to the common good is so natural that a pure intelligence could not turn away from it in a state of pure nature. In fact, the angel, elevated to the supernatural order, turned away from the most divine common good there is, namely, supernatural beatitude, and it was only as a consequence that he lost his natural common good. The fallen angel ignores, with a practical ignorance (*ignorantia electionis*), the common good of grace. As for us, we are now ignoring every common good, even speculatively.[73] The common good, and not the person and liberty, being the very principle of all law, of all right, of all justice, and of all liberty, a speculative error on this subject entails fatally the most execrable practical consequences.

PART TWO

The Principle of the New Order

This is not the wisdom that descends from above.
It is earthly, sensual, devilish.
—James 3:15

But the good angels, in knowing a creature, do not fix on it, since this would be to darken and become night, but refer it to the praise of God, in whom as in their cause they know all things.
—Ia, q. 85, a. 5, ad 2

And the angel, in knowing himself, does not dwell on himself, as if enjoying and placing his end in himself—for thus would night come to be, as with the angels who sinned—but refers his knowledge to the praise of God.
—*Q.D. de veritate,* q. 8, a. 16, a. 6

The program lists "Philosophy and Order in International Relations" as the subject of this paper.[74] Actually I was asked to present to you as a subject for discussion the problem of "Metaphysics and International Order."

Although the problem of international order properly concerns political science and prudence—of the speculative sciences, philosophy of nature would be closer to it than metaphysics—it is significant that the most radical philosophy of international revolution has always been careful to point out that traditional metaphysics is its absolute contrary.[75] The philosophy of revolution has well understood that metaphysics takes upon itself the defense of first principles and that it is the science which leads us most properly to things naturally more noble and more divine than man. And this clearly has implications for a world order among nations. The aim of revolutionary philosophy, on the contrary, is not international order in the strict sense of the word, but rather a universal order in which man in a state of privation, both material and spiritual, is the radical and necessary principle. The present discussion will be confined to this fundamental, although restricted, aspect of the problem.

Negation of the Primacy of the Speculative

Modern philosophers, despite their apparent differences, generally agree that metaphysics or speculative wisdom, inasmuch as it chiefly considers things better than man, estranges man from his true self and power and consequently is destructive of human nature and among the great enemies of mankind. Because it is in its way superhuman, it is thought anti-human, distracting man from the total effort in which he should be engaged to conquer the earth and respond to his desire to live.[76] It is destructive of human nature and must consequently be counted among mankind's greatest enemies.

And, indeed, as Aristotle says in the *Ethics,* if man were the best thing in the universe, political science and prudence, not speculative wisdom, would be the best knowledge.[77] Let us, then, consider the hypothesis that political science and prudence are the best knowledge and see what follows rigorously from it.

The first and more general consequence is that things would be merely what we should wish them to be. Political science and prudence are practical in that they direct toward an end, according to right reason. But this presupposes that we know the nature of that which is to be directed and of the end, in other words, the rectitude of practical regulation presupposes the rectification of the speculative intellect.[78] Therefore, if, *per impossibile,* practical regulation were independent of speculative truth, then what things are or should be, such as man, society, and the human good, would be merely what we wished them to be. Practical science would no longer even be a science. Simple practical knowledge would no longer be truly practical. All direction would be by chance and would no longer be direction.

The hypothesis also implies specifically the negation of prudence itself. One might argue that, since in practical things the end is the principle, and since the artisan chooses the end he desires to realize—this kind of house, and this one rather than that one—we are free to choose the end. But art and prudence differ in this respect, for prudence does not choose the end but only the means. If prudence chose the end, like art it could not choose the means, and thus it would be one with art. And if this were so, the truth of the prudential judgment would not depend upon the rectitude of the appetite with respect to the good but upon the intellect only, that is, upon its conformity with the chosen end.[79] And since art regards truth only, and not like prudence both the true and the good, the judgment of the morally corrupt could be as sound as that of the virtuous man (which is commonly accepted in the practice of politics), and all defects in moral action would be due to defects in knowledge only. Furthermore, since art is about contraries, e.g., health and disease in medicine, if prudence were like art in this respect prudence would be indifferent to good and evil.[80] Right and wrong would be determined only by success in accomplishing a chosen end. It would always be absurd to defend oneself by saying that one acted according to conscience and with good intention. Every concrete lapse from the chosen end, whether due to chance or will, would be a fault.

On this hypothesis, then, man would be the measure of all things, and there could be no other measure. But the proposition that man is the measure of all things remains abstract. To be consistent, we must ask, "Which man?" or "Which men?" We cannot ask "Which man or which men have the right to impose their measure?" He will be right who has the power to impose it. To be consistent, we can only wait to see what happens.

This would be the emancipation of man as pure *artifex,* and this emancipation answers to a quite characteristic desire of man. There is in man a tendency to grant primacy to the practical over the speculative, to art over prudence. And this tendency arises from man's intellectual weakness, as the following arguments show.

Aristotle says toward the beginning of the *Metaphysics,*[81] that "the possession of [wisdom] might be justly regarded as beyond human nature, for in many ways human nature is in bondage." The contemplative life is not properly human, but rather superhuman, whereas the active life is most proportionate to human nature.[82] Man's best part, the speculative, is the feeblest in him. Will he submit himself to the pitiless demands of the object of that part of him which is at once most noble and weakest?

In speculative knowledge, the intellect is measured by the object and in speculative wisdom we are mainly concerned with things better than we are. In practical knowledge, insofar as it is practical, the intellect itself is the measure, and we ourselves are in a way the end of all works of art.[83] It is impossible to think of the objects of the speculative without at the same time feeling our condition of inferiority, both as to our nature and as to our mode of knowing.[84]

We might be inclined to prefer art to prudence because truth in art is not conditioned by the conformity of appetite with the good but only with the chosen work, whether it be good or bad. And the end of art is this particular work, this machine, this statue; whereas the goodness of this prudential act depends upon the conformity of this act with the good life as a whole.[85]

Furthermore, art, because it imitates nature, succeeds for the most part, and the artisan does not have to deliberate about the means, but in acts which depend upon the conformity of appetite with the good, we fail for the most part.[86] And the reason for this is man's dual nature, and the contrariety of sense and reason.[87] St. Thomas explains why this contrariety entails that human action is for the most part evil, for a reason which may be drawn from the very nature of man. Man is not perfected by nature: his "second perfections" are not inborn, but either acquired or infused. Until it is perfected by virtue, human nature is not determined *ad unum* and risks going wrong more often than not.[88]

It is always because of the weakness of his speculative intelligence that man will be tempted to exalt his capacity to construct delightful imitations; he will be tempted to dominate all the imitable originals, those above us as well as those inferior to us. The fine arts, in fact, are the most human means of making objects better than ourselves proportioned to us.

In the Beginning, the Word of Man

The history of modern philosophy has actually worked out the various conclusions we have deduced from the hypothesis that man is the best thing in the universe. I would now like to show very briefly that by progressively ignoring and denying those things which are better than man, and consequently wisdom itself, modern thought simply relinquished what is better in man himself: actually it has endowed that which is both spiritually and materially most inferior in man with quasi-divine attributes.

The *Encyclopedia Britannica* defines humanism, as

> in general any system of thought or action which assigns a predominant interest to the affairs of men as compared with the supernatural or the abstract (from Lat. *Humanus*, human, connected with *homo*, mankind). The term is specially applied to that movement of thought which in western Europe in the 15th century broke through the medieval traditions of scholastic theology and philosophy, and devoted itself to the rediscovery and direct study of the ancient classics. This movement was essentially a revolt against intellectual, and especially ecclesiastical authority, and is the parent of modern developments whether intellectual, scientific or social.[89]

We would never be able to subscribe to this tentative definition of humanism if it were thought to be applicable to all those called humanists. When Saint Robert Bellarmine and Saint Peter Canisius are called humanists, I do not think the term has the same meaning as applied to Pico della Mirandola,[90] Erasmus, or Rabelais. The humanism of these latter implies a humanist conception of man.[91] And it should also be said that, in a Rabelais, contrary to what is customarily said, this humanism is far more an attitude than a doctrine.

Let us look at a text that we call humanist in the philosophical sense of the word—and that is the sense we will have in mind from now on. It is drawn from the *Discourse* on the dignity of man by Pico della Mirandola.

> Finally, the best of workmen decided that that to which nothing of its very own would be given should be, in compositive fashion, whatsoever had belonged individually to each and every thing. Therefore He took up man, a work of indeterminate form; and, placing him at the midpoint of the world, He spoke to him as follows:
>
> "We have given to thee, Adam, no fixed seat, no form of thy very own, no gift peculiarly thine, that thou mayest feel as thine own, have as thine own, possess as thine own the seat, the form, the gifts which thou thyself shalt desire. A limited nature in other creatures is confined within the laws written down by Us. In conformity with thy free judgment, in whose hands I have placed thee, thou art confined by no bounds;

> and thou wilt fix limits of nature for thyself. I have placed thee at the center of the world, that from there thou mayest more conveniently look around and see whatsoever is in the world. Neither heavenly nor earthly, neither mortal nor immortal have We made thee. Thou, like a judge appointed for being honorable, art the molder and maker of thyself; thou mayest sculpt thyself into whatever shape thou dost prefer. Thou canst grow downward into the lower natures which are brutes. Thou canst again grow upward from thy soul's reason into the highest natures which are divine."
>
> O great liberality of God the Father! O great and wonderful happiness of man! It is given to him to have that which he chooses and to be that which he wills. As soon as brutes are born, they bring with them, "from their dam's bag," as Lucilius says, what they are going to possess. Highest spirits have been, either from the beginning or soon after, that which they are going to be throughout eternity.[92]

We will not analyze this text in detail, but notice its insistence on formlessness. It is true that because of his capacity to have the form of another and to be other things thanks to knowledge, man is at the center of the cosmos, whereas other cosmic creatures are limited, either to their individual form or to sensible singular forms. Moreover, when we consider formally this formlessness, this unlimited potentiality, we grasp rational nature in its characteristic non-being and, far from occupying the center of creation because of that, man is then the lowest rung in the hierarchy of intelligent creatures.

For all that, Mirandola does not consider this formlessness only in the line of knowledge; this formlessness will be most admirable because it can expand the field of liberty. It is not a question of liberty of understanding, but "deciding for yourself the limits of your nature in accord with the free will which is proper to you"; it is a question of a faculty to establish its own rule of conduct and to direct itself pushed to the point of making it equivalent to a sharing in the knowledge of good and evil.

There is here an exaltation of formlessness, of the indetermination proper to man's rational nature, that will reach fruition in the idealism of Hegel and, in a more acute manner, in the materialism of Feuerbach and Marx.

To attribute man's perfection to this formlessness itself and to the subjective power to activate it is to make material and efficient causes primary.

The desire to experience in a very tangible way the infinity of this power as the first principle and his most proper activity pushes man to the adoration of the infinite in his hands and in his language, the organs of practical reason. The infinite which underwrites technical progress that the *homo faber* of today erects into an end then becomes a horrible thing. Looked at as such, this infinite when made into an end will become an object of hellish despair.

Liberty of contrariety before the natural end also bears the mark of a properly human imperfection. It can be thought a perfection only by comparison with beings without will. It cannot exist in a perfect intellectual nature. Would man then be the masterpiece of creation because he can go wrong and deviate even from his purely natural end? Because he is composed of contrary natures? That is, because he is defective in the very notion of an intelligence and free nature?

Since he can accept or reject his end, since he can direct himself to his natural end, is not man more of a *causa sui* than an intellectual nature created in the possession of his end? That is the sophism that underlies the rhetoric of della Mirandola.

This perversion is properly human. The fallen angel was satisfied beyond measure with the actual perfection that was his in virtue of his creation. Man, by contrast, is here satisfied with his potentiality in a disordered manner, with the fact that he was not established in possession of his end. I say "in a disordered manner" since man can rejoice that he was not fixed by nature in the manner of irrational creatures. But it is forbidden for him to look backward—*Nemo respiciens retro, aptus. . . .*

The exaltation of this poetic activity in which he makes for himself objects, imitations which have a terminal note in the line of knowledge and suffice unto themselves, was a deliberate return to the time when divinities were, in all their concrete determinations, works of man; when these divinities were in large measure in the image of man, subjected to human conditions, and on which the poet could exercise his dominion. It was not a return to classical art considered in all its amplitude; that was, in many respects, truly religious, that is, subject to originals recognized as superior. Rather it was a deliberate return to classical poetry insofar as it could be profane even when confronted with divine originals. One wants, in short, a profane poetry whose realm is universal, religious at the most by extrinsic denomination. One wants the emancipation of pure poetry "which has for its objects things which, because of their defect of truth, cannot be grasped by reason."[93] Every imitable original must be reduced, when confronted by human genius, to the condition of operable matter. But this is in effect to accord primacy to the *infima doctrina*.[94]

Descartes speaks expressly of this philosophy which would have as its end not primarily knowledge for its own sake, but the transformation of all things for the benefit of man. Marx is the faithful echo of this passage from the *Discourse on Method* (Part VI): ". . . in place of that speculative philosophy taught in the Schools one should find a practical one by which, knowing the force of the activities of fire, water, air, the stars, the heavens and all other bodies which surround us as distinctly as we know the different metiers of our artisans, we would in the same way be able to use them for all uses proper to them, and thus we would make ourselves the masters and possessors of nature. This is to be desired not only for the discovery of an infinity of artifacts which would enable us to enjoy painlessly the fruits of the earth and all the commodities found there, but chiefly for the preservation of health. . . ."[95]

In order to grasp the full import of this text, recall what Descartes had to say on the subject of theology. "I revere our theology and hope as much as anyone to gain heaven, but, having learned as a most assured thing that the road is no less open to the most ignorant as to the learned, and that revealed truths are above our understanding, I did not dare to subject them to the weakness of my reasoning, and I think that, in order to undertake to examine them, one would need an extraordinary assistance from Heaven and to be more than a man" (Part I).

Even speculative philosophy is too difficult, too uncertain, or insufficiently adjusted to the level of reason. What then is left to us if not that practical philosophy which ends moreover by setting aside morals and substituting medicine and hygiene for the cure and prevention of all spiritual evils.

Hume's skepticism serves even more to ground the negations which condition a philosophy fully and openly humanist. The following passage from Hume's *Enquiry Concerning Human Understanding* is a kind of premonition of his skepticism: "It seems, then, that nature has pointed out a mixed kind of life as most suitable to the human race, and secretly admonished them to allow none of these biases to *draw* too much, so as to incapacitate them for other occupations and entertainments. Indulge your passion for science, says she, but let your science be human, and such as may have a direct reference to action and society. Abstruse thought and profound researches I prohibit, and will severely punish, by the pensive melancholy which they introduce, by the endless uncertainty in which they involve you, and by the cold reception which your pretended discoveries shall meet with, when communicated. Be a philosopher; but, amidst all your philosophy, be still a man."[96]

Man turns away from research into and contemplation of things better than man,which is to say that he turns away as well from that which is best in himself. He falls back on the powers that are most properly his. Among these powers there is one, in a way the most profound, which touches on all the principles absolutely first for us, the power of properly human language. One can say and write things that one cannot think. One can say, "It is possible to be and not to be at the same time and in the same respect," or "The part is greater than the whole," even though one cannot think them. And yet, these phrases are grammatically correct. The transcendent power of language: one can utter both the thinkable and the unthinkable. A power to use the purely irrational. I can say, "I do not exist." And, lo, I can ground my "I exist" on pure non-being. I say it. Who can stop me? Let him try. I say it again. The self. Selves. Soon a society of selves.

The liberty of speech is discovered, speech freed from intelligence. The "external utterance" is emancipated. Thought is made subject to language. In the beginning, the word of man.

"I say unto you, on the day of judgment, men will give an account of every vain word that they have spoken. For you will be justified by your words and you will be condemned by your words."

We should not forget the gradual emancipation of the historian who finds in the critical method a substitute for prudence that allows him to judge historical events objectively no matter what his subjective disposition may be. Henceforth, the historian need no longer be a prudent man whose judgments of human actions would be conditioned not by knowledge alone, but by the rectitude of his own appetite—*qualis unusquisque est, talis ei finis videtur*—it will be enough to apply the scientific method which, it is said, is wholly impersonal. We are in the end freed from this terrible text: "As you have judged others, so you will be judged, and in the same way you have measured others you will be measured." The truth permits adulterous men to cry out in a public place, this woman has been caught committing adultery *in flagrante delictu.* Does the beam in your eye prevent you from seeing the speck in another's? Is this not a case of impersonal truth? Is not such truth the right of everyone? Why shouldn't the historian be as free as the physicist? Facts are facts.

But the fear of God. . . .

The attitude of philosophers toward their reader has completely changed. Now it is not so much the truth that they express; rather, the reader and writer become the principal objects of preoccupation. Philosophical writers always hope, for their own advantage, they say, that the reader will approve their opinions. Even more important, the reader for whom they write is no longer the philosopher but rather that vague individual sometimes called the man of good sense, sometimes the cultivated man, sometimes the reader in general. Compare this procedure with that of Aristotle or Saint Thomas. The *Discourse on Method* is essentially a rhetorical work. It was also one of the first appeals to the very formlessness of the unformed man which will burst forth one day in an appeal to the unformed mass precisely as unformed.

Philosophical works take on a form that puts them more and more outside the possibility of refutation. They are rooted in attitudes. More and more, philosophy becomes an expression of the personality of philosophers. It becomes a literary activity. And who can refute a poem? Who can refute the thought of an author?

Have philosophers truly become more critical? The critical spirit is one of the greatest delusions of history. Never before have philosophers postulated more evidences and things "supposed to be known." There can be no modern philosopher who has succeeded better in getting accepted impossible evidences, carefully couched in intuitions and conceded as supposedly known, than the austere critic of Königsberg.

Nonetheless, beneath the infinite diversity of systems there is hidden a profound unity which will one day come to light in Marxism—unity of purpose, of final cause: the emancipation of man purely as man envisaged in his formlessness taken as a sufficient principle for all that man can be, the power of his powerlessness, the fecundity of the non-being of man.

Kant's efforts to free the speculative intellect from the fetters of metaphysics by confining it to the logical order[97] was the most decisive step toward that philosophy of revolution—the "armed critique"—which today openly menaces the entire world. Perhaps we ourselves, succumbing to this very tradition, have lost faith in the human intellect to such a degree that we are reluctant to admit that what men think could be of consequence.[98]

Given the kind of emancipation of the human intellect Kant had in mind, the choice of logic as the instrument of emancipation was most appropriate. The power of its abuse can be shown from what we ourselves hold on the nature of logic. The necessity of logic flows from the natural imperfection of our intellect which is originally in a state of potentiality.[99] In this sense it is profoundly and properly human. It is nevertheless the most perfect of arts; its matter is necessary; it is both a speculative art and a speculative science—both speculative and directive; both instrumental and transcendent. Remaining entirely within the speculative intellect, it is the most liberal of arts, but at the same time the most subservient: it is for use only.

This very art which has its root in the potentiality of our intellect will become the omnipotent method of Hegel: "Method is the absolute, unique, supreme, infinite force, which no object can resist; it is the tendency of reason to find itself again, it recognizes itself in all things." All things are now in the image of our mind, and logical reason has become the principle which posits all things.[100]

And in turn Hegel will lay stress on that part of logic which by its very nature can serve his purpose most appropriately—dialectic, not merely the dialectic of the *Topics,* but the dialectic which consists in using principles of *logica docens* to reach reality.[101] However this use of logic could not of itself adequately attain reality unless the logical and the real were identical, and this could not be unless contradiction were possible. But this is precisely what Hegel did maintain. Contradiction is to him a simple fact which he illustrates with an example drawn from geometry. "A notion, which possesses neither or both of two mutually contradictory marks, e.g. a quadrangular circle and a rectilinear arc, no less contradicts this maxim, geometers never hesitate to treat the circle as a polygon with rectilinear sides."[102] Now the principle of contradiction is a rather important matter. And it is most relevant to our subject since the negation of it is the first principle of the modern philosophy of revolution.[103] Marx, Engels, and Lenin worried much about the disrespect and neglect of Hegelian logic in their disciples, and orthodox Marxism has continued to emphasize its importance. So let us see briefly in Aristotelian terms what Hegel does to get around contradiction and to proclaim it the very principle of fecundity.

A remote genus is predicable of the species with identity. Thus the circle and the polygon are the same figure. This predication with identity is possible because the remote genus is not divided by the species.[104] But Hegel identifies the properties of the remote genus with those of the proximate genus. It then follows that the circle and the polygon are the same plane figure, i.e., a plane figure is identical with the differences that divide it. This procedure might seem plausible because the circle may be defined dialectically as the limit of a regular inscribed polygon whose sides increase indefinitely in number, in which there is an apparent tendency of one species to pass continuously into the other, through a purely quantitative change.

In this way we can see how Hegel's "dialectic of speculative reason," from the confusion proper to the logical order, from the genus as potential, pretends to derive all things in their difference instead of merely using this logical device, in which one species is considered the limit of the other, as a human means of imitating *in obliquo* divine wisdom which attains all diversity in one without negating the diversity. Thus the logic of contradiction and negation becomes infinitely fruitful. It is not that we would deny this dialectical process. We only ask that it be recognized as only dialectical. The process is legitimate and fecund provided that one sees it as a purely logical expedient to surmount in a tentative way the multiplicity of our means of knowing where our knowledge is defective given the very notion of wisdom.

It is true that the dialectical reduction of volume to surface and of surface to line and of line to point makes our knowledge more perfect and more like the divine knowledge which in a unique species, in the universal means of knowing, knows all things in what is most proper to them. We understand human intelligence better when we see it as the limit of a falling away in the very nature of intelligence. But, lest we destroy the very term of this reduction, we must take into account that it is purely dialectical, that the movement impressed on things is only a movement of reason projected onto the objects, and that this reduction remains in a state of tending. This movement does not aim at the reduction of the natures known: that reduction is had in strictly scientific knowledge where one nature is known as the reason for another, the one and the other remaining radically distinct. Rather it aims at the reduction of the means of knowing, a reduction that can only be tentative. If it were completed it would result in the destruction of the natures we want to grasp in their differences. Victim of an emancipated language, Hegel thought to engender in this way a new and richer object—the square circle, for example.

Is it then only a scholastic subtlety, one of those precisions of the school, that separates us from these dialecticians? Perhaps. But let us not underestimate the precisions of the school. Hegel here abuses one of the most powerful instruments of metaphysics in order to imitate the divine wisdom. The same is true of that other instrument—also purely human—the negation of the negation, the fecundity of which is manifest in mystical theology.[105] Here then is the movement of pure reason, of reason considered formally in its pure ratiocination, and the negation, another characteristic of human intelligence, becomes perfectly emancipated and takes on attributes which are properly divine.

Et facta est nox

This perversion of human thought at its very root was to bear its fruits in Marxism, which, not content to consider this process as a philosophical game, will put it into practice even down to "Herr Krug's pen."[106] To be sure, the Hegelian dialectic was already of a fundamentally compositive and practical mode, but it remained practically sterile. Marxism will identify Hegel's dialectical process with things envisaged in their ultimate concretion. But among the things which surround us, it is in fact matter which is the proper principle of their ultimate concretion. Matter itself will become the primordial principle, the "first reason." Do you think you are governed by a perfect intelligence and an infinitely good will? You are exclusively determined by material conditions. Finality? Scholasticism!

And just as in Hegel the movement of reason arises from the contradiction inherent in being, so for the Marxists the contradiction of matter will become clear in the movement of matter, itself perfectly contradictory, they think: contradiction and the movement of contradiction from which all things are born. In contradiction, that is, in birth by destruction, the fecundity of privation is made clear. It is what you call being, but in truth it is not. In short, what is *is* what is not. "For the dialectical method," Stalin says, "what above all is important is not that which for a moment appears stable, but already begins to disappear; what is above all important is what comes to be and develops, even if the thing seems at a given moment unstable, since for the dialectical method, the only invincible is that which becomes and develops."[107] Applied to society, this means that salvation comes from the revolt of the deprived, that is, from the deprived class. Power resides in it because it is what is not. "Social reforms are never accomplished by the weakness of the strong but by the strength of the weak."[108] "Feudalism too had its proletariat—the serfs contained all the seeds of the bourgeoisie. Feudal production also had two antagonistic elements, that one designates as the good and bad side of feudalism, considering that it is always the bad side which ends by carrying off the good. It is the evil side which produces the movement that makes history by provoking struggle." "The human essence had to fall into this state of absolute poverty in order to give birth to its internal richness." Once man is rescued from the things that he had believed better than himself, "he can move around himself—his veritable sun."

Do you feel pity at human misery and the lot of the dispossessed? Does the egoism and wickedness of the rich make you indignant? Bourgeois! Then you don't see that you want to kill the goose that lays the golden eggs! "If it is true," Stalin goes on, "that development comes about by exposing internal contradictions, by the conflict of contrary forces, a conflict destined to surmount them, it is clear that the struggle of the proletarian class is a perfectly natural phenomenon, inevitable."[109] Far from wanting to assuage conflict by a just distribution of goods, far from appealing to an "eternal justice"[110] to which every man should conform himself, it is necessary on the contrary to encourage the struggle and push the conflict to the point of exasperation. The ways to the emancipation of non-being must be opened. "Consequently, in order not to be deceived in politics, one must follow a proletarian politics of class, intransigently, and not a politics reshaping the harmony of the interests of the proletariat and the bourgeoisie, not a conciliating politics of 'integration' of capitalism in socialism."[111] Have an understanding with the adversary, so long as this is a means of crushing him. You can count on his weakness. In the integrity of his cowardice, he will not dare plumb your cynicism.

Let your cynicism be universal. Let it extend to the entirety of being. Let Yes mean No, and No, Yes.[112]

Sit autem sermo vester, est, est: non, non: quod autem his abundantius est, a malo est—Let your speech be yes, yes, no, no: that which exceeds this comes from the evil one.

In what does this process of despoilment even to absolute privation end? "The human essence," says Marx, must fall into absolute poverty in order that its inner richness can arise from itself."[113] Once man has broken all his links with no matter what, he will be able to move "around himself, as his true sun."[114]

This is the principle of the new order.

The pure self. The self with all that it has most from itself as pure subject, willed, this time, as end. The self boastful of that which, in him, is not. For whom then does he wish to ravish himself?

"The destruction of religion," says Marx, "as the illusory happiness of the people, is a requirement of real happiness. . . ."

"Religion is only the illusory sun that moves around man, as long as he does not move around himself."

"The religious hypocrisy that gives to another what it has taken from me, to give it to God. . . ."

"Every critique must proceed from the critique of religion."

"The critique of religion arrives at the doctrine that man is the supreme being for man."

"Philosophy does not conceal it. The profession of Prometheus, 'in a word, I hate all the gods . . .' is its own profession, which it holds and will hold against all the gods of the heaven and of the earth which do not recognize human conscience as the highest divinity. This divinity will suffer no rival. . . . [Philosophy] repeats what Prometheus says to Hermes, servant of the gods: 'Be sure of this, I will never exchange my miserable lot for your servitude. I prefer being bound to this rock than to be the faithful valet, the messenger of Zeus the father. . . .'"[115]

That is what Marx following Feuerbach says; Feuerbach issued from Hegel; Hegel issued from Fichte and Kant; Kant issued from. . . .

Non Serviam!

"Now," it is said in *Mystical Theology*, "we remove by negations that which is above all that one could remove and take away, we ought first to remove and subtract that which is most distant and remote from him. For would not one rather say that God is life and goodness than to say that he is not air or a stone?"[116] Marxism too has its way of negation to arrive at the term it thinks the most perfect: man purely man in his most complete despoilment from which will emerge his hidden inner richness. It too begins by denying what is most distant and remote from its term. Its first negation is the negation of God. The order is reversed.

What is this human essence that the Marxist tends to appropriate, what is the object of the "joy that man gives to himself"?[117] What is this inner richness? The question elicits indignation. Is it not at once evident and ineffable? Ineffable. Doesn't the scaffolding of negations say enough? The Marxist says nothing on this score, and he could say nothing. The perversion is therefore achieved. "As also at present," continues the *Mystical Theology*, "we are going to enter into that obscure mist which is above all understanding, not only will we find there a lessening of words, but an entire privation of words and thoughts. . . . For now that our discourse proceeds by mounting from low to high, to the degree that it rises, it is restricted and lessened, and when it will have passed everything that can be climbed, it unites entirely with him who cannot be explained nor declared in language."

Who could explain these modern positions in the light of philosophy alone? We can indeed indicate some "technical" errors. No doubt there would remain the weight of "systems" singularly increased by the disappearance of authors and the liberty that engenders them. But to make only such criticisms would not get at the root of such philosophies. It is not a matter here of merely accidental errors of thought in its evolution toward an always fuller truth, as was the case with the ancients. These errors have their root in appetite.

The practical force with which these authors and their disciples adhere to their errors can only be explained by a love for these errors which are *powerful as death.* I say powerful as death because the Marxist must sacrifice his whole being, he must face total death even to the annihilation of his self. He must coldly nourish himself on the most absolute despair. His every action tending always to violence can only result in the total destruction of his self. Dead, he will be, he thinks, as if he had never existed. No recompense, no justice, no pity. He who existed only for himself, exists in order not to be. Are his pains compensated for by some heritage that he will leave? But who will be his heir? Humanity? But humanity is made up of a multitude of selves and all await the same end. For each individual it will soon be as if he had never existed. Whether or not he has acted, acted well or badly, matters not.

But it is important, comes the cry. Even so, it is important to act! Haven't we here the essential condition of an absolutely gratuitous human act? Does not a man owe to himself this absolute generosity? The true Marxist can only live in a total abnegation, the power and weakness of negation. He cannot destroy everything. He consoles himself by living, he lives this life insofar as it allows him to deny it. May there always be things in order that he might live negation. He perpetuates himself in death by transmitting this negation from generation to generation. This is a generosity born from hatred and distrust. A heroism issued from a supreme capitulation. In the *Ethics*, this kind of heroism is the excess contrary to heroism—it is called bestiality.[118]

Negation of what? What does one want?

Amen, Amen, I say to you, unless the seed of grain fallen on the earth does not die, it alone remains, but if it dies, it bears much fruit. He who loves his life will lose it, and he who hates his life in this world will keep it for eternal life.

Why is the guile of the wise of this world turned on the masses? What do they seek in the crowd? What do they want of the miserable? The question is opportune for never before have the wise of this world manifested a more profound contempt for this same mass, even for its purely material good. And why not? A man dies in the same way as a dog. What does it matter to him whether he existed or not? Will you weep for the death of men? Then weep as well for dogs.

Marx dares to quote these sacred words, "*Let the dead bury their dead* and weep for them."

What is there in the mass that attracts the wisdom of this world? Has this guile chosen the most apt victim for its revenge? This wisdom lusts for power, but what is the power of the miserable crowd? It is true that it harbors a power of material crushing that one has hardly begun to exploit. There is as well another, one whose destruction is wanted first: the power of the weak. For the Omnipotent, the Lord of Mercy, has said: *I have compassion for this crowd. That which the world considers as nothing, God has chosen in order to confound the strong; and God has chosen that which in the world is beneath consideration and without power, that which is nothing, in order to reduce to nothing that which is.*

We are now witnessing a supreme effort to attack the work of God. They want to take Him away from the humble who are the most powerful with the Omnipotent—the true power of the weak. One would instill pride in them for who is less worthy of mercy than a proud wretch? One would inculcate in them the philosophy of the wise of this world. "Theory," Marx says, "also becomes a material force when it penetrates the masses. Theory is capable of penetrating the masses as soon as it makes demonstrations *ad hominem*, and it makes demonstrations *ad hominem* when it becomes radical. To be radical, it must take things by the root. And the root of man is man himself." "Just as philosophy finds in the proletariat its material arms, so the proletariat finds in philosophy its spiritual arms once clarity of thought has penetrated with naive terror into the depths of the people. . . ."[119] Thus the wise of this world want to seduce the *parvuli* by means of nocturnal knowledge. They will avenge themselves because God has held in disdain their wisdom. *Has not God convicted the wisdom of this world of folly?*

I bless you, Father, Lord of heaven and earth, that you have hidden these things from the wise and prudent, and have revealed them to the little ones. Yes, Father, I bless you because of what it has pleased you to do.

The intrigues of human intelligence and will, supposedly emancipated from the supernatural, are only intelligible as simulations of properly divine truths.[120] How can we explain this exaltation of the unformed if not as a perversion of obediential potency and the quite particular elevation of the least perfect of intelligent creatures. Why this attempt to free our words from our thoughts? In fact, we rank the divine names to the degree that He surpasses our conception of Him. How to explain this deification of motion, whether it is a matter of real motion, the most imperfect of acts, or the ratiocinative motion of reason, the most extrinsic and tenuous form of thought? In the light of revealed doctrine, this deification is only a seductive perversion of *the wisdom which is more mobile than all other mobile things.* The very idea of struggle and universal combat is a simulation of a state of affairs which in a certain way has its principle in the supernatural order. Indeed, without grace, the pure spirits, entirely determined in their nature and indefectible, would all have remained from the morning of their existence in a state of perpetual peace. Their order had been absolutely imperturbable. But was it not the elevation to the supernatural order by grace the principle of merit and the exercise of a liberty of contrariety which was followed by the fall and a combat which invaded the whole of creation, a combat in which God Himself takes part by the sacrifice of His only son? And has not God placed an enmity between the naturally most intelligent and powerful creature and the most humble of human creatures, creatures at the lowest degree of immortal creation? That victory should be the work of the weak is a monstrous caricature of the Woman who, from the beginning, was destined to crush the head of all pride.

Purely philosophical wisdom is impotent to judge modern philosophies. Christian philosophy should know that. The moderns have rejected the possibility of philosophy being the servant of a higher science. That is in a stroke to deny any principle higher than those which are first for us. This rejection cannot be without consequences. It implies the negation of all true wisdom. Man will have to deny even nature. And what natural truth has he in fact not denied?

In other words, modern philosophy has developed outside natural truth, that is to say, outside philosophy. But it has not been able to escape from the more universal order that faith and theology make known to us. The divine light alone can sound the depths of the night in which the wisdom of the serpent has sought refuge. This night is the counterfeit of the obscure and caliginous depth of the Inaccessible Light.

Appendix I
The Flourishing of the Person

We uphold the necessity of taking into account the characteristic traits of an individual, either to encourage his aptitudes and natural bents or to repress them depending on whether they are good or bad. But it is necessary to see that it is the end that is the reason for this necessity of taking note of individuating notes, and to see that this end is the measure and criterion of what is needed to attain it well. In short, it is a question of a hypothetical necessity and not of a necessity such "that what is necessary should be so as an end, for the necessary is derived from matter—*ponitur ex parte materiae*—whereas it is in the end that one finds the reason for the necessity. We do not say, in effect, that it is necessary that there should be such an end because the matter is such, but quite the contrary, it is because the end and the form are such that it is necessary that the matter be such. It is thus that necessity is in the matter—*ponitur ad materiam*—whereas the reason for the necessity is taken from the end."[121]

Provided that one understands it in this sense, we admit the necessity to develop, in their order to the end, the innate good traits of the person. And the same should be said of the family and the nation. There is a certain flourishing there, since these traits come from nature which is the intrinsic principle of operation. This flourishing can only better proportion the subject to his end; the end requires this proportion; it is the first principle of it.

But the humanists who accord primacy to the material and efficient causes do not understand it this way. By reason of his characteristic notes, the individual person himself will be the measure of his end: the end, first principle of the ordination of the person to the end, will be identical to the very order inscribed in the person. The accomplishment of his end will consist, for the person, in rejoining himself, in finding himself, and recognizing himself in his inner richness wholly characteristic and sealed with individuating notes. He will be himself the first principle of respect and of liberty, which are owed him with respect to this "personality."

Whence as well that radical multiplication of ends that humanism teaches and the primacy that it accords art. It is entirely in the line of humanism to see the prime root, the most fundamental reason, of the social character of man, not in the common good, but in the poetic nature of the individual, in the need to express oneself and tell others of oneself under the pressure of an inner superabundance of the pure self. Every object then becomes an original means of a work which will have its first true principle in the self. Understand from that that the person of the other is necessary because I feel the need to make myself heard, because I have to have someone to appreciate me, I need a person-subject. In short, with respect to me, your reason for being is to participate in my personal life. Is it indeed a man who speaks thus? And is this not the excuse given by practically practical personalists for their paradoxical horror of solitude and their irrepressible desire to meddle? That is why the humanist doctor has a greater desire to teach than a desire to know. His knowledge has for its end the expression of his self: the need to speak is the very principle of his knowledge. And that is quite logical. Is not his liberty prior to knowledge? Isn't it the most profound thing in the self?

As we have indicated, the nation understood in the Thomist sense of *patria* also has its right to flourish from certain of its proper characters. The common good of civil society requires that the proper character of the nation or of nations be respected, for which it must truly be a common good. The common good does not require a homogeneity of subjects, rather the opposite. Moreover, if the common good of civil society is for the nation, it is not for all that taken as end, it is not a pure means for the flourishing of the nation. The good of civil society should be conformed to the nation, it ought to be 'its' good. From which it does not follow that the former is subordinated to the latter. To subordinate the good of civil society to the good of the nation is to subordinate reason to nature. One then reverts to the irrational and voluntarist nationalism of the *Discourse to the German Nation.* Civil society would be for the nation only a pure means of rejoining itself as nation, whereas in truth the good of civil society is more divine than that of the nation. The flourishing of it is not even the proper end of the nation, it remains in the order of dispositions and means.

Appendix II
Every Person Desires His Good

Every person desires his good insofar as he desires his perfection. We have seen that "his good" is distinguished from the alien good, from the good of another taken purely and simply as such. The good of man, "his good," comprises not simply the proper good of the single person; "his good" includes as the most worthy and divine good the common good. When we restrict "his good" to the proper good of his single person, we deprive man of that which is for him his greatest good. The person would be reduced to the condition of the brute. He would not be able to defend the common good under the very note of common good. Egoism would be in perfect conformity with reason. The sacrifice of the individual person for the common good would have for principle and term the self-love of man purely as man.

And yet, certain personalists, more naive than others, have not hesitated to make theirs this very logical and perfectly ignoble conclusion. See "The Theory of Democracy" by Mortimer J. Adler and Father Walter Farrell in *The Thomist* 4, no. 2 (1942). "In short," they say, "every act of justice implies a relation to the common good, and, what seems paradoxical, is by that very fact selfish, because the common good is not an end in itself; it is a means to the individual happiness that every man pursues, but which can only be attained and possessed by virtue, justice included. From which it follows that no obligation founded on justice can turn man away from the pursuit of his proper good to turn him toward some alien good, unless that obligation is part of his individual happiness, or, a means to attain it" (pp. 323–24).

> As we have seen, the intentions of natural justice are selfish. They do not look to the good of another *as other,* but only as part of the community which should be conserved for the good proper to himself. On the other hand, just as natural justice and natural love are selfish, so neither of

> them is heroic. Neither the one nor the other leads a man to martyrdom. Although natural love is less selfish than justice because it involves a kind of forgetfulness of the true self, and although natural love, unlike justice, pushes men to the generosity of sacrifice, it does not the less remain on the level of imperfect action, by which the agent always seeks to perfect himself at the same time as the other, and in fact considers the others as an extension of himself, as an *alter ego*. In this sense, the impulses of natural love never arise from the fundamental tendencies of natural desire—*which consists, for every thing, to seek its proper perfection* (pp. 329–30).

And in a note (256), they add,

> One might object that heroism is an undeniable fact in pagan societies—that the literature of Greece and Rome, for example, are full of examples of men who sincerely sacrificed their life for their country in military action. Such a heroism can moreover be explained by pagan beliefs in the immortality of the soul and the recompense reserved for heroes in the Elysian Fields of a future life. Today we could cite the Japanese as an example of a people among whom heroes are encountered—men who almost commit suicide for the well-being of their country, acting out of a 'religious' belief in the Emperor. But examining it more closely, one will see, we believe, that such a heroism is counterfeit and that it does not involve sacrifice because it does not involve forgetfulness of oneself; the exploit is done in expectation of recompense, whether it be a higher rank among the shades of the dead, or of the lasting renown of one's name in the memories of men. The predominant motive among the ancients was not the privileges and joys accorded the brave in the Elysian Fields. Even setting aside the myths of a future life and these slight 'beliefs' in an immortal soul, the pagan 'hero' would have been motivated by concern for his renown—pride in self, pride in his family, this pride having to be satisfied by this sort of 'immortality' that a man can enjoy when he is honored in the annals of his people.

This opinion, which does not deserve refutation, will be, for the future, a witness of the depths to which we have fallen. We can oppose to deplorable opinion a certainty, also timely and of perfect practical rectitude. It is taken from a letter written in the last hours of the fall of Bataan and appeared in the *Washington Daily News.*

> I have seen horrible things but I have also seen admirable acts of courage, of sacrifice, of loyalty. Finally, I have found what I have sought all my life—a cause and a task in which I can lose myself completely and to which I can give the last ounce of my strength and thought. I have mentally and spiritually conquered the fear of death. My prayer, morning and evening, is that God who suffered so for me will grant me his strength and his peace. In the last two months I have taken part in one of the most disinterested cooperative efforts ever accomplished by any group of individuals. Some errors have been made, but that has nothing to do with the manner in which my comrades of Bataan, both Filipinos and Americans, have reacted to their baptism of fire. If the same ardor were devoted to the betterment of the world in times of peace, what a wonderful world we would have. (*The Reader's Digest* [Sept. 1942]: 14)

That is love for the common good.

Appendix III
Nabuchodonosor, My Servant

> But the evil princes themselves are the ministers of God, for it is by a divine disposition that they are princes, to inflict chastisements, even though that is not their intention, according to *Isaiah* 10, 5–7: *Woe to the Assyrian: he is the rod and the staff of my anger, and my indignation is in their hands, I will send him to a deceitful nation and I will give him a charge against my wrath to take away the spoils, and to lay hold on the prey, and to tread them down like the mire of the streets. But he shall not take it so, and his heart shall not think so; but his heart shall be set to destroy and to cut off nations not a few.* And *Jeremiah* 25, 9: *Behold I will send, and take all the tribes of the north, saith the Lord, and Nabuchodonosor the king of Babylon my servant: and I will bring them against this land and against the inhabitants thereof and against all the nations that*

> *are round about it: and I will destroy them and make them an astonishment and a hissing and perpetual desolation.* And also because these evil princes, from time to time, God permitting, afflict the good, which turns to the benefit of the latter, according to these words *We know that all things work together for the good for those who love God.* (Saint Thomas, *On the Epistle to the Romans*, 13:3, lectio 1)

> The will to harm comes from man himself, but the power to harm comes from God who permits it, *a Deo permittente.* God does not permit the wicked one to harm as much as he might want to, but imposes a limit on him. *And I said: Hitherto thou shalt come, and shalt go no farther* (*Job* 38, 11). And thus the devil has not harmed Job save in the measure that God permitted. And so too Arius would not have been able to harm the Church except insofar as God permitted. In the *Apocalypse* 7, the angel says to the four angels he has asked to harm the land and sea *Do not harm the earth or the sea or the trees, till we have sealed the servants of our God on their foreheads.* (Saint Thomas, *On 2 Timothy*, 3, lectio 2)

Appendix IV
Ludwig Feuerbach Interprets Saint Thomas

Feuerbach, from whom Marx and Engels borrowed their absolute humanism, considered authentically Christian thought as a thought evolving toward his own *anthropotheismus.* In *The Essence of Christianity,* he opposes the conception of the Christians to those of the ancients concerning the relations of the individual man to the whole of his species, to society, to the universe.

"The ancients," he writes, "sacrificed the individual to the species (*Gattung*); the Christians, the species to the individual. Or: paganism conceived and estimated the individual only evisaged in his distinctness from the whole of the species: Christianity, however, conceived the individual in his unique immediacy and indistinction from the whole."

Feuerbach took care to make use of Saint Thomas when he could and to base himself on him—in order to surpass him, of course. Thus he had to explain the doctrine of Ia, q. 60, a. 5. There, Thomas seems in complete agreement with Aristotle: the good of the whole is better than the good of the part alone. But, says Feuerbach, it is otherwise when Saint Thomas takes the supernatural point of view and speaks as a theologian. Then the person is not only an individual: he is a whole, an absolute. Here is his commentary.

> It is known that Aristotle expressly says in the *Politics* that the individual (der Einzelne), as he does not suffice to himself, is, in his relation to the state, as a part to the whole. . . . It is true that Christians too 'sacrifice the individual', which means here the singular as part of the whole, the genus, common being (Gemeinwesen). "The part," says Saint Thomas, one of the greatest thinkers and Christian theologians, "sacrifices himself instinctively for the preservation of the whole. Every part naturally loves the whole more than itself. And each individual naturally loves the good of his species more than his singular good or well-being. Every being, thus, in its own fashion and naturally, loves God, as the universal good, more than itself" (*Summa*, Ia, q. 60, a. 5). In this perspective, Christians think, therefore, like the ancients. Saint Thomas, in the *De regimine principum*, III, 4, praises the Romans because they put their fatherland before all else and sacrificed their proper good to the well-being of the fatherland. And yet, all such thoughts and opinions count for the Christian only on earth, not in heaven; in ethics, not in dogmatics; in anthropology, not in theology. As object of theology, the individual is a single supernatural entity, immortal, self-sufficient, absolute, a divine being. The pagan thinker Aristotle (*Ethics*, IX, 9) declared that friendship was necessary for happiness; the Christian thinker, Saint Thomas Aquinas does not think so. "The society of friends is not necessarily required for happiness, for man finds the fullness of his perfection in God." "So much so, that if there should be but one soul enjoying the possession of God, he would still be happy, although he had no neighbor to love" (IaIIae, q. 4, a. 8). Thus, the pagan considers himself as an individual, even in happiness, and consequently as having need of another being like himself, of the same species; the Christian, on the contrary, has no need of another self, for the individual is not only an individual, but equal to the whole (Gattung), a general being (allgemeines Wesen), since he possesses "the fullness of his perfection in God" and thus in himself.

There are things to find fault with in this presentation, but we will stick to the essential. Is there an opposition between the point of view St. Thomas takes in Ia, q. 60 and that of IaIIae, q. 4? It would be ridiculous to say that in the first case Saint Thomas puts himself in a purely natural point of view or that he does not envisage the created person except insofar as he has the note of part of the universe, whereas in the *IaIIae* he considers things from a supernatural point of view where the person would have, on the contrary, the note of a whole. This presupposes a strange conception of the subject of the *Summa* and of the order of its treatises.

Feuerbach is obliged to invoke this distinction because he does not see that it is quite another thing to be under the dependence of the whole and of its parts in order to attain the good of the whole. The fundamental reason for which we call every created person a part is that his greatest good is incommensurable with the good of the single person taken as such; it is rather as an individual that the human person is a whole. No created person is a nature proportioned, or proportionable, to the purely and simply universal good as its proper good as a single person. Otherwise, every person would be God. That is why for Feuerbach man is God.

How did this philosopher arrive at the divinization of man? Romantic philosophy divinized the confused universal, and what we call the universal *in causando* would only be a manifestation of that. The concept 'animal' would be richer than the concepts of 'man' and 'brute', because it includes them and is higher than them. The anteriority according to the order of potentiality is converted into absolute priority. That is why man will substitute himself for God.

For Hegel, as it was no doubt for David of Dinant, being is the *summum genus*, and that is the primary reason of all things. In fact, this Hegelian being is nothing other than what we call 'the first known,' that is, being as the most common predicate, the most undetermined, the most confused, the most superficial concept that can be conceived, the more purely potential concept, which better reflects the pure potentiality of the most imperfect intellect possible, which signifies most proximately the original pure subjectivity of our intelligent self. Thanks to the movement of reason, Hegelian being takes on the nature of the potency that is act. Dialectics has for its function to make explicit the infinite richness of being. Pure potentiality presents itself as a substitute for pure actuality. It is the pure undetermined which will have the fecundity that we attribute to pure act.

After all, what is this *summum genus?* The question is opportune. Feuerbach will expressly identify the infinity of the genus (die Unendlichkeit der Gattung) with being as the most common predicate: he will identify the pure community of it with being the subject of metaphysics; the being which is the subject of metaphysics with the fullness of being, with God in whom thought is identical with being; and, since we are what we know, the fullness of being will be nothing else than the proper being of man. God is nothing other than man. Each individual human is at once part and whole, individual alone and God. As individual, man is limited; as a properly conscious being, he is unlimited, infinite. "Consciousness in the proper and rigorous sense, and consciousness of the infinite are inseparable; limited consciousness is not consciousness; consciousness is essentially a nature all comprehensive and infinite. The consciousness of the infinite is nothing other than the infinity of consciousness. Or: in the consciousness of the infinite, the consciousness of the infinity of the proper being (of self) is object."

And Feuerbach attempts to indicate the historical roots of this conception. He cites Saint Thomas for each of his most fundamental assertions. Let us admit that once granted this crude total adequation of two kinds of universality, nothing is easier than to twist certain texts of St. Thomas in favor of his anthropotheism. Is not the knower the known? Is not the soul in a certain way all things? Does not understanding comprise the entire order of being? Isn't it a *virtus infinita?* Is not the object of this *virtus* the *universale bonum?* How could a man, so considered, have the note of a part?

Feuerbach recognizes as well that "man is nothing without an object." He is nothing so long as he is not grasped as unlimited *Gattung*, as long as he is not taken in his pure universality. Is he then under a dependence to an object? Surely. He must make the conquest of the object, he must make the conquest of himself. As long as the object of man is conceived as external to man, man conceives himself as limited, he is only an individual, he is not part of the whole, he alienates himself in a strange God, the God of religion. But it is necessary that God be at the very center of man, that man be the center of man, that he rejoin himself as the principle of himself. "The object to which a subject relates essentially and necessarily is nothing other than the proper being of the subject, envisaged this time as object being (gegenständliche Wesen)." The absolute being, the God of man, is the proper being of man. The power of the object over man is consequently the power of his own being.

The German philosopher believes that he can surmount the antinomy between Catholicism and Protestantism by pushing the latter to its ultimate conclusion. "In Catholicism, man exists for God; in Protestantism, God exists for man. . . . The history of Christianity has had for its principal result the revelation of this mystery: the realization and the knowledge of theology as anthropology."

The doctrine of Feuerbach is not humanist in the sense that it would accord primacy to human affairs understood in the ordinary sense. He objects equally to vulgar atheism. His God is the God who for the Jews, for philosophers, for Christians was only a dream. "I in no wise say—this would be too simplistic—that God is not, the Trinity is not, the Word of God is not, etc. I simply say that they are not the illusions that theology creates—that they are not strange mysteries, but that they are mysteries in us (einheimische), mysteries of human nature." The God of religion is an external God to whom man submits by being limited, he is the infinity of alienated man. In religion, man does not yet have direct consciousness of himself (sich direct bewusst); religion is the condition of the infancy (kindliche Wesen) of humanity. By contrast, the God of anthropotheism has become perfectly commensurable with man. He is man emancipated from the limits of his individuality. He is the very heart of man.

In theology become openly anthropology, the Pelagian and the Augustinian ought no longer speak occultly. The one and the other have their qualities and their faults. At bottom, the difference between them is only a "pious delusion."

> The distinction between Augustinianism and Pelagianism consists uniquely in this that the first expresses in the religious mode what the second expresses in the rationalist mode. Both say the same thing, one and the other appropriate the good to man—Pelagianism does it directly in a rationalist, moralist manner; Augustinianism does it indirectly, in a mystical, that is, religious manner. Pelagianism denies God, denies religion—*isti tantam tribuunt potestatem voluntati, ut pietati auferant rationem* (Augustine, *On Nature and Grace against Pelagius*, ch. 58)—it has for foundation the Creator alone, hence nature, and not the Redeemer . . . in short, it denies God, it erects man into God insofar as it makes of man a being who has no need of God, who is sufficient unto

> himself and who is independent. . . . Augustinianism is only Pelagianism reversed; what the one posits as subject, the other posits as object.

The anthropotheism of Feuerbach indeed leads us beyond Pelagianism. The latter sustained the integrity of human nature and its sufficiency; it denied the ascendancy of evil. The former on the contrary incorporates evil and seeks in it a profundity which makes man commensurable with God. "Human misery is the triumph of divine mercy; contrition for sin causes the intimate joy of divine sanctity."[122] Feuerbach found the clearest proof that his philosophy was already precontained in religion in the love of God for man, a love which is expressed in the Incarnation. Here is another of those passages where the most sublime truth is engaged in the most revolting sophism.

> The clearest and most incontestable proof is that, in religion, man regards himself as a divine object, as a divine end, and thus in religion he is related uniquely to himself—the clearest and most incontestable proof of all is the love of God for man, the foundation and central point of religion. For man, God puts off his divinity. Notice in what the elevating fact of the Incarnation consists: the highest being, who knows no need, humiliates and abases himself for man. Thus the vision of my proper being appears to me in God; I have value for God; the divine sense of my own being is thus revealed to me. How to express in a more elevated way man's value: God becomes man for man, man is the end, the object of the divine love? God's love for man is an essential determination of the divine being. God is a God who loves me, who loves man above everything. That is where the accent falls, in this consists the deep emotion of religion. God's love makes me loving; God's love for man is the foundation of man's love for God; the divine love causes, awakes human love. *We love God then because He has first loved us* (1 John 4.19). What is it that I love in and about God? It is love and, indeed, love for man. But when I love and adore the love with which God loves man, isn't it man that I love, is not my love at least indirectly love of man? Moreover, is not man the content of God when God loves man? And is not what I love what is deepest in me. Have I a heart when I do not love? No. Love alone is the heart of man. But what is that love without the very thing that I love? What I thus love is my heart, what I contain, my essence.

One cannot read such blasphemies without a shudder. But we must confront them. So, man does not have his true grandeur from this that God has abased Himself for him; God will have abased Himself and stripped away His divinity because of the goodness of man; He will first be drawn by the goodness of this creature that He made; man will have remained at bottom so loveable that God could not leave him in this condition of misery into which the pernicious indulgence of Adam has plunged him: that would have been incompatible with the dignity of His creature, and notice who would have been unjust. God has loved us first means only that God discovered us first; if His love is the foundation of ours, this is formally insofar as it is love for man. The Incarnation would have as its point to aid man to become aware of his own grandeur and powers. It would be the rending of the veil that separated man from himself. The things that God has chosen would only appear to be things that are not—*ea quae non sunt.* A merciful raising up? At bottom, it would have been the pity man has for himself that saved him; God was only an instrument of man's mercy toward himself. In truth, the wretch would have delivered himself, he would have raised himself by the power of his powerlessnes, by the strength of his weakness, as Marx will repeat.

These authors have for us the advantage of not speaking *in angulis.* This is the upshot of the perverse turning to the self, the effort to enjoy the pure self in his most radical subjectivity. Since man is chosen for divine love, what does he have that is attractive for the very creator of man? Certainly not the things *that are* of man. Does not man's grandeur reside in that which is not in him? Is it not his quite particular formlessness, his non-being, that seized God? That which is in man, is it not a defect of his non-being? Here is how evil, that positivity rooted in privation, only occurs in order to open man to his power. Passibility causes power to spring up. Does it therefore make us commensurable with God? Then it is in our non-being that we encounter being as such. Man's true being is identified with his non-being.

"The Passion," Feuerbach goes on, "is an essential condition of God become man, or, what amounts to the same, of the human God, of Christ. Love declares itself in suffering. All the thoughts and all the sentiments which attach first to Christ are linked to the idea of suffering. God as God is the sum of human perfection, God as Christ is the sum of human misery. Pagan philosophers celebrated activity, particularly the immanent activity (Selbsttätigkeit) of intelligence as the highest activity, the divine activity; Christians celebrated suffering, and located suffering even in God. While God as *Actus Purus,* as pure activity, is the God of abstract philosophy, Christ, on the contrary, the God of Christians, is *Passio Pura,* pure suffering—the most elevated metaphysical thought, the supreme being of the heart."

Would anyone have thought man would go so far in order to have his soul without losing it, to possess it from himself and for himself? And is that not the product of a desire to put oneself as it were behind oneself where man would possess his own liberty, where he would have it from his own hands, where he would have it as God has it, where he would have the knowledge of good and evil? "The First Man chiefly sinned," Saint Thomas says, "by desiring to resemble God by means of the knowledge of good and evil which the serpent promised him and which would render him capable of fixing for himself moral good and evil, or even to foresee the good or evil which could befall him. He sins secondarily in desiring to resemble God in his own power of acting, in order to attain happiness by virtue of his own nature, by that personal power of which Eve had the love in her soul, as Augustine says."[123]

Man will set himself up as an absolute, even at the price of identifying himself with He Who Is, with that which is most distant from him. One is astonished that Feuerbach is also a materialist. But it should be well noted that the antinomy between modern idealism and materialism is all on the surface. The absolute idealism of Hegel is, to tell the truth, more materialist than Marxist materialism. In effect, Hegelian being, being an extreme in the genus of indetermination, has more the character of matter than matter in the physical order; it is infinitely poorer than prime matter. And the speculative reason of Hegel is fundamentally practical, bearing on a being transcendentally *factibile.* So-called Hegelian speculation is in truth a revolt against practical truth, against the condition of this truth that is rectitude of appetite.[124]

Here we are then squarely in the path mapped out by David of Dinant "who posited in the most stupid manner that God is prime matter," and whom Saint Thomas accused of being a consummate ass.

Appendix V
The Revolt of the Philosophers of Nature

In his *Ludwig Feuerbach,* F. Engels, comparing the German revolution to the French revolution, writes:

> Just as in France in the 18th century, so it will be in the 20th in Germany, the philosophical revolution preceding the political one. But what a difference between the two! The French in open struggle against all official science, against the Church, often even against the State, their writings printed across the border, in Holland or in England, and they themselves often in danger of spending time in the Bastille. The Germans, on the contrary, the professors, those masters of youth named by the State, whose works are recognized as teaching manuals, with the system that crowns the whole development, that of Hegel, raised almost to the rank of official philosophy of the Prussian royalty! The revolution had to hide behind these professors, behind their pedantic and obscure phrases, their ponderous boring sentences. Those who at the time passed as the representatives of the revolution, weren't they precisely the fiercest adversaries of this philosophy which cast trouble on minds? But what neither the government nor the liberals could see, one man saw in 1833. And that man was Henrich Heine.[125]

Without any doubt Engels is referring to Heine's *Zur Geschichte der Religion und Philosophie in Deutschland* in which he ends his reflections on the destructive character of the *Kritik* of Kant and the apparently detached and inoffensive system of Hegel.[126]

> When one sees such sorrowful leaves burgeon on the philosophical tree which would soon bloom into poisoned flower; when one notices above all that German youth, immersed in metaphysical abstractions, forgot the most pressing interests of the time, and became unfit for practical life, patriots and the friends of liberty came to feel a just resentment against philosophy, and some even went so far as to break away from it with a frivolous joy whose results were sterile.
>
> We are not stupid enough to refute seriously these malcontents. German philosophy is an important matter which concerns the whole of humanity, and our grand nephews alone will decide if we deserve blame or praise for having done our philosophy first and our revolution second. It seems to me that a methodical people like ourselves ought to begin with reform and later occupy ourselves with philosophy, and to come to the political revolution after having passed through these

phases. I find that a quite reasonable order. The heads that philosophy used for meditation can be broken at pleasure by the revolution, but philosophy would never have been able to use heads that the revolution had already cracked. So fear not, compatriots, the German revolution will be neither more debonair nor more soft because Kant's *Kritik*, the transcendental idealism of Fichte and the philosophy of nature will have preceded it. These doctrines developed revolutionary forces which only await the moment to explode and fill the world with fright and admiration. Then will appear Kantians who no longer wish to hear of piety either in the world of facts or in that of ideas, the overturning without mercy, with axe and sword, the soil of our European life to extirpate the last roots of the past. There will appear on the same scene armed Fichteans, whose fantasy will no longer be mastered either by fear or interest; for they live in the spirit and disdain matter like the first Christians whom one could not sway either by bodily punishment or terrestrial joys. Yes, such transcendental idealists, in a social overturning, will be even more inflexible than the first Christians, for the latter endured martyrdom to arrive at heavenly bliss, whereas the transcendental idealist regards even martyrs as pure appearance and remains in the fortress of thought. But the most frightening of all in the German revolution will be the philosophers of nature who will identify themselves with the work of destruction, for if the hand of the Kantian strikes a strong sure blow since his heart is not moved by any traditional respect, if the Fichtean scorns heedlessly all dangers because for him they do not exist in reality, the philosopher of nature will be terrible because he puts himself in communication with the original powers of the earth from which he conjures the forces hidden by tradition and can invoke the whole of Germanic pantheism, which awakes in him that ardor of combat we find in the ancient Germans and who wishes to fight, not to destroy, nor even to vanquish, but simply to fight. Christianity has softened to a point this brutal battle ardor of the Germans, but it has not been able to destroy it, and when the cross, that talisman which fetters it, will have been broken, then will begin anew the ferocity of the old warriors, of the frenetic exaltation of the Berserkers that the poets of the North sing still today. Then, and alas this day will come, the old warrior divinities will rise from their fabled tombs and brush from their eyes the dust of centuries; Thor will take up his gigantic hammer and demolish the gothic cathedrals. . . . When you hear the din and the tumult, be on your guard, dear neighbors in France, and do not meddle in the business we are engaged in in Germany—evil might come to you. Don't try to blow out the flame and extinguish it, for you could easily burn your fingers. Don't laugh at such warnings, even though they come from a dreamer who invites you to mistrust Kantians, Fichteans, philosophers of nature; don't laugh at the fantastic poet who awaits in the world of facts the same revolution that has been accomplished in the realm of spirit. Thought precedes action as lightning does thunder. The thunder in Germany is in truth German indeed: it is not gentle, a slow rumble, but it will come and when you hear a crack like no crack ever before heard in the history of the world, know that German thunder will finally have reached its goal. At this noise, eagles will fall dead from the heights, and lions in the most remote deserts of Africa will turn tail and slink to even more remote regions. A drama will be performed in Germany next to which the French revolution will be an innocent idyll. It is true that today all is calm, and if you see here and there men gesticulate in a lively way, do not imagine that these are the actors who will one day be put on stage. They are only dogs that run in an empty field, barking and baring their teeth, before the troop of gladiators enter who will fight to the death.

And the hour will sound. The peoples gather as in the seats of an amphitheater around Germany to see great and terrible games. I warn you, Frenchmen, to be very quiet, do not applaud. We might very well misinterpret your intentions and deal with you a little brutally following our impolite nature. For if once, in our condition of indolence and servitude, we have been able to measure ourselves by you, even more so will we in the arrogant drunkenness of our youthful liberty. You yourselves know what one can do in such a state, a state you no longer are in. . . . Beware! I have only good intentions and I tell you bitter truths. You have more to fear from a Germany delivered than from the entire Holy Alliance with Croats and Cossacks. Above all, you are not loved in Germany, something almost unthinkable for you are indeed loveable, and you have taken, during your stay in Germany, much trouble to please at least the better and the most beautiful half of the German people; but it is just that half that loves you that will not take up arms and whose

> friendship will help you little. What is had against you, I have never been able to grasp. One day, at Goettingen, in a beer hall, a young Old-German said that we must avenge ourselves with the blood of the French for Konradin Hohenstaufen whom you decapitated in Naples. That you have certainly long ago forgotten, but we forget nothing. You see that when the desire takes us to cross swords with you motives will not be lacking in Germany. In any case I warn you to be on your guard, whatever happens in Germany, whether the prince royal of Prussia or Doctor Wirth comes to the dictatorship, hold yourselves in arms, remain calm at your posts, rifles at the ready. For you I have only good intentions, and I have been almost frightened when I recently heard that your ministers have the project of disarming France. . . .
>
> Since, despite your present romanticism, you were born classic, you recognize our Olympus. Among the joyful divinities that there regale themselves with nectar and ambrosia, you see a goddess who, among her soft delights, retains always a cuirass, helmet on her head, and a lance in her hand.
>
> She is the goddess of wisdom.[127]

Assuredly, to judge by the concrete character of this prophesy, the German poet had his demon. Nonetheless, let us try to single out one of the most common reasons for this fury of the philosophers of nature.

At the beginning of Book Two of the *Physics* nature is defined as "the principle and cause of movement and repose in the thing in which it resides primarily, per se and not accidentally." In the course of this same book it is demonstrated that nature acts for an end, the first principle and first cause of nature itself. In the light of this demonstration, Saint Thomas defines nature as "a reason (*ratio, logos*) put in things by the divine art in order that they might act for an end" (lectio 14; see too *Metaphysics* XII, lectio 12). In effect, action for an end presupposes intelligence, or at least a participation in intelligence. Nature properly speaking is therefore a substitute for intelligence. *Ratio indita rebus ab arte divina,* the most unreasonable nature is always a divine logos. Even the purely material principle, the passive principle of all things, being as well properly nature, is as it were a divine word.[128]

The aim of the philosophy of nature is to know, in their ultimate specific concretion, these divine *logoi* and the end that specifies them and attracts them; to know perfectly the natural being whose form is separable and the term of all other natures, as Aristotle says in the same book of the *Physics* (ch. 2) as well as in *On the Parts of Animals* (ch. 5). Nonetheless, this aim is only a dialectical limit of the study of nature, a term that we can ceaselessly approach but which we can never adequately attain.

Let us take note that the role of hypotheses increases to the degree that we approach things in their concretion. In the hypothesis, there is not only the aspect which calls for experimental confirmation, there is also the more profound tendency to get ahead of experience and to deduce it by way of a conclusion. Given the method we must employ on the road to that ultimate concretion, it would suffice to isolate this tendency in order that, at the limit, there would arise a world entirely of our making. Seen in this respect, the limit toward which experimental science tends is the condition of the demiurge.

The method of the discovery of the reasons which anticipate experience is a method of reconstruction. Always in this precise relation taken abstractly, to reconstruct the universe is in some fashion to construct it.[129] And if, *per impossible,* this limit could be reached, the universe would bc only a projection of our own *logoi.* But to attain that limit it would be necessary that we have practical knowledge of natural things; it would be necessary that natures are things makeable by us.[130] It should be said that the Renaissance had a lively awareness of this role of the hypothesis, even though the most eminent savants did not have a very exact notion of a scientific hypothesis, but one seized on the aspect of anticipation, the creative aspect. The fecundity of the constructive intelligence was exalted, a fecundity matched by a practical power over things. It is also in this light that one should see, I think, the primacy of the Cartesian cogito. Enthusiasm increased in the measure that the application of the method of limits issued from Platonism and secularized by Nicholas of Cusa expanded. At bottom, this method is the very basis of every hypothesis.

We have already said that the attempt to see the entire cosmos as a great flow, as an immense torrent arising always from a unique logos, from a first reason where natures are like whirls in the flux, is very laudable, even essential to a sapiential view, provided that one takes into account the limits and conditions of this method. But naturalism—I mean naturalism in the profound sense as opposed to a vulgar naturalism of mechanistic materialism, for example—wants to push this method to the point of substitution of our reasons for natures, that is, to the expulsion of the divine *logoi.* That is indeed what Hegel tried to accomplish.

"Thus," says Marx, "the metaphysicians who, in making abstractions, imagined themselves to be making an analysis, and who, to the extent that they detached themselves more and more from objects, imagined that they approached the point of penetrating them, these metaphysicians had their reason for saying that the things here below are embroideries, of which logical categories form the cloth. That is what distinguishes the philosopher from the Christian. The Christian has only one incarnation of the Logos, despite logic; the philosopher never finishes with incarnations" (*Morceaux choisis,* p. 64).

Hegel does not take into account that for the deduction of each species that species must absolutely be presupposed, as in the case of the straight line the notion of which is absolutely anterior to its character as limit, which is never more than phenomenal. It is true that dialectical Reason presupposes Understanding, but the latter is always the root of the former.[131]

When one expects from this method the results that the Hegelian hopes to obtain, it shows itself as sterile, however fecund when one understands it well. Marx indeed has seen this sterility. The study of nature can never renounce the primacy of sensible experience. The pretensions of idealist deduction are only "hypocritical leaps of speculations, which it constructs a priori" (*Morceaux choisis,* 47). He has also seen well that all these constructions of our thought are but means only. Hegel surmounted natures as such only in a purely apparent manner.

For us, these intermediary constructions have for their limit natures, the divine logoi, seminal reasons, which are not operable by us, although to the degree that we approach them, our practical empire over the world ceaselessly expands. Marx is as intent on nature in itself as Hegel, but he is not content with a phenomenal conquest: he wants a practical conquest. And in truth there can be no other. Nature as a thing in itself, as the object which escapes our grasp, represents for Marx accordingly a foreign power. Everything that is properly nature is an obstacle, but a useful, necessary obstacle. The thing in itself must be converted into a thing by and for us. The idealist dissolution is not bad because it is dissolution; it is bad because it allows objects to subsist under the pretext that they are from us at least as objects. This is an illusion. The idealist dialectic hesitates before practical, concrete destruction which demands the conquest of foreign powers.

"In its mystifying form," says Marx, "dialectic was a German mode, because it appeared to transfigure existing things. In its notional form, it is a scandal and an abomination for the bourgeoisie and its doctrinaire spokesmen, because in the positive understanding of existing things it implies at the same time the understanding of their negation, of their necessary destruction, because it conceives all the forms in the course of movement, consequently, in their ephemeral side, because it does not permit itself to be imposed on by anything, because it is essentially critical and revolutionary" (*Morceaux choisis,* 68).

That is what Marxism owes to Hegelian philosophy: the power of dissolution, but pushed to the limit. "There is nothing definitive, absolute, sacred for it," says Engels, "it shows the fragility of all things and in all things, and nothing exists for it but the uninterrupted process of becoming, of the transitory, of the endless rising of the lower to the higher, which itself is only a reflection in the thinking head. It has, to be sure, a conservative side as well; it recognizes the justification of certain steps of development of knowledge and society for their time and conditions, but only in this measure. The conservatism of this way of seeing is relative, its character revolutionary and absolute—the only absolute, moreover, that it allows to prevail" (*Ludwig Feuerbach,* 13).

Confronted with this intelligence in revolt, the world of natures must be converted into operable matter, and the resistance of natures serves to lever an action turned against them. And all that which tends to take on the stability of a nature, all that which perfects or fulfills it in the line of nature, also becomes a constraint on our liberty, an obstacle to be overcome; therefore, not only the completely natural society that is the family, but even political society whose roots are natural, have to be exterminated.

The Word must be sought in every word that He has made, to the most remote limits of creation. Every word of God will come to trouble the silence of our night—like a thunderbolt.

We say, to the most remote limits of creation. The philosopher of nature thus will seize upon the weakest cause, the undetermined causes, chance and fortune, causes "without reason" (paralogon),[132] to rationalize them in order that the world might be ours and nothing escape our empire. Above all, in order that the ineffable Providence which appears most in chance and fortuitous events, of which it is the only determined cause, might be denied.

Profane reason will substitute itself for the Reason that governs history. Marxism accordingly will be a historical materialism. The judgment of history will supplant the Judgment of God.

It is a terrible idea. But the revolt of the philosophers of nature is also terrible. Those who don't think so are its most certain instruments—the lukewarm who will be vomited from the mouth of God.

NOTES

1. Karl Marx, *Morceaux choisis,* edition of Nouvelle Revue Francais, p. 37.

2. "... because seeing their proper dignity, they desired singularity, which is especially the mark of the proud ... [the devil refused supernatural beatitude] to be had without his own singularity, but common with men; as a consequence he wanted to have a special prevalence over them rather than communion, as St. Thomas observes in this question 58, a. 3, toward the end. He invokes for this the authority of Pope St. Gregory... 'The Angels desired to lose a shared elevation because they preferred a private, that is, they refused celestial beatitude because it was participated and common to many, and wanted only a private one, namely as private and proper, and thus the two conditions of pride, namely, singularity, or as having nothing in common with inferiors, which seemed to them vulgar, even though it would have been a supernatural glory, and not to have it as by a special benefit and grace, and as it were precarious: for this is what the proud specially refuse, and the angels specially refused. The parable from Luke 14 is pertinent here, that which concerns the man who gave a great feast to which he called many, and when he summoned the invited they began to excuse themselves, for perhaps they then refused to come to that feast because it was great and for many, disdaining to consort with such a number, and chose rather their private accommodations, although quite inferior, as of the natural order, this one because he had bought a house, that one a yoke of oxen, another because he had taken a wife, each offering his own excuse, and for a private good because it was their proper good, refusing the feast because it was large and common to many. This is most proper to the spirit of pride'" (John of St. Thomas, *Cursus theologicus* [Vivès edition], t. IV, d. 23, a. 3, nn. 34–35, pp. 950–51). "... he who judges that he does not have his natural and proper excellence by a special grace and as a benefit from God, but by right of creation, and not as common to many, but singularly his own" (ibid., n. 40, p. 955). "The angel in his first sin inordinately loving a spiritual good, namely his own proper existence, and his own proper perfection or natural beatitude ... so willed, that at the same time from the side of the mode of willing, although not on the side of the thing willed, he per se willed aversion from God, and not to be subject to his rule in the pursuit of his altitude" (Salmanticenses, *Cursus theologicus* [Palmé edition], t. IV, d. 10, dub. 1, p. 559b).

3. St. Thomas, *Summa theologiae,* IIIa, q. 8, a. 7, c.

4. John of St. Thomas, *Cursus theologicus,* n. 39, p. 954.

5. "... quanto aliqua causa est altior, tanto ejus causalitas ad plura se extendit. Habet enim causa altior proprium causatum altius quod est communius et in pluribus inventum" (St. Thomas, *In VI Metaphysicorum,* lectio 3, n. 1205).

6. "Manifestum est enim, quod unaquaeque causa tanto prior est et potior quanto ad plura se extendit. Unde et bonum, quod habet rationem causae finalis, tanto potius est quanto ad plura se extendit. Et ideo, si idem bonum est uni homini et toti civitati: multo videtur majus et perfectius suscipere, idest procurare et salvare illud quod est bonum totius civitatis, quam id quod est bonum unius hominis. Pertinet quidem ad amorem, qui debet esse inter homines, quod homo conservet bonum etiam uni soli homini. Sed multo melius et divinius est, quod hoc exhibeatur toti genti et civitatibus. Vel aliquando amabile quidem est, quod exhibeatur uni soli civitati, sed multo divinius est, quod hoc exhibeatur toti genti, in qua multae civitates continentur. Dicitur hoc autem esse *divinius,* eo quod magis pertinet ad Dei similitudinem, qui est ultima causa omnium bonorum. Hoc autem bonum, scilicet quod est commune uni vel pluribus civitatibus, intendit methodus, idest quaedam ars, quae vocatur civilis. Unde ad ipsam maxime pertinet considerare finem ultimum humanae vitae, tanquam ad principalissimam" (*In I Ethicorum,* lectio 2, n. 30).

Compare this text with the following passage from the *De Voluptate* of Lorenzo Valla, where he responds to the question *An moriendum sit pro aliis* (Ought one to die for others?) (L. II, c. 2). "I have no obligation to die for one citizen, nor for two, nor for three, and so on to infinity. How then can I be obliged to die for the Fatherland which is the sum of all these? Does the fact that one adds one more change the nature of the obligation?" (in P. Monnier, *Le Quattrocento* [Paris, 1924], t. I, p. 46). "The humanists, says Cino Rinuccini, understand nothing of domestic economy. They live foolishly without concern for paternal honor and the benefit of children. They ignore what is the best government, whether of one or of several, whether of many or of few. They flee fatigue, saying that he who serves the commune serves no one, they do not defend the republic in war nor defend it with arms. Finally they forget that the more common the good, the more divine it is. (*Ne si ricordano che quanto il bene è più commune, tanto a più del divino*)" (ibid., 332).

7. "Bonum suum cujuslibet rei potest accipi multipliciter: Uno quidem modo, secundum quod est ejus proprium *ratione individui.* Et sic appetit animal suum bonum cum appetit cibum, quo in esse conservatur. Alio modo, secundum quod est ejus *ratione speciei.* Et sic appetit proprium bonum animal inquantum appetit generationem prolis et ejus nutritionem, vel quicquid aliud operetur ad conservationem vel defensionem individuorum suae speciei. Tertio vero modo, *ratione generis.* Et sic appetit proprium bonum in causando agens aequivocum, sicut caelum. Quarto autem modo, *ratione similitudinis analogiae* principiatorum ad suum principium. Et sic Deus, qui est extra genus, propter suum bonum omnibus rebus dat esse" (*III Summa contra gentes,* c. 24).

8. "Quodlibet singulare naturaliter diligit plus bonum suae speciei quam bonum suum singulare" (Ia, q. 60, a. 5, ad 1).

9. "Nor does the argument of P. Suarez count, namely that nutrition is ordered to one's proper conservation and generation to an alien individual. A thing is more inclined to its proper good than to an alien good, because friendship with others arises from friendship toward oneself. In response, a thing is more inclined to its proper good as distinguished from the alien good, but not from the common good. There is a greater inclination (*ponderatio*) to this than to the proper good, because the proper good is contained within the common good and depends on it, and thus friendship toward another arises from friendship toward oneself, when the other is wholly alien, not when the other is as a common and higher good, with respect to which this reasoning has no force" (John of St. Thomas, *Cursus Philosophicus* [ed. Reiser], t. III , p. 87a).

10. ". . . natura reflectitur in seipsam non solum quantum ad id quod est ei singulare, sed multa magis quantum ad commune: inclinatur enum unumquodque ad conservandum non solum suum individuum, sed etiam suam speciem. Et multo magis habet naturalem inclinationem unumquodque in id quod est bonum universale simpliciter" (Ia, q. 60, a. 5, ad 3).

11. *III Summa contra gentes,* c. 112.

12. *Q.D. de veritate,* q. 2, a. 2, c.

13. "Quanto aliquid est perfectioris virtutis et eminentius in gradu bonitatis, tanto appetitum boni communiorem habet et magis in distantibus a se bonum quaerit et operatur. Nam imperfecta ad solum bonum proprii individui tendunt; perfecta vero ad bonum speciei; perfectiora vero ad bonum generis; Deus autem, qui est perfectissimus in bonitate, ad bonum totius entis. Unde non immerito dicitur a quibusdam quod bonum, inquantum hujusmodi, est diffusivum: quia quanto aliquid invenitur melius, tanto ad remotiora bonitatem suam diffundit. Et quia in quolibet genere quod est perfectissimum est exemplar et mensura omnium quae sunt illius generis, oportet quod Deus, qui est in bonitate perfectissimus et bonitatem communissime diffundens, in sua diffusione sit exemplar omnium bonitatem diffundentium" (*III Summa contra gentes,* c. 24).

14. "Cum affectio sequatur cognitionem, quanto cognitio est universalior, tanto affectio eam sequens magis respicit bonum commune; et quanto cognitio est magis particularis, tanto affectio ipsam sequens magis respicit privatum bonum; unde et in nobis privata delectio ex cognitione sensitiva exoritur, dilectio vero communis et absoluti boni ex cognitione intellectiva. Quia ergo angeli quanto sunt altiores, tanto habet scientiam magis universalem . . . ideo eorum dilectio maxime respicit commune bonum" (*Q.D. de spiritualibus creaturis,* a. 8, ad 5).

15. "Magis ergo diligunt se invicem, si specie differunt, quod magis pertinet ad perfectionem universi . . . quam si specie convenirent, quod pertineret ad bonum privatum unius species" (ibid.).

16. *XII Metaphysics,* c. 10, 1075a15.

17. See Marx, *Morceaux choisis,* 203.

18. "Amare bonum alicujus civitatis ut habeatur et possideatur, non facit bonum politicum; quia etiam aliquis tyrannus amat bonum alicujus civitatis ut ei dominetur; quod est amare seipsum magis quam civitatem; sibi enim ipsi hoc bonum concupiscit,

non civitati. Sed amare bonum civitatis ut conservetur et defendatur, hoc est vere amare civitatem; quod bonum politicum facit; in tantum quod aliqui propter bonum civitatis conservandum vel ampliandum, se pericula mortis exponant et negligant privatum bonum. Sic igitur amare bonum quod a beatis participatur ut habeatur vel possideatur, non facit hominem bene se habentem ad beatitudinem, quia etiam mali illud bonum concupiscunt; sed amare illud bonum secundum se, ut permaneat et diffundatur, et ut nihil contra illud bonum agatur, hoc facit hominem bene se habentem ad illam societatem beatorum; et haec est caritas, quae Deum per se diligit, et proximos qui sunt capaces beatitudinis, sicut seipsos" (*Q.D. de caritate,* a. 2, c.1).

19. ". . . sic enim et populus totus erit quasi unus tyrannus" (*de Regno,* c. 1).

20. "Proprium autem bonum hominis oportet diversimode accipi, secundum quod homo diversimode accipitur. Nam, proprium bonum hominis inquantum homo, est bonum rationis, eo quod homini esse est rationale esse. Bonum autem hominis secundum quod est artifex est bonum artis; et sic etiam secundum quod est politicus, est bonum ejus bonum commune civitatis" (*Q.D. de caritate,* a. 2, c.).

21. ". . . Ad hoc quod aliquis sit bonus politicus, requiritur quod amat bonum civitatis. Si autem homo, inquantum admittitur ad participandum bonum alicujus civitatis, et efficitur civis alicujus civitatis, competunt ei virtutes quaedam ad operandum ea qua sunt civium, et ad amandum bonum civitatis, ita, cum homo per divinam gratiam admittitur in participationem coelestis beatitudinis, quae in visione et fruitione consistit, fit quasi civis et socius illius beatae societatis, quae vocatur coelestis Jerusalem secundum illud, *Ephes.* 2, 19: *Estis cives sanctorum et domestici Dei*" (ibid.).

22. "Unde homini sic ad caelestia adscripto, competunt quaedam virtutes gratuitae, quae sunt virtutes infusae; ad quarum debitam operationem praexigitur amor boni communis toti societati, quod est bonum divinum, prout est beatitudinis objectum" (ibid.).

23. ". . . cum nullum meritum sit sine caritate, actus virtutis acquisitae non potest esse meritorius sine caritate. . . . Nam virtus ordinata in finem inferiorem non facit actum ordinatum ad finem superiorem, nisi mediante virtute superiori; sicut fortitudo, quae est virtus hominis qua homo, non ordinat actum suum ad bonum politicum, nisi mediante fortitudine quae est virtus hominis in quantum est civis" (*Q.D. de virtutibus,* a. 10, ad 4). "Dicit ergo primo [Philosophus], quod neque etiam fortitudo est circa mortem quam aliquis sustinet in quocumque casu vel negotio, sicut in mari vel in aegritudine, sed circa mortem quam quis sustinet pro optimis rebus, sicut contingit cum aliquis moritur in bello propter patriae defensionem . . . quia mors quae est in bello est in maximo periculo, quia de facili ibi moritur homo; etiam est in periculo optimo, quia homo pericula sustinet hic propter bonum commune, quod est optimum. . . . Virtus autem est circa maximum et optimum" (*In III Ethicorum,* lectio 14, nn. 537–38).

24. ". . . eminentem obtinebunt coelestis beatitudinis gradum, qui officium regium digne et laudabiliter exequuntur. Si enim beatitudo virtutis est praemium, consequens est ut majori virtuti major gradus beatitudinis debeatur. Est autem praecipua virtus qua homo aliquis non solum seipsum, sed etiam alios dirigere potest; et tanto magis,

quanto plurium est regitiva; quia et secundum virtutem corporalem tanto aliqua virtuosior reputatur, quanto plures vincere potest, aut pondera levare. Sic igitur major virtus requiritur ad regendum domesticam familiam quam ad regendum seipsum, multoque major ad regimen civitatis et regni. . . . Tanto autem est aliquid Deo acceptius, quanto magis ad ejus imitationem accedit: unde et Apostolus monet *Ephesios*, 5, 1: *Estote imitatores Dei, sicut filii charissimi.* Sed si secundum Sapientis sententiam: Omne animal diligit simile sibi, secundum quod causae aliqualiter similitudinem habent causati, consequens igitur est bonos reges esse acceptissimos, et ab eo maxime praemiandos" (*de Regno*, c. 9).

25. ". . . cum amor respiciat bonum, secundum diversitatem boni est diversitas amoris. Est autem quoddam bonum proprium alicujus hominis in quantum est singulare persona; et quantum ad dilectionem respicientem hoc bonum, unusquisque est sibi principale objectum dilectionis. Est autem quoddam bonum commune quod pertinet ad hunc vel ad illum inquantum est pars alicujus totius, sicut ad militem, inquantum est pars exercitus, et ad civem, inquantum est pars civitatis; et quantum ad dilectionem respicientem hoc bonum, principale objectum dilectionis est illud in quod principaliter illud bonum consistit, sicut bonum exercitus in duce, et bonum civitatis in rege; unde ad officium boni militis pertinet ut etiam salutem suam negligat ad conservandum bonum ducis, sicut etiam homo naturaliter ad conservandum caput, brachium exponit" (*Q.D. de caritate*, a. 4, ad 2).

26. ". . . et hoc modo caritas respicit sicut principale objectum, bonum divinum, quod pertinet ad unumquemque, secundum quod esse potest particeps beatitudinis" (ibid.).

27. ". . . bonitas cujuslibet partis consideratur in proportione ad suum totum: unde et Augustinus dicit . . . quod *turpis est omnis pars quae suo toti non congruit.* Cum igitur quilibet homo sit pars civitatis, impossible est quod aliquis homo sit bonus, nisi sit bene proportionatus bono communi; nec totum potest bene existere nisi ex partibus sibi proportionatus" (IaIIae, q. 47, a. 10, ad 2).

28. "Primo quidem, quia bonum proprium non potest esse sine boni communi rei familiae vel civitatis vel regni. Unde et Maximus Valerius dicit de antiquis Romanis quod *malebant esse pauperes in divite imperio quam divites in paupere imperio.*—Secundo quia, cum homo sit pars domus et civitatis, oportet quod homo consideret quid sit sibi bonum ex hoc quod est prudens circa bonum multitudinis: bona enim dispositio partis accipitur secundum habitudinem ad totum: quia ut Augustinus dicit . . . *turpis est omnis pars suo toti non congruens*" (IIaIIae, q. 47, a. 10, ad 2).

29. *Q.D. de caritate*, a. 2, c.

30. "Actus . . . rationalis creaturae a divina providentia diriguntur, non solum ea ratione quo ad speciem pertinent, sed etiam in quantum sunt personales actus" (*III Summa contra gentes*, c. 64).

31. "Unumquodque intendens aliquem finem, magis curat de eo quod est propinquius fini ultimo: quia hoc etiam est finis aliorum. Ultimus autem finis divinae voluntatis est bonitas ipsius, cui propinquissimum in rebus creatis est bonum ordinis totius universi: cum ad ipsum ordinetur sicut ad finem, omne particulare bonum hujus vel

illius rei, sicut minus perfectum ordinatur ad id quod est perfectius; unde et quaelibet pars invenitur esse propter suum totum. Id igitur quod maxime curat Deus in rebus creatis, est ordo universi" (ibid., c. 64).

32. "Distinctio rerum et multitudo est ex intentione primi agentis, quod est Deus. Produxit enim res in esse propter suam bonitatem communicandam creaturis, et per eas repraesentandam; et quia per unam creaturam sufficienter repraesentari non potest, produxit multas creaturas et diversas; ut quod deest uni ad repraesentanda, divinam bonitatem, suppleatur ex alia. Nam bonitas quae in Deo est simplicter et uniformiter, in creaturis est multipliciter et divisim; unde perfectius participat divinam bonitatem et repraesentat eam totum universum, quam alia quaecumque creatura" (Ia, q. 47, a. 1, c.).

33. In quolibet effectu illud quod est ultimus finis, proprie est intentum a principali agente sicut ordo exercitus a duce. Illud autem quod est optimum in rebus existens, est bonum ordinis universi. . . . Ordo igitur universi est proprie a Deo intentus, et non per accidens proveniens secundum succesionem agentium. . . . Sed . . . ipse ordo universi est per se creatus ab eo, et intentus ab ipso" (Ia, q. 15, a. 2, c.).

34. "Id quod est bonum et optimum in effectu, est finis productionis ipsius. Sed bonum et optimum universi consistit in ordine partium ipsius ad invicem, qui sine distinctione esse non potest; per hunc enim ordinem universum in sua totalitate constituitur, quae est optimum ipsius. Ipse igitur ordo partium universo et distinctio earum est finis productionis universi" (*II Summa contra gentes,* c. 39).

35. IIIa, q. 8, a.7.

36. See Appendix I.

37. "Manifestum est enim quod duplex est bonum universi: quoddam separatum, scilicet Deus, qui est sicut dux in exercitu; et quoddam in ipsis rebus, et hoc est ordo partium universi, sicut ordo partium exercitus est bonum exercitus. Unde Apostolus dicit *Rom.* 8, 1: *Quae a Deo sunt, ordinata sunt.* Oportet autem quod superiores universi partes magis de bono universi participent, quod est ordo. Perfectius autem participant ordinem ea in quibus est ordo per se, quam ea in quibus est ordo per accidens tantum" (*Q.D. de spiritualibus creaturis,* a. 8, c.).

38. Ia, q. 47, a. 2.

39. "Per hoc autem quod dicimus substantias intellectuales propter se a divina providentia ordinari, non intelligimus quod ipsae ulterius non referantur in Deum et ad perfectionem universi. Sic igitur propter se procurari dicuntur et alia propter ipsas, quia bona quae propter divinam providentiam sortiuntur, non eis sunt data propter alterius utilitatem; quae vero aliis dantur, in earum usum ex divina ordinatione cedunt" (*III Summa contra gentes,* c. 112).

40. *In III Sententiarum,* d. 35, q. 1, a. 4, sol. 1.

41. *In I Sententiarum,* d. 7, q. 2, a. 2, q. 3, ad 4.

42. *III Summa contra gentes,* c. 113.

43. "Praecellunt enim [intellectuales et rationales naturae] alias creaturas et in perfectione naturae et in dignitate finis. In perfectione quidem naturae, quia sola creatura rationabilis habet dominium sui actus, libere se agens ad operandum; dum caeterae

vero creaturae ad opera magis aguntur quam agant. . . . In dignitate autem finis, quia sola creatura intellectualis ad ipsum finem ultimum universi sua operatione pertingit, scilicet cognoscendo et amando Deum: aliae vero creaturae ad finem ultimum pertingere non possunt nisi per aliqualem similitudinis ipsius participatione" (ibid., c. 111).

44. ". . . homo peccando ab ordine rationis recedit; et ideo decidit a dignitate humana, prout scilicet homo est naturaliter liber, et propter seipsum existens, et incidit quodammodo in servitutem bestiarum. . . . Pejor enim est malus homo quam bestia" (IIaIIae, q. 64, a. 2, ad 3).

45. "Paternitas igitur est dignitas Patris, sicut et essentia Patris: nam dignitas absolutum est, et ad essentiam pertinet. Sicut igitur eadem essentia quae in Patre est paternitas, in Filio est filiatio; ita eadem dignitas quae in Patre est paternitas, in Filio est filiatio" (Ia, q. 42, a. 4, ad 2).

46. "Sicut autem subsistentia, quando est modus proprius, subordinatur naturae: subsistentia autem divina, assumens naturam creatam terminando, potius subordinat illam sibi" (John of St. Thomas, *Cursus theologicus* [Solesmes edition], t. II, p. 159, n. 12).

47. ". . . sicut est ordo in causis agentibus, ita etiam in causis finalibus: ut scilicet secundarius finis a principali dependeat, sicut secundarium agens a principali dependet. Accidit autem peccatum in causis agentibus quando secundarium agens exit ab ordine principalis agentis: sicut, cum tibia deficit propter suam curvitatem ad executione motus quem virtus appetitiva imperabat, sequitur claudicatio. Sic igitur et in causis finalibus, cum finis secundarium non continetur sub ordine principalis finis, est peccatum voluntatis, cujus objectum est bonum et finis.—Quaelibet autem voluntas naturaliter vult illud quod est proprium volentis bonum, scilicet ipsum esse perfectum, nec potest contrarium hujus velle. In illo igitur volente nullum potest voluntatis peccatum accidere cujus proprium bonum est ultimus finis, quod non continetur sub alterius finis ordine, sed sub ejus ordine omnes alii fines continentur. Hujusmodi autem volens est Deus, cujus esse est summa bonitas, quae est ultimus finis. In Deo igitur peccatum voluntatis esse non potest.—In quocumque autem alio volente, cujus propium bonum necesse est sub ordine alterius boni contineri, potest peccatum accidere voluntatis, si in sua natura consideratur. Licet enim naturalis inclinatio voluntatis insit unicuique volenti ad volendum et amandum sui ipsius perfectionem, ita quod contrarium hujus velle non possit; non tamen sic est inditum ei naturaliter ut ita ordinet suam perfectionem in aliud finem quod ab eo deficere non possit: cum finis superior non sit suae naturae proprius, sed superioris naturae. Relinquitur igitur suo arbitrio quod propriam perfectionem in superiorem ordinat finem. In hoc enim differunt voluntatem habentia ab his quae voluntate carent, quod habentia voluntatem ordinant se et sua in finem, unde et liberi arbitrii esse dicuntur: quae autem voluntate carent, non ordinant se in finem, sed ordinantur a superiori agente, quasi ab alio acta in finem, non autem a seipsis" (*III Summa contra gentes,* c. 109).

48. "Hoc autem differt inter hominem et substantiam separatam, quod in uno homine sunt plures appetitivae virtutes, quarum una sub altera ordinatur. Quod quidem in substantiis separatis non contingit: una tamen earum est sub altera. Peccatum autem in voluntate contingit qualitercumque appetitus inferior deflectatur. Sicut igitur

peccatum in substantiis separatis esset vel per hoc quod deflecteretur ab ordine divino, vel per hoc quod aliqua earum inferior deflecteretur ab ordine alicujus superioris sub ordine divino manentis; ita in homine uno contingit peccatum dupliciter. Uno modo, per hoc quod voluntas humana bonum proprium non ordinat in Deum: quod quidem peccatum est commune et sibi et substantiae separatae. Alio modo, per hoc quod bonum inferioris appetitus non regulatur secundum superiorem: puta quando delectabilia carnis, in quae concupiscibilis tendit, volumus non secundum ordinem rationis. Hujusmodi autem peccatum non contingit in substantiis separatis esse" (ibid.).

49. "Considerandum est etiam quod, cum proprium alicujus bonum habet ordinem ad plura superiora, liberum est volenti ut ab ordine alicujus superiorum recedat et alterius ordinem non derelinquat, sive sit superior sive inferior" (ibid.).

50. "Sicut miles, qui ordinatur sub rege et sub duce exercitus, potest voluntatem suam ordinare in bonum ducis et non regis, aut e converso. Sed si dux ab ordine regis recedat, bona erit voluntas militis recedentis a voluntate ducis et dirigentis voluntatem suam in regem, mala autem voluntas militis sequentis voluntatem ducis contra voluntatem regis: ordo enim inferioris principii dependet ab ordine superioris" (ibid.).

51. "Quamvis igitur multa, quae videntur esse per accidens reducendo ipsa ad causas particulares, inveniantur non esse per accidens reducendo ipsa ad causam communem universalem, scilicet virtutem coelestem, tamen etiam hac reductione facta, inveniuntur aliqua esse per accidens" (*In VI Metaphysicorum,* lectio 3, n. 1212).

52. ". . . manifestatio eorum quae dependent ex voluntate intelligentis, non potest dici illuminatio, sed locutio tantum; puta si aliquis alteri dicat, Volo hoc addiscere, Volo hoc vel illud facere. Cujus ratio est, quia voluntas creata non est lux, nec regula veritatis, sed participans lucem: unde communicare ea quae sunt a voluntate creata, inquantum hujusmodi, non est illuminare. Non enim pertinet ad perfectionem intellectus mei, quid tu velis, vel quid tu intelligas cognoscere: sed solum quid rei veritas se habeat" (Ia, q. 107, a. 2, c.).

53. "Per suum esse substantiale dicitur unumquodque ens simpliciter; per actus autem superadditos dicitur aliquod esse secundum quid. . . . Sed bonum dicit rationem perfecti, quod est appetibile, et per consequens dicit rationem ultimi; unde id quod est ultimo perfectum, dicitur bonum simpliciter; quod autem non habet ultimam perfectionem quam debet habere, quamvis habeat aliquam perfectionem, in quantum est in actu, non tamen dicitur perfectum simpliciter, nec bonum simpliciter, sed secundum quid" (Ia, q. 5, a. 1, ad 1).

54. "Unumquodque autem in rebus naturalibus, quod secundum naturam hoc ipsum quod est, alterius est, principalius et magis inclinatio in id cujus est, quam in seipsum. Et haec inclinatio naturalis demonstratur ex his quae naturaliter aguntur: quia unumquodque, sicut agit naturaliter, sic aptum natum est agi, ut dicitur in II Physic. Videmus enim quod naturaliter pars se exponit ad conservationem totius: sicut manus exponitur ictui, absque deliberatione, ad conservationem totius corporis. Et quia ratio imitatur naturam, hujusmodi inclinationem invenimus in virtutibus politicis: est enim virtuosi civis, ut se exponat mortis periculo pro totius reipublicae conservatione; et si homo esset naturalis pars hujus civitatis, haec inclinatio esset ei naturalis" (Ia, q. 60, a. 5, c.).

55. "... Natura et substantia partis, hoc ipsum quod est, essentialiter et primo propter totum et totius esse est. Quod convenire cuilibet creaturae respectu Dei, patet. Quia quaelibet creatura, secundum suam naturam, est naturalis pars universi: ac per hoc naturaliter diligit plus universum quam seipsam, juxta primum fundamentum. Ergo, a fortiori, magis diliget ipsum bonum universale: tum quia est eminentius totum universum, tum quia est omne bonum, tum quia bonum ipsum universale quod est Deus gloriosus, est finis et bonum ipsius universi, et consequenter a quocumque magis amatur universum, ab eo magis amabitur ipse; ut patet de exercitu et duce, juxta doctrinam XII *Metaphy.* 10" (Cajetan, ibid., n. 5). Also, *In III Sententiarum*, d. 29, q. 1, a. 3, c.

56. "Dicendum quod Philosophus loquitur de amicabilibus quae sunt ad alterum in quo bonum quod est objectum amicitiae invenitur secundum aliquem particularem modum: non autem de amicabilibus quae sunt ad alterum in quo bonum praedictam invenitur secundum rationem totius" (ibid., ad 1). See as well the commentary of Cajetan.

57. Ia, q. 60, a. 5, c.

58. "... felicitas est operatio hominis secundum intellectum. In intellectu autem est considerare speculativum cujus finis est cognitio veritatis, et practicum cujus finis est operatio. Et secundum hoc duplex felicitas assignatur hominis. Una speculativa quae est operatio hominis secundum virtutem perfectam contemplativam quae est sapientia. Alia autem practica quae est perfectio hominis secundum perfectam virtutem hominis practicam quae est prudentia. Est autem quaedam operatio secundum prudentiam et speculatio secundum sapientiam hominis secundum seipsam solum. Et est quaedam operatio prudentiae et speculatio totius civitatis; et ideo est quaedam felicitas practica et speculativa quaedam hominis secundum seipsum, et est quaedam felicitas practica totius civitatis et quaedam contemplativa totius civitatis. Felicitas autem speculativa secundum unum hominem melior est practica quae est secundum unum hominem, sicut evidenter docet Aristoteles in X *Ethicorum*; quoniam illa perfectio intellectus eligibilior est quae est respectu objecti magis intelligibilis, quia ratio perfectionis sumitur ex objecto; talis autem est speculativa. Felicitas enim est perfectio intellectus respectu primi et maxime intelligibilis. Felicitas autem practica est perfectio intellectus respectu agibilis ab homine quod multo deficit a ratione intelligibilis primi; ergo felicitas contemplativa unius eligibilior est quam felicitas practica; et iterum magis est continuus et sufficiens et delectabilis haec quam illa. Et eadem ratione contemplativa totius civitatis eligibilior est quam politica seu civilis, et contemplativa totius civitatis simpliciter eligibilior est contemplativa quae est secundum unum. Et hoc est quod intendebat dicere Aristoteles in I *Ethicorum*: si idem est uni et civitati, majusque et perfectius quod civitati videtur et suscipere et salvare. Amabile enim est uni: melius vero et divinius genti et civitati. Et ratio hujus potest esse, quia contemplativa et civilis civitatis comparantur ad contemplativam secundum unum, sicut totum ad partem: totum autem rationem magis perfecti et majoris boni habet quam pars; et ideo ista quam illa" (*In VII Politicorum*, lectio 2; Thomas's commentary as completed by Petrus de Alvernia).

59. "... bonum gratiae unius majus est quam bonum naturae totius universi" (IaIIae, q. 113, a. 9, ad 2).

60. M.H. Doms, the author of a much read book, *Du sens et de la fin du mariage* (Desclée de Brouwer), in which he maintains a personalist and profoundly perverse conception of the great sacrament, wants, basing himself on Scheeben, to teach the opposite (pp. 28–32; 69). St. Augustine and St. Thomas expressly reject this teaching (*De Trinitate,* XII, ch. 5; Ia, q. 93, a. 6). On marriage and its mystical sense, one should read Cornelius a Lapide, *In Epistolam ad Ephesios,* 5:32, and Denis the Carthusian, *Enarratio in Canticum Canticorum,* a. 2. The need to make known the writings of St. Augustine against the Pelagian exaltation of man and of liberty, as well as his writings on marriage, has become ever more urgent.

61. "Cum dicitur quod Deus diligitur ab Angelo inquantum est ei bonus, si ly inquantum dicat finem, sic falsum est: non enim diligit naturaliter Deum propter bonum suum, sed propter ipsum Deum. Si vero dicat rationem amoris ex parte amantis, sic verum est: non enim esset in natura alicujus quod amaret Deum nisi ex eo quod unumquodque dependet a bono quod est Deus" (Ia, q. 60, a. 5, ad 2).

62. ". . . homo non ordinatur ad societatem politicam secundum se totum, et secundum omnia sua" (IaIIae, q. 21, a. 4, ad 3).

63. See *Divini Redemptoris,* Acta Apostolicae Sedis, March 31, 1937, p. 79.

64. It is true that in brutes knowledge is ordered to something less than knowledge, namely, nutrition and generation. Under this strict aspect, the knowledge of the brutes, purely instrumental, is a sort of anomaly. This anomaly disappears when one considers the brutes as ordered to man where knowledge has the note of term, and where the cognitive senses (as opposed to the natural senses) are not simply useful. Knowledge cannot be of itself ordered to something lower than knowledge. Its condition is a certain anomaly when 'to be another' is subordinated to the obscure 'being itself'.

65. Marx, *Morceaux choisis,* 232, 229, 217.

66. Since the moral person is not properly an individual substance, one cannot apply to it the definition *rationalis naturae substantia individua.* The moral person is essentially common, such as the person of the chief as chief, or the common personality that is a society (Salmanticenses, *Cursus theologicus* [Palmé], t. VIII, d. 14, dub. 1, p. 23B). The term 'person' that we encounter in the two cases—physical person, moral person—is neither univocal nor analogous, but properly equivocal. The jurist who does not occupy himself formally with natures can bring them together in a quasi-genus: subject of right. Note, in passing, the important distinction to be made between 'subject of right' and 'foundation of right', that moderns tend to confuse. The right is defined by law and law by the common good.

67. ". . . ut pro fine habeatur ratio status communis, quae est pernicies reipublicae bene ordinate" (John of St. Thomas, *Cursus theologicus,* t. VII, d. 19, a. 6, a. 12, p. 694). "Aliud habet justitia legalis ex parte boni communis, quatenus illi debet princeps bonum gubernationem, et sic oportet, quod respiciat altiorem finem, quam ipsum bonum commune, scilicet Deum quod nisi respiciat gubernatio boni communis, declinabit in ratione status" (ibid., n. 16, p. 696).

68. Ibid., n. 13, p. 695.

69. Marx has very well seen this tyrannical and alienating power of the State, but he sought the solution in the very logical application of Kantian personalism. According to Kant, man is an end to himself. The ultimate end that God proposed in the creation of rational beings is the persons themselves in their proper dignity. This dignity does not derive from the fact that the person can himself attain to the ultimate end of the universe, an end other than the person; the person derives his dignity from himself in that he has himself for his end and accomplishes it himself in his liberty and autonomy (see *Foundations of the Metaphysics of Morals,* section two). According to Marx, any ordination to something other than the self wounds man's dignity which requires that man himself should be the root of man. "To be radical is to take things by their root. And the root of man is man himself"; ". . . man is the supreme essence of man" (*Morceaux choisis,* pp. 186–87). "Philosophy no longer conceals it. The profession of Promethus: 'in a word, I hate all the gods . . .' is its own profession, the discourse that contends and will always contend against all the gods of heaven and earth, who do not recognize human conscience as the highest divinity. This divinity suffers no rival" (p. 37). "Human emancipation will only be realized when the real individual man will have absorbed the abstract citizen, when as individual man in his empirical life, in his individual work, in his individual relations, he will have become a generic being and thus will have recognized his proper powers as social powers and will have organized them himself as such, and, by consequence, will no longer separate from himself the social power under the form of political power" (p. 217). "Communism is the positive abolition of private property considered as the separation of man from himself, therefore communism as the real appropriation of the human essence by man and for man, is the return of man to himself as social man, that is, the human man, complete, conscious returns, with all the richness of the previous development. This communism, being an achieved naturalism, coincides with humanism; it is the true end of the quarrel between existence and essence, between objectification and the affirmation of self, between liberty and necessity, between the individual and the species" (p. 229). "It is beyond this reign of necessity that the development of the powers of man begins, who is himself his own end, which is the true reign of liberty, but which can only come about in basing itself on this reign of necessity" (p. 234). Immortality, which would put man in dependence on something other than himself, which consequently is contrary to his dignity, will be itself 'courageously' denied. Here is something quite conformed to the Marxist dialectic, as it is, this time, with the truth: thus dignity implies its own negation.

70. See Appendix III below

71. Cajetan, *In IIamIIae,* q. 81, a. 6; John of St. Thomas, op. cit., t. VII, d. 19, a. 6, nn. 9–18.

72. Antoine de Saint Exupéry, *Pilote de guerre* (New York: Editions de la Maison Française), 212.

73. Even the sin of Adam was without speculative ignorance. "*Adam non est seductus, sed mulier.* Seductio autem duplex est, scilicet in universali, et in particulari eligibili, quae est ignoratio electionis. Quicumque ergo peccat, seducitur ignorantia electionis in particulari eligibili. *Mulier autem fuit seducta,* ignorantia in universali,

quando scilicet credidit quod serpens dixit; sed vir non credidit hoc, sed deceptus fuit in particulari, scilicet quod gerendus esset mos uxori, et cum ea comedere deberet, et inexpertus divinae severitatis credidit quod facile ei remitteretur" (*In I ad Tim.*, c. 2, lectio 3). See too *In II ad Tim.*, c. 3, lectio 2, on the words *semper discentes et numquam veritatem invenientes.*

74. This is an expanded version of an address that appeared in the proceedings of the American Catholic Philosophical Association, 1941, pp. 1–15.

75. Joseph Stalin, in *History of the Communist Party* (Moscow, 1939) ch. 4, section 2. The same section was issued separately by International Publishers, New York, under the title *Historical and Dialectical Materialism.*

76. "A sophista vero differt philosophus 'prohaeresi,' idest electione vel voluptate, idest *desiderio vitae*" (*In IV Metaphysicorum*, lectio 4, n. 575).

77. "It would be strange to think that political science or prudence should be the best knowledge, since man is not the best thing in the world" (*Nicomachean Ethics*, VI, 7, 1141a20).

78. John of St. Thomas, *Cursus theologicus* (Solesmes edition), t. I, p. 395.

79. Cajetan, *IaIIae*, q. 58, a. 5.

80. Cajetan, *IIaIIae*, q. 47, a.1.

81. I, ch. 2, 982b28.

82. *Q.D. de virtutibus cardinalibus*, a. 1, c. ". . . vita autem humana est quae est homini proportionata. In hoc homine autem invenitur primo quidem natura sensitiva, in qua convenit cum brutis; ratio practica quae est homini propria secundum suum gradum; et intellectus speculativus, qui non perfecte in homine invenitur sicut invenitur in angelis, sed secundum quamdam participationem animae. Ideo vita contemplativa non est proprie humana, sed superhumana; vita autem voluptuosa, quae inhaeret sensibilibus bonis, non est humana, sed bestialis. Vita ergo proprie humana, est vita activa, quae consistit in exercitio virtutum moralium."

83. *Physics*, II, ch. 2, 194a30–35. –*In II Physicorum*, lectio 4, n. 8: ". . . nos utimur omnibus quae sunt secundum artem facta, sicut propter nos existentibus. Nos enim sumus quodammodo finis omnium artificialium." This proposition is verified even in the case of religious works of art for these imitations are made with a view of representing the originals in a way more proportioned to us.

84. ". . . prudentia est circa bona humana, sapientia autem circa ea quae sunt homini meliora" (*In VI Ethicorum*, lectio 6, n. 1191).

85. *Nicomachean Ethics*, VI, ch. 5; St. Thomas, lectio 4, nn. 1162–63: "Et dicit (philosophus) quod ad prudentem videtur pertinere, quod sit potens ex facultate habitus bene consiliari circa propria bona et utilia, non quidem in aliquo particulari negotio, puta qualia sint bona vel utilia ad sanitatem vel fortitudinem corporalem; sed circa ea quae sunt bona et utilia ad hoc quod tota humana vita sit bona . . . quia scilicet illi quidem dicuntur prudentes non simpliciter, sed circa aliquid determinatum, qui possunt bene ratiocinari quae sunt bona vel utilia ad aliquem finem determinatum, dummodo ille finis sit bonus; quia ratiocinari de his quae pertinet ad malum finem est contrarium prudentiae: dummodo hoc sit circa ea quorum non est ars; quia bene

ratiocinari de hoc non pertinet ad prudentiam, sed ad artem. Si ergo ille qui est bene consiliativus ad aliquid particulare est prudens particulariter in aliquo negotio; consequens est, quod ille sit totaliter et simpliciter etiam prudens, qui est bene consiliativus de his quae pertinent ad totam vitam."

86. *Q.D. de potentia,* q. 3, a. 6, ad 5.

87. Ia, q. 49, a. 3, ad 5; q. 63, a. 9, ad 1; q. 23, a. 7, ad 3; IaIIae, q. 71, a. 2, ad 3.

88. *In II Ethicorum,* lectio 7; *In I Sententarium,* d. 39, a. 2, ad 4.

89. And in the article on the Renaissance, humanism "denotes a specific bias which the forces liberated in the Renaissance took from contact with the ancient world—the particular form assumed by human self-esteem at that epoch—the ideal of life and civilization evolved by the modern nations. It indicates the endeavor of man to reconstitute himself as a free being, not as the thrall of theological despotism, and the peculiar assistance he derived in this effort from Greek and Roman literature, the *litterae humaniores,* letters being rather on the side of man than of divinity.—In this article the Renaissance will be considered as implying a comprehensive movement of the European intellect and will toward self emancipation, toward reassertion of natural rights of reason and the senses, toward the conquest of this planet as a place for human occupation, and toward the formation of regulative theories both for states and individuals differing from those of medieval times."

90. "Statuit tandem optimus opifex. . . ." See the pages taken from the *Theologica Platonica* of his master and friend Marsilius Ficinus.

91. Some authors use the term "humanist" to mean a very elevated conception of man's natural faculties. This usage is the cause of many purely verbal misunderstandings. When one gives so large a meaning to the term, it must be said that Saint Thomas is infinitely more humanist than Erasmus, indeed that he is opposed to Erasmus as to the destroyer of that which is best in man. What is nowadays called the "vulgar" conception of humanism, and which is based on the work of Burckhardt, Monnier, and Symonds, cannot avoid this trick of imposition. Moreover, this imposition can find a justification in the vulgar conception of humanism, which also turns on the meaning of man's "natural powers." One need not concede this slide of terminology in every domain. In St. Thomas, "essence" is not a fuel and, in reading him, we ignore this new imposition which is not devoid of foundation. One would not weaken the thesis of the movement that includes men who are called humanists according to the vulgar conception of humanism because of their ideas, citing passages from Erasmus, for example, against a Mirandola, rising against the so-called rationalization of the Gospels and against the philosophical hellenism of the Middle Ages. Erasmus is profoundly humanist when he wants to reject Aristotle, and even more so when he attacks Scholastic theology on the pretext of defending Christian wisdom. One diminishes his genius by wishing to excuse him by citing the abuses into which decadent Scholasticism fell. One diminishes the mastery of a master even more when one maintains that his work, isolated from infinitely complex historical circumstances, is not truly intelligible. One who attacks the great Scholastics of the Middle Ages and ignores his own greatest contemporaries ought also, in our opinion, to attack what is most profound in Greek wisdom,

that is, that by which man can best gain Christian speculative wisdom and moral science, since that is as well what is most divine in man. The bare evangelism that Erasmus preached is what is most humanist in him in our sense of the term. His doctrine is no less humanist for being called the "philosophy of Christ," or for having called the use of philosophy in theology paganism. We do not deny as quite well-founded the enthusiasm later manifested for the great discoveries of science. What we see as humanist in that is the hope with which this new power is invested. Nor do we deny the strength of the temptation; we call humanist the way in which certain persons reacted to it, and we count them among our adversaries. No doubt words are conventional in meaning—*ad placitum.* But that need not lead us to fail to follow the advice of Saint Thomas: "We ought not have terms in common with the gentiles, lest the sharing of terms become an occasion of error; the faithful should avoid the use of the term 'fate', lest they seem to approve those who employ it in a bad sense" (*III Summa contra gentes,* c. 93).

92. Pico della Mirandola, *On the Dignity of Man,* trans. Charles Glen Wallis and Paul J.W. Miller (Indianapolis: Hackett, 1998), 4–5.

93. ". . . poetica scientia est de his quae propter defectum veritatis non possunt a ratione capi; unde oportet quod quasi quibusdam similitudinibus ratio seducatur" (*In I Sententiarum,* Prol., q. 1, a. 5, ad 3; Ia, q. 1, a. 11, obj. 1 and ad 1).

94. Ia, q. 1, a. 9.

95. *Discours de la méthode,* ed. Gilson (Paris: Vrin, 1930), part 6, pp. 61–62.

96. *The English Philosophers from Bacon to Mill,* ed. Burtt (New York: Modern Library, 1939), 587.

97. Even Kant's conception of physics is dialectical insofar as its method is interrogative.

98. See Appendix V below.

99. *In I Posteriorum analyticorum,* lectio 1.

100. Compare with F.C.S. Schiller, *Hypothesis,* in *Studies in the History and Method of Science,* ed. Charles Singer (Oxford, 1921), vol. 2, pp. 429–30.

101. St. Thomas, *In Boethii de trinitate,* q. 6, a. 1; *In IV Metaphysicorum,* lectio 4; *In I Posteriorum analyticorum,* lectio 20, n. 5: "Pars autem logicae, quae demonstrativa est, etsi circa communes intentiones versetur docendo, tamen usus demonstrativae scientiae non est in procedendo ex his communibus intentionibus ad aliquid ostendendum de rebus, quae sunt subiecta aliarum scientiarum. Sed hoc dialectica facit, quia ex communibus intentionibus procedit arguendo dialecticus ad ea quae sunt aliarum scientiarum, sive sint propria sive communia, maxime tamen ad communia." See also John of St. Thomas, *Cursus philosophicus* (ed. Reiser), t. I, p. 278.

102. *The Logic of Hegel,* translated from the *Encyclopaedia of the Philosophical Sciences* by W. Wallace (London, 1892), 221. Engels finds a similar confirmation of this verbal negation in the calculus of his time. "When the mathematics of straight and curved lines has thus pretty well reached exhaustion a new almost infinite field is opened up by the mathematics that *conceives curved as straight* (differential triangle) and *straight as curved* (curve of the first order with infinitely small curvature). O metaphysics!" (*Dialectics of Nature,* trans. Wallace [New York: International Publishers, 1940], 201).

103. G.V. Plekhanov, *Les questions fondamentales du marxisme,* Dialectique et Logique (Paris: E.S.I., n.d.). 97 ff.

104. *Metaphysics,* V, ch. 6, 1016a25 (lectio 7, n. 863); *Physics* IV, ch. 14, 224a (lectio 23, n. 13); St. Albert, *IV Physicorum,* tract. 3, c. 17.

105. On the negation of negation, see F. Engels, *Anti-Dühring,* t. 1, ch. 13, entitled "Dialectic. Negation of the Negation" (Paris: Alfred Costes, 1931), 198 ff.

106. "It appears that a certain Herr Krug, supposing Hegel to be attempting in the philosophy of nature to deduce all actual existent objects from the pure Idea, enquired whether Hegel could deduce the pen with which he, Herr Krug, was writing. Hegel demolishes the unfortunate Krug in a contemptuous and sarcastic footnote, in which he states that philosophy has more important matters to concern itself with than Krug's pen. And the general position he takes up is that the philosophy of nature cannot and should not attempt to deduce particular facts and things, but only universals. It cannot deduce *this* plant, but only plant in general, and so on. The details of nature, he says, are governed by contingency and caprice, not by reason. They are irrational. And the irrational is just what cannot be deduced. It is most improper, he tells us, to demand of philosophy that it should deduce *this* particular thing, *this* particular man, and so forth.... In my opinion Hegel was wrong, and Krug right, as regards the question of the pen. And Hegel's ill-tempered petulance is possibly the outcome of an uneasy feeling that Krug's attack was not without reason. If we are to have an idealistic monism it must explain everything from its first principle, thought. And that means that it must deduce everything. To leave anything outside the network of deduction, to declare anything utterly undecidable, is simply dualism" (W.T. Stace, *The Philosophy of Hegel* [London: Macmillan, 1924], 308). Lest the context of this citation mislead, it should be said that Professor Stace is not a Marxist.

107. Stalin, *Historical and Dialectical Materialism,* 100.

108. Marx, *Morceaux choisis,* 156.

109. Stalin, *Historical and Dialectical Materialism,* 197.

110. Ibid., 103.

111. Ibid., 156.

112. See Plekhanov, *Les questions fondamentales,* 97 ff.

113. Marx, *Morceaux choisis,* 233.

114. Ibid., 222.

115. Ibid., 222, 221, 223, 37.

116. Denys l'Aréopagite, *Traité de la théologie mystique,* trans. Du R.P. Dom Jean de S. Francois, in *Oeuvres de S. Denys Aréopagite* (Paris: Nicolas Buon, 1629), ch. 3, pp. 546, 545.

117. Marx, *Morceaux choisis,* 233.

118. "Then I saw rise from the sea a beast..." (Apoc. 13:1). In the *Expositio II* on the Apocalypse, found in the edition of Saint Thomas's works (Vivès, t. 32), we find the following commentary: "*And I saw,* that is, interiorly, *a beast,* that is, a body, a crowd of perverse men living in a bestial manner and cruelly devouring other men, namely by causing them both spiritual and corporeal damage.... *From the sea,* that is,

from a world shaken by the tempest of tribulations and temptations and made bitter by these transgressions, for this beast will be made up of the different nations of the world" (p. 298).

In the Desclée edition of Crampon's translation, we read, in the same place, in a note, "The four beasts of Daniel represent each an empire (7:17, 23), that of the Apocalypse which unites in itself the traits of all the others (5:2) should necessarily represent the ensemble of empires and be the *powerful political symbol* of the material power of states, *put at the service of the dragon to oppress the servants of God.* It rises from the sea, like the four beasts of Daniel (7:1) because empires arise ordinarily from wars and the troubles which agitate the peoples.—*Then I saw rise from the earth another beast . . .* (Apoc. 8:11). "*From the earth:* the first beast came from the sea, that is from the agitation and overturning of peoples; this one comes out of the earth, the calmest element: it is born in a tranquil social state, in the bosom of civilization.—*Another beast:* all the traits that follow make up the symbol of *false science,* of the wisdom of this world *in the service of impiety.* It is also designated later as 'the false prophet'" (ibid.).

119. Marx, *Morceaux choisis,* 186, 187.

120. The Encyclical *Divini Redemptoris* denounces modern communism as a doctrine of false redemption: *fucata tenuiorum redemptionis specie profertur*" (Acta Apostolicae Sedis, March 31, 1937, p. 69). See the remarks of Father Alphonse-Marie Parent in his essay entitled "Autour du racisme," in *L'Academie canadienne Saint-Thomas d'Aquin* (1941).

121. ". . . non ita quod id quod est necessarium, sit sicut finis; quia id quod necessarium est, ponitur ex parte materiae; sed ex parte finis ponitur ratio necessitatis. Non enim dicimus quod necessarium sit esse talem finem, quia materia talis est; sed potius e converso, quia finis et forma talia futura est, necesse est materiam talem esse. Et sic necessitas ponitur ad materiam, sed ratio necessitatis ad finem" (*In II Physicorum,* lectio 15, n. 4).

122. In *Ego Sapientia,* Part Two, I insisted on the perverse interpretation to which one could subject the doctrine of the power of the weak. I did not know, not yet having read this book of Feuerbach in its complete text, that modern philosophy had really done it in so elaborate a manner.

123. "Sed primus homo peccavit principaliter appetendo similitudinem Dei quantum ad *scientiam boni et mali,* sicut serpens ei suggessit: ut scilicet per virtutem propriae naturae determinaret sibi quid esset bonum et quid malum ad agendum; vel etiam ut per seipsum praecognosceret quid sibi boni vel mali esset futurum. Et secundario peccavit appetendo similitudinem Dei quantum ad propriam potestatem operandi, ut scilicet virtute propriae naturae operaretur ad beatitudinem consequendam: unde Augustinus dixit xi super Gen. Ad litt., quod *menti mulieris inhaesit amor propriae postestatis*" (IIaIIae, q. 163, a. 2, c.).

124. "To what does the immense scaffolding of *The Essence of Christianity* lead? To the exaltation of sexual sensuality, where man-species achieves Man-Gattung in a physical and concrete way" (op. cit., ch. 18). "Religion is, according to Feuerbach, the sentimental relation of man to man who . . . now find reality directly and without

intermediary in the love between me and thee. And it is thus that sexual love becomes, at the end of the day, for Feuerbach, one of the most, if not the most, elevated forms of the exercise of his new religion" (Engels, *Feuerbach*). But Marxism too, despite the protests of Engels who found all this dispiriting, will lead to an analogous *Dämmerung*. "What are goods? Material goods. What are material goods. Nourishment, clothes, shoes, housing, fuel, instruments of production, etc." (Stalin, *Historical and Dialectical Materialism,* 113).

125. Friedrich Engels, *Ludwig Feuerbach,* in *Études philosophiques* (Paris: Editions sociales internationales, 1935), 10–11.

126. This text, which Professor A. Viatte made known to me several years ago, has been cited many times, but always incompletely. One leaves out just the passages which, from my point of view, are the most important, that is, those which directly incriminate philosophy.

127. Heinrich Heine, *De l'Allemagne* (1834) (Paris: Callmann Levy, 1878), t. I, pp. 179–83. [Ed. I have translated from De Koninck's text. A doubtless more elegant translation can be found in *The Works of Heinrich Heine,* in two volumes, trans. Charles Godfrey Leland (London: William Heinemann, 1892), vol. 1, pp. 205–11.]

128. We do not mean by that that only natures, even natures in a large sense, are the works of the divine art. Every work of God, everything of which He is the cause, is a work of the divine art. *Omnia per ipsum facta sunt: et sine ipso factum est nihil, quod factum est.*

129. Think of the discussion of Marsilio Ficino above.

130. It is to be noted that in the treatises most advanced in the way of concretion, Aristotle opposes natural doctrine to the speculative sciences: "The mode of necessity, however, and the mode of demonstration are different in natural science from what they are in the theoretical sciences" (*On the Parts of Animals,* I, 1, 640a1). He opposes this treatise to those composed "according to philosophy" (642a5). In some respects, natural doctrine, art, and prudence agree in a quasi-genre opposed to metaphysics and mathematics, according to St. Thomas.

131. It is also true that the relative anteriority of the absolute Idea presents difficulties of interpretation, but it remains that each category inferior to the first, and each species, is transcendentally the fruit of pure becoming, of the movement of reason, by means of contradiction. The impossibility of making clear the relation between the first reason and the absolute Idea makes sufficiently clear the impossibility that bothers Hegel himself.

132. *Physics II,* ch. 5, 197a18; lectio 9, n. 4. See as well *III Summa contra gentes,* c. 99: "Ordo enim inditus . . . etc."

On the Common Good

Review of

The Primacy of the Common Good

Yves R. Simon

1944

Of all the philosophical investigations which may throw light on our political, social and moral problems, none is more badly needed and eagerly demanded than a thorough study of the concept of common good. To achieve such a contribution, nobody is better qualified than the young and profound philosopher who heads the Laval School of Philosophy. On the other hand, there is little doubt that the various "personalistic" schools, which have been noisy in the last twelve years or so, shelter quite a few trends of such a nature as to jeopardize the primacy of the common good. Cardinal Villeneuve, in his preface to *De la primauté du bien commun contre les personalistes,* stresses felicitously both the timeliness of a sound doctrine of the common good and the urgency of a criticism of those "personalistic" trends.

It cannot be said that De Koninck's book meets our expectation. It cannot be said, either, that it brings about disappointment. The truth is that the present volume is by no means a book on the common good. It contains only a short essay on that topic (79 pages). The rest of the volume is made up of an essay on *The Principle of the New Order* and a few notes in which the writer describes philosophical and theological errors that he perceives at the root of the calamities of our time.

Considering, accordingly, that we have to do, not with a book, but with a short essay, let us say that De Koninck has outlined, with unusual profundity and accuracy, the main aspects of a theory of the common good. It would be unfair to blame such a brief treatment for what we do not find in it. We do find in it a most valuable contribution to the definition of the common good and to the vindication of its primacy. Here is a survey of the main doctrinal points treated by De Koninck.

1. If the common good were merely a collection of private goods, its excellence would be merely material. The genuine principle of its excellence is its *communicability.* "The common good is greater [than the private good] for every being which participates in it, inasmuch as it is communicable to other particular beings" (p. 8).

2. The common good of a multitude is the good of every member of the multitude. If it were merely the good of the multitude itself, considered

as a kind of individual entity, it would not really be common. Let us not say, for instance, that the species seeks its own good against the natural desire of the individual; let us say, rather, that the individual itself naturally seeks the good of the species more than its private good. Accordingly, the common good is not by any means a *bonum alienum.*

3. Rational creatures, on the ground of their ability to grasp the all-embracing concept of being, are capable of relating themselves actively, through knowledge and love, to the common good of the whole universe. Rational natures are distinguished by the incomparably greater intimacy, as well as by the loftiness, of their relation to the common good (pp. 12–16).

4. Loving the common good in order to possess it is not loving the common good as such. That kind of love for the common good characterizes tyrants. A society made of people who all love the common good that way would be a society of tyrants (p. 17).

5. The subordination of the temporal common good to the supernatural good should not be mistaken for the subordination of a good that is common to a good that is private; the higher good to which all temporal good is subordinated is itself a common good (p. 19). "The supernatural good of the individual person is essentially subordinated to the supernatural common good, in such a way that it is impossible to distinguish between the supernatural virtue of a man and the supernatural virtue of the same man considered as part of the heavenly city" (p. 63).

So much for the doctrine outlined by De Koninck. It calls for many specifications and further developments, but it constitutes a very sound foundation for any further development of the theory of the common good.

Turning to the polemical side of the essay, we realize at once that the writer was confronted by a great difficulty. De Koninck's purpose is to vindicate the primacy of the common good *against the personalists.* It is a hard job, for the obvious reason that the term personalism covers a great variety of ill-defined doctrines and attitudes. We shall examine two questions: First, what are the characteristics of personalism according to the writer's judgment? Second, who are the "personalists" against whom he vindicates the primacy of the common good?

The first question is not too clearly answered. It is a pity that the writer did not sum up, in orderly fashion, the elements of what he calls personalism. He gives us only piece-meal information about his own conception of the philosophy which he fights with such ardent conviction.

P. 14. We find an allusion to people who like the individual good of their own person better than the common good, and erect their personality into a universal standard of values. The reader is not explicitly told whether the personalists do the rationalizing for those disgusting fellows. He strongly suspects that they do.

P. 16. Allusion is made to "authors" who rank the common good "second."

Pp. 22 and 77. There is severe criticism of the "people" (in French: *on*) who think that a man can be a good political leader without being good as a man (*bonus vir*). This is surely a dreadful error, but so old and so widespread that we wonder whether it has any particular connection with personalism. If it does not, why should it be mentioned here? The reader is led to believe that it does.

P. 41. "If the human person were really what the personalists say it is . . . the self would be the principle of his destiny; it would be also the term of its destiny; it would not subordinate itself to an end distinct from itself except in order to subordinate that end to itself. . . ."

P. 67. "Personalism reverses that order of the goods; it attributes the greatest good to the formality which is the lowest in man. What the personalists understand by person is, in truth, what we understand by mere individual—a material and substantial whole confined in itself, and they reduce the rational nature to the sensuous nature, whose object is the private good."

P. 73. He quotes Jacques de Monléon: "The so-called personalists who put the person above the common good no longer can see in the latter the bond of the persons."

P. 75. "In fact, personalism adopts the totalitarian concept of the state."

The reader cannot help being very anxious to identify the men who profess all those vicious stupidities; on the other hand, since the writer describes personalism as a redoubtable threat, and speaks with a high sense of duty, of his task as critic of the personalists (p. 4), he could be expected to complete his warnings, and make them practical, by giving the names of at least a few personalistic leaders, and the titles of the main works in which personalistic errors are set forth. The discretion of De Koninck, in this connection, is one of the striking features of his essay. A careful examination reveals only four names. Karl Marx is often referred to, but he is not an immediate threat in the Province of Quebec, where Bolshevism never was very popular. In Appendix II, p. 130, we find a quotation from Mortimer Adler and the Rev. Walter Farrell, accompanied by a severe and fully deserved criticism. Lastly, in a footnote, p. 181, De Koninck mentions H. Doms, whose ideas on marriage are called "personalistic and deeply perverse." That is all, and that is too little. Nobody would believe that De Koninck's polemics are directed exclusively against Adler, Farrell, Doms and . . . Karl Marx. The question *who are those personalists?* is left unanswered.

There are many personalistic groups in the world today, and it is arbitrary to speak of "personalism" as if it were one doctrine, and of "the personalists" as if they were one unified school of thought. Let us ask ourselves who are the men for whom the expression "the personalists" is likely to stand in the imagination of the average reader of a book published by the Laval University Press, Quebec, and the Editions Fides, Montreal, in 1943. He may think of the editors and contributors of the magazine *The Personalist* (Los Angeles); of Mounier and *Esprit* (Paris); of Aron, Dandieu, and the magazine *L'Ordre Nouveau* (Paris); of Scheler, Doms, Dietrich von Hildebrand; perhaps of Adler and Father Farrell; of a few French Canadian writers, the best known of whom is François Hertel, author of a book entitled *Pour un ordre personaliste* (Montreal, 1942); finally, on account of the very important part played by the concept of person in the work of Maritain, there is no reason why he should not believe that the expression "the personalists" stands for Jacques Maritain.

The last interpretation was favored by the circumstances; it was bound to prevail. To my knowledge, everybody has believed, and still believes that Maritain is the real target of De Koninck's vindication of the primacy of the common good, against the personalists.

Now, re-read the collection of personalistic stupidities and monstrosities that we abstracted from De Koninck's essay: insofar as he trusts the writer, and believes that his criticism is aimed at Maritain, the reader cannot help taking for granted that those stupidities and monstrosities are really found in Maritain's work. The net effect of the essay, insofar as Maritain is concerned, resembles that which could have been brought about—perhaps not so successfully—by plain calumny. A charitable guess is that De Koninck was so unaware of the situation as not to foresee what the reaction of his readers would be.

I do not think for a moment that De Koninck intended to libel Maritain, anxious though he may be to oppose his influence. I rather suspect that he has chosen the line of least resistance. As we have already remarked, a thorough discussion of contemporary personalism would be an extremely difficult job, owing to the fact that there are many personalistic doctrines and that most of them are very ill-defined. It was much easier to pick a few ideas which are felt to be in the air, some gossip, a few pieces of personalistic oratory, and to leave it up to the reader, who bears in mind the great name of Maritain, to ascertain whether Maritain is a personalist or not. In this connection, it must be pointed out that De Koninck's essay presents other evidences of a preference for the line of least resistance. Even its purely doctrinal parts, which we praised so highly, are spoiled by what seems to be hasty composition. Admirable texts of St. Thomas are translated into a dialect which becomes intelligible only after you have re-translated it into Latin.

It looks very much as though De Koninck had not read the ethical and political writings of Maritain. Such an ignorance can hardly be involuntary. Seeking reasons for that voluntary ignorance is none of our business.

To conclude: Insofar as De Koninck's essay vindicates the primacy of the common good and carries out the criticism of definite positions, it is entirely praiseworthy. Insofar as it seems to be aimed at books that the writer never read, at doctrines that he does not expound, and men whose names he does not mention, the essay *against the personalists,* is a purely private affair, which concerns exclusively De Koninck, and is of no concern whatsoever to the reader.

In Defense of Jacques Maritain

Attack on

The Primacy of the Common Good

I. Thomas Eschmann, O.P.

Pontifical Institute of Mediaeval Studies, Toronto, Ontario

Université de Montréal, Institut d'études médiévales
Albert-le-Grand, Montréal, Québec

1945

. . . *utpote personae sunt.*
—Pius XII, Encyclica *Mystici Corporis*

I. On Censures, Insinuations, and Citations

The problem of Person and Society in the philosophy of St. Thomas, for many years past a favorite topic among European Thomists, has recently become an acute question on the continent of North America, owing, in no small measure, to the publication by the Dean of the Faculty of Philosophy at Laval University, Québec, Professor Charles De Koninck, of a book entitled *De la primauté du bien commun contre les personnalistes. Le principe de l'ordre nouveau.*[1]

Whether the author of this book was aware of it or not—it makes no difference to the fact—this title pointedly assails the eminent French philosopher, Jacques Maritain. One need only call to mind Maritain's *La primauté du spirituel* and the well-known fact that throughout all his writings this doctrine of the primacy of the spiritual is crystallized and, as it were, concretized in the primacy of the personal.

This also explains the fact that the reading public generally connected Professor De Koninck's thesis with Maritain and his doctrine [184] of the person.[2] Such a view of the impact of Professor De Koninck's book was at once inevitable and, given the circumstances, perfectly correct. For, books comprise more than their objective, abstract content, more than the mere words in which they are written. They embrace all the circumstances of time, place, and occasion with which their publication is surrounded. Books are qualified *actus humani*, public documents burdened with all the references and relations accompanying their appearance in print. What point could there be, then, in any possible denial by Professor De Koninck of the charge that he meant to challenge Maritain's teaching or of any possible assertion on his part that he knew nothing about such teaching? His book is not only a polemic against "the" personalists—an all too convenient anonymity which permits every attack, and leaves every avenue of retreat wide open—but it is also, and primarily, a polemic against personalism, i.e. that doctrine on the person which right here and now, in our own day and age, in this very country and among Thomists, is represented most prominently by Jacques Maritain.[3]

This is the personalism which is at issue in a passage on page three of Professor De Koninck's book, a passage which recalls by its style and bearing the "heroic" ages of baroque-Scholastic controversy:

> Le péché des anges fut une erreur pratiquement personnaliste: ils ont préféré la dignité de leur propre personne à la dignité qui leur serait venue dans la subordination à un bien supérieur mais commun dans sa supériorité même. L'hérésie pélagienne, dit Jean de Saint Thomas, peut être considérée comme une étincelle de ce péché des anges. Elle n'en est qu'une étincelle, car, alors que l'erreur des anges fut purement pratique, l'erreur des pélagiens était en même temps spéculative. Nous croyons que le personnalisme moderne n'est qu'une réflexion de cette étincelle, spéculativement encore plus faible. Il érige en doctrine spéculative une erreur qui fut à l'origine seulement pratique. . . .
>
> Nous n'entendons pas soutenir ici que l'erreur de tous ceux qui se disent aujourd'hui personnalistes est plus que spéculative. Qu'il n'y ait là-dessus aucune ambiguité. Sans doute notre insistance pourra-t-elle blesser ceux des personnalistes qui ont identifié cette doctrine à leur personne. C'est la leur responsabilité très personnelle. Mais il y aussi la nôtre—nous jugeons cette doctrine pernicieuse à l'extrême.

There is a proper and profound Thomistic doctrine of the relative superiority, within definite orders, of *their* respective common goods over the particular goods contained in those orders. It is this doctrine which Professor De Koninck has distorted into the contradictory [185] and unintelligible position of the absolute superiority of "the" common good over all and everything. This will be shown later in detail. Father Jules A. Baisnée in a recent article in *The Modern Schoolman*[4] has chosen to defend and emphatically to recommend this distortion, and to defend and recommend it against Jacques Maritain. Father Baisnée reveals himself to be much impressed by the weight of the authorities which, according to him, Professor De Koninck's antipersonalist position commands. He, therefore, seems to yearn for an official condemnation of Personalism (Jacques Maritain's doctrine on the person not only not excluded but unmistakably included). He imputes to "the" Personalists (Jacques Maritain's doctrine not only not excluded but unmistakably included) that "freedom, autonomy, dignity are words which come frequently under [their] pen . . . and they insist that subordination of man to any general good but the good of God, . . . to any authority but the divine authority would mean a denial of man's very personality."[5] This statement is not supported by any reference to any text of any "Personalist" and, therefore, embraces every text of every man who ever, by whomsoever, has been brought in whatever connection with "Personalism." If it be applied to Jacques Maritain, it will be—*sit venia verbo*—utter nonsense, and there is no point here in carrying coals to Newcastle. One is almost tempted to repeat St. Thomas' challenge: ". . . non loquatur in angulis" (*De unitate intellectus*)! The above statement may be taken for one of the "reasons" why Father Baisnée fears that "Personalism" (Jacques Maritain's doctrine not only not excluded but unmistakably included) may be "a revival of the polycephalous monster of Pelagianism."[6] A strange fate, surely, for Jacques Maritain to be now a Pelagian after having been accused, not many years ago, of opening the door to Lutheranism!

Pacem, amici! Would it not be better for us to stop short on the road of censuring and adding-up authorities, before it is too late and Catholic scholarship is once more made a laughing stock?

[186] To the extent, however, to which authoritative pronouncements might enter into consideration, it would be well to bear in mind that the language at least of recent papal utterances favours the personalist ideas as expounded by Maritain. At any rate, these utterances should deter Catholic writers from attempting to make a crusade of their opposition to these ideas. I am far from affirming or even insinuating that Maritain's specific set of conceptions which is labelled, *personalism*, has received any official sanction whatsoever. But I do affirm that, in the light of the documents quoted below, it is rather surprising to see Catholic writers indiscriminately indulging in subtle insinuations, strong language, and even specific and grave censures, without having previously made the necessary distinctions with the utmost care and an absolutely unequivocal clarity, and without having given specific references to explicit statements of individual writers.

Pius XI says in the Encyclical *Divini Redemptoris*:

> Societas . . . ex divini Creatoris consilio naturale praesidium est, quo quilibet civis possit ac debeat ad propositam sibi metam assequendam

> uti; quandoquidem Civitas homini, non homo Civitati exsistit. Id tamen non ita intellegendum est, quemadmodum ob suam *individualismi* doctrinam *Liberales,* quos vocant, asseverant; qui quidem communitatem immoderatis singulorum commodis inservire jubent: sed ita potius ut omnes, ex eo quod cum societate composito ordine copulantur, terrenam possint, per mutuam navitatis conspirationem, veri nominis prosperitatem attingere . . . Iamvero, quemadmodum homo officia illa repudiare non potest, quibus Dei iussu civili societate obstringitur, . . . ita pari modo societas iis iuribus civem spoliare non potest a Creatore Deo eidem impertitis . . . neque eorundem usum impossibilem reddere. . . . Dum *communistarum* effata personam ita extenuant, ut civium cum societate necessitudines praepostere subvertant, humana mens, contra, ac divina revelatio eam . . . sublime extollunt.[7]

Pius XII says in the Encyclical *Mystici Corporis Christi*—and this is, indeed, a very remarkable passage which I do not hesitate to call the *Magna Charta* of the Christian doctrine of person:

> Dum enim in naturali corpore unitatis principium ita partes iungit, ut propria, quam vocant, subsistentia singulae prorsus careant; contra in mystico Corpore mutuae coniunctionis vis, etiamsi intima, membra ita inter se copulat, ut singula omnino fruantur persona propria. Accedit quod, si totius et singulorum mutua inter se rationem consideramus, in physico quolibet viventi corpore totius concretionis emolumento membra singula universa postremum unice destinantur, dum socialis quaelibet hominum compages, si modo ultimum utilitatis finem inspicimus, ad omnium et uniuscuiusque membri profectum, utpote personae sunt, postremum ordinantur.[8]

It is, no doubt, an exaggeration to say that Pope Pius XII "makes a plea for personalism." Popes do not make pleas, they pronounce. [187] It is rather Jacques Maritain who makes the plea for the full and intelligible acceptance of the Holy Father's defense of the human person. His doctrine is indeed a serious and, on the whole, successful attempt to give expression, within the framework of a true philosophy and a correctly interpreted Thomism, to the ideas which are put forward in the above quoted documents and in numerous other papal pronouncements.[9]

II. Professor De Koninck on Part and Whole

On page 27 Professor De Koninck states that even the personalists will not have great difficulty in admitting, with him, that individual persons are subordinated to that ultimate separate and extrinsic good of the universe which is God, nor that this subordination is formally motivated by the fact that God is the common good. But this will not suffice. It must be stressed, indeed—such is the author's thesis—that persons are subordinated to the intrinsic common good of the universe, i.e. its order. And they are thus subordinated because they are material parts materially composing and materially constituting that order and common good. For, is not the ultimate reason why God has created the intellectual beings or persons none other than exactly the order and common good of the universe?

> Si l'on concéde que les personnes singulières sont ordonées au bien ultime séparé en tant que celui-ci a raison de bien commun, on ne concédera pas si volontiers que, dans l'univers même, les personnes ne sont voulues que pour le bien de l'ordre de l'univers, bien commun intrinsèque meilleur que les personnes singulières qui le constituent matériellement.

Let us here for the moment consider the second part of this thesis, viz. the statement regarding the intrinsic common good of the universe and its relation to the intellectual beings or persons. Even Professor De K. somehow seems to feel that his is a "revolting" statement (cf. p. 30). He, therefore, makes every effort to be very [188] careful in establishing a Thomistic proof of it. In fact, he asks, is not the same statement repeatedly implied in St. Thomas' discussion of the question: What is the end God has proposed to Himself in the production of all things? Four texts are cited by the author.[10] Let us here reproduce, in Latin, the first two, taken from *Contra Gentiles*, III, 64; they will sufficiently show in what specific set of Thomistic texts Professor De K. has found a proof, satisfying to his mind, of his assertion. The italicized sentences are not held worthy of quotation, by the author:

> Deus res omnes in esse produxit, non ex necessitate naturae, sed per intellectum et voluntatem. Intellectus autem et voluntatis ipsius non potest esse alius finis ultimus nisi bonitas eius, ut scilicet eam rebus

communicaret. . . . Res autem participant divinam bonitatem per modum similitudinis. Inquantum ipsae sunt bonae. Id autem quod est maxime bonum in rebus causatis, est bonum ordinis universi, quod est maxime perfectum, ut Philosophus dicit:[11] cui etiam consonat Scriptura divina *Gen.* I, cum dicitur (vers. 31), "Vidit Deus cuncta quae fecerat, et erant valdc bona," cum de singulis operibus dixisset simpliciter quod "erant bona." Bonum igitur ordinis rerum causatarum a Deo est id quod est praecipue volitum et causatum a Deo. Nihil autem aliud est gubernare aliqua quam eis ordinem imponere. *Ipse igitur Deus omnia suo intellectu et voluntate gubernat.*

Amplius. Unumquodque intendens aliquem finem, magis curat de eo quod est propinquius fini ultimo: quia hoc etiam est finis aliorum. Ultimus autem finis divinae voluntatis est bonitas ipsius, cui propinquissimum in rebus creatis est bonum ordinis totius universi: cum ad ipsum ordinetur sicut ad finem omne particulare bonum huius vel illius rei, sicut minus perfectum ordinatur ad id quod est perfectius; unde et quaelibet pars invenitur esse propter suum totum. Id igitur quod maxime curat Deus in rebus creatis, est ordo universi. *Est igitur gubenator ipsius.*

To be sure (Professor De K. continues arguing) in chapter 112 of the same work and book Aquinas seems to have made a statement contrary to the author's own. The intellectual creatures, St. Thomas says, are governed for themselves. How, then, can they still be a material part of the cosmos? How can it still be true that God's intention, in the creation of such beings, is "nothing but the order of the universe"? Let nobody be alarmed, Professor De K. assures the troubled reader! For, in this same chapter, if you read it to the end, the doctrine of the intellectual creatures as material parts of the universe shines forth even with greater clarity and splendor: "Le fait que les parties principales constituant matériellement l'univers [189] sont ordonnées et gouvernées pour elles-mêmes ne peut que faire éclater davantage la suréminente perfection de l'ensemble qui est la raison intrinsèque première de la perfection des parties." For, attention must be paid to St. Thomas' words, ibid., "Per hoc autem quod dicimus substantias intellectuales propter se a divina providentia ordinari, non intelligimus quod ipsae ulterius non referantur in Deum et ad perfectionem universi."

Such, then, are Professor De K.'s doctrine and arguments at this juncture. If they were true, then the personalists, and with them all the Christian Fathers and theologians and philosophers, should close their shops, go home and do penance, *in cinere et cilicio,* for having grossly erred and misled the Christian world throughout almost two thousand years. That is to say, they should do so provided the Divine Cosmos leaves to them such a "home" in which to do penance! For, being material parts of the cosmos and subordinated, as material parts, to the stars and the spheres, they will have just as much responsibility, just as much choice, as the pistons in a steam engine. Let it be said, at once, that we simply refuse even to discuss this, Professor De K.'s own, private doctrine and thesis which is most patently erroneous. Let us be charitable and forget that such a statement ("Les parties principales constituant matériellement l'univers . . .") has ever been made in a work which pretends to exhibit the pure wisdom of St. Thomas Aquinas.[12]

For the sake of the proper understanding of the present issue only one point remains to be set in relief: What is the true meaning of St. Thomas' texts quoted above? What is his authentic doctrine on the intellectual substances as parts, i.e. principal or *formal* parts, of the universe?

Will it be granted that it is inadmissible to read St. Thomas with scissors and paste, by cutting the texts out of their literary and historical context and just quoting what, in a particular instance, seems to be suitable? Will it be granted that, if St. Thomas has explicitly stated and solved a given problem, a Thomist worthy of [190] that name is obliged to take account of this fact and can not afford to refer to some other texts which either have nothing to do with the problem or, at best, refer to it in a distant and mediate fashion?

Here is the problem as stated by St. Thomas:

> Videtur quod imago Dei inveniatur in irrationabilibus creaturis . . . [for, and this is the third *argumentum in contrarium*] quanto aliquid est magis perfectum in bonitate, tanto magis est Deo simile. Sed totum universum est perfectius in bonitate quam homo, quia etsi bona sint singula, tamen simul omnia dicuntur "valde bona," *Gen.* 1 (St. Augustine). Ergo totum universum est ad imaginem Dei et non solum homo.[13]

The reader will not fail to remember the texts from *Contra Gentiles,* III, 64, quoted above. At this place in the *Summa* St. Thomas makes use of the same doctrinal and traditional material but applies it to a different problem, viz. that of comparing the universe and intellectual substance, the *Imago Dei,* with regard to their respective likeness with God. This problem has often occupied the medieval mind which was so sensitive to everything concerning the great and old controversy between Greek and Christian thought. See also St. Bonaventure, 3 *Sent.,* d. 2, 1. 1 (ed. Quaracchi, p. 37*a* and *b*); St. Thomas, 3 *Sent.,* d. 2, 1. 1, q. 3; 3 *Sent.,* d. 16, 1. 2; 3 *Sent.,* d. 32, 5, q. 4 ad 2; *De Caritate,* 7 ad 5; *ST,* III, 4, 1 ad 4.

The following is St. Thomas' answer:

> Universum est perfectius in bonitate quam intellectualis creatura: extensive et diffusive. Sed intensive et collective similitudo divinae perfectionis magis invenitur in intellectuali creatura, quae est capax summi boni.—Vel dicendum, quod pars non dividitur contra totum, sed contra aliam partem. Unde cum dicitur quod sola natura intellectualis est ad imaginem Dei, non excluditur quin universum secundum aliquam sui partem sit ad imaginem Dei; sed excluduntur aliae partes universi.

St. Thomas' solution of the problem is so clear, so complete, and so perfectly balanced that it needs no explanation. Let us however try to paraphrase: Which is more like God, i.e. more to the image of God, the whole universe, or one single intellectual creature? The whole universe is more like God "extensively and diffusively." That is, if you consider God as the cause and fountain-head of the whole universe and of every creature pertaining to it, you will judge that there is quantitatively more likeness in the whole than in the parts. But before you consider God as cause, you must first look at Him as He is in Himself the supreme good by His essence. In this way a single intellectual creature is more perfectly likened to Him, because only the intellectual substance (every single intellectual substance) is capable of being, by knowledge and love, united with God as God is in Himself. "Intensively," thus, and "collectively," i.e. [191] considering the fact that the essentially most perfect likeness is gathered together in one single point, a single intellectual substance by far surpasses everything that might, in a certain sense, be said to be like God. The intellectual substance is, indeed, the only proper image of God.—Thus far, St. Thomas has proposed the same solution of the problem which already can be read in St. Bonaventure (loco cit.). The Angelic Doctor then continues, not by proposing another solution, but by stressing a certain aspect of the same solution which in the foregoing has been left aside. Are not the intellectual substances parts, i.e. of course, principal, formal, constitutive, primary, parts of the universe? Are they not, as it were, the sons of that great family or economy of the universe of which God is the *paterfamilias?*[14] Are they not, just as sons are, very deeply interested in the vicissitudes of that which is their possession and heredity—and the possession and heredity of each one of them, according to Holy Scripture (Matth. 24:47): "Super omnia bona sua constituet eum"?[15] The statement, therefore, that the intellectual substance alone is *ad imaginem Dei,* might be expanded by saying that the universe in one of its parts, and precisely in its first and foremost constitutive parts, is *ad imaginem Dei.* In this way a solution of the problem is obtained which is most properly "Thomistic" in that it takes account of every possible aspect of the problem.

Professor De K. has not remembered that there are two entirely different problems in St. Thomas' cosmology.

First. Against Greco-Arabian necessitarianism St. Thomas states that there exists an intelligent and loving Creator, i.e. a personal God and a divine and all-embracing Providence. Were this not so, he argues, the universe would fall apart into so many unconnected and unconnectable bits, and it would be impossible to maintain the fact of the order of the universe on whose existence and sublime beauty both the Greeks, and especially the Christian Fathers, have so energetically insisted. In this group of texts—it is a very large one—St. Thomas frequently, and with obvious enjoyment, avails himself of two quotations from Aristotle, viz., (*a*) *bonum commune est divinius . . . ,* and (*b*) *quod est optimum in rebus existens est bonum universi.*[16] By these citations no proper doctrine on the common good is taught; and still less is anything said about the relations between the common good and the proper good of the intellectual [192] substances. Their impact is clearly to show, against a Greek heresy, that, even in the Greek thinkers themselves, and above all in Aristotle, who was so fondly cherished in the Arabian world, there are principles upon which one may proceed to prove the fact of divine Providence.

This is the group of texts Professor De K. argues from. He should not have done so, because they do not properly and immediately belong to the question he undertook to treat.

Second. Another and entirely different problem in St. Thomas' cosmology is the question: What is the position and rank, within the universe, of the intellectual or rational substances, among them, human souls? The treatment of this problem originates in, and is directed against, another set of Greek and Arabian errors, viz., the Greek (Stoic) divinsation of the cosmos, the Platonic world soul, the Plotinian theory of emanations, the Arabian unity of the intellect, and so on. In St. Thomas' discussion of these problems, a doctrine is set forth which may well be called Thomistic personalism (and which, by the way, is one of the major sources of Jacques Maritain's personalism).

This is the group of texts—an immensely rich one, as every Thomist knows—which Professor De K. should have taken into account. But he did not.

When St. Thomas says:

> [substantiae intellectuales] ulterius . . . referuntur ad Deum et ad ordinem universi [*Contra Gent.*, III, 112],

he very exactly circumscribes the situation of those who are like the sons of God in the universal economy. They are, first, through a personal relation, ordained to God as He is in Himself. Only then, and second—since God is also the Creator of a universe—they are parts, i.e. formal, constitutive parts of that whole to which these substances, each one in its proper way, will bring the divinely appointed order.

The most essential and the dearest aim of Thomism is to make sure that the personal contact of all intellectual creatures with God, as well as their personal subordination to God, be in no way interrupted. Everything else—the whole universe and every social institution—must ultimately minister to this purpose; everything must foster and strengthen and protect the conversation of the soul, every soul, with God. It is characteristically Greek and pagan to interpose the universe between God and intellectual creatures. Is it necessary to remind Thomists that they should not, in any way whatever, revive the old pagan blasphemy of a divine cosmos?

[193] III. Professor De Koninck's Notion of God

Let us further examine the remarkable passage of Professor De Koninck quoted at the beginning of section II. For, if I am not mistaken, it presents still another very interesting feature.

The author supposes that his personalist opponents would not be unwilling to agree with him in saying that individual persons are ordered to the last "separate" good *en tant que celui-ci a raison de bien commun.* It is thus his own conviction that we are ordered to God *because* God is a common good. The precise and formal *ratio* why we find in God our last end and beatitude, why we must love God and obey His laws, is that God is a common good.

Taken, for the moment, in its purely verbal expression, this seems to be a very surprising affirmation, indeed.

For, up to now, according to common theological and philosophical language, we have held that our ordination to God is based upon the fact that God is the *most perfect* and *supreme* good, the *bonum per se,* or, as we also say, the *universal* good.

Surely, this language is not unfamiliar to Professor De K. It is, therefore, at once clear that in his mind the two notions, universal good and common good, are completely identified. They mean one and the same thing considered under one and the same aspect.

That this is indeed the case, we may find confirmed in the following passage. A few lines before the above-quoted passage Professor De K. speaks of "le bien commun qu'est la béatitude" and, describing its contents, he says:

> L'universalité même du bien est principe de béatitude pour la personne singulière. C'est, en effet, en raison de son universalité qu'il peut béatifier la personne singulière. Et cette communication au bien commun fonde la communication des personnes singulières entre elles *extra verbum:* le bien commun en tant que bien commun est la racine de cette communication qui ne serait pas possible si le bien divin n'était déjà aimé dans sa communicabilité aux autres: "praeexigitur amor boni communis toti societati, quod est bonum divinum, prout est beatitudinis obiectum."[17]

If I am not mistaken, this whole passage means that, according to Professor De K., the analysis of the essential structure of the order of man to God reveals four consecutive moments: (1) God is the universal or common good, *and this means,* a good communicable to others; (2) we immediately reach the divine good in the light (*sub ratione*) of this very communicability; (3) in this consists our beatitude which is, indeed, *formally* (cf. "rapport très formel," p. 21) a common good; (4) thus our communication with that common good is the basis for the communications among ourselves.

To prove his assertion by a text of St. Thomas, Professor De K. [194] extracts a few words from the *Q.D. De Caritate* (art. 2). The content of these words is so important that we ask the patient reader to excuse us for transcribing the relevant text in its entirety. We shall italicize the words to which Professor De K. draws attention.

> Si [? sicut] autem homo, inquantum admittitur ad participandum bonum alicuius civitatis et efficitur civis illius civitatis: Competunt ei virtutes quaedam ad operandum ea quae sunt civium, et amandum bonum illius civitatis, ita, cum homo per divinam gratiam admittatur in participationem caelestiae beatitudinis, quae in visione et fruitione Dei consistit, fit quasi civis et socius illius beatae societatis, quae vocatur Caelestis Jerusalem, secundum illud *Eph.* 2, 19: "Estis cives sanctorum et domestici Dei." Unde homini sic ad caelestia adscripto competunt quaedam virtutes gratuitae, quae sunt virtutes infusae, ad quarum debitam operationem *praeexigitur amor boni communis toti societati, quod est bonum divinum, prout est beatitudinis objectum.*

To begin our criticism of this whole position, let us first say a few words with regard to the author's exegetical methods. The word "praeexigitur," extracted from St. Thomas' text, is at once commandeered by Professor De K. to supplement the arsenal of his own ammunition. Whereas, according to St. Thomas' text, there is something prerequisite for the exercise of the infused virtues, according to Professor De K. this something is made a prerequisite for a moral philosophy and a social metaphysics. A facile device to support one's own assertions by authority! The solemn gravity of an apparently authentic quotation, given in Latin, turns out to be an empty show. Was this quotation intended to impress the reader or is it possible that the author himself was impressed by his pseudo-discovery?[18] Strictly speaking, the disclosure of such an inept method of dealing with a text would authorize us in taking no further account whatsoever either of this excerpt or of the teaching based upon it.

Is it true that St. Thomas taught, as Professor De K. would have us to believe, that the object of our beatitude, the very first and essential element of our ordination to God, is the divine good, insofar as this good is a common good, constituting, first and foremost, a society ("amor boni communis toti societati, quod est bonum divinum, prout est beatitudinis obiectum")? By no means! This interpretation is false. St. Thomas' argument in the *De Caritate,* loco cit., proceeds *a simili,* i.e. by comparing two highest goods, each taken in its own order, not, properly speaking, two common goods. The highest good of the earthly city is called a common good. No description [195] or definition of it is given in this text. St. Thomas is here not lecturing on social metaphysics or political philosophy, but on charity; and the example of the city is only used as an *argumentum ad hominem.* To the earthly city, referred to in the example, the Heavenly City corresponds as the thing exemplified; and, through the words "quasi [!] civis" (to which corresponds in the parallel text, *Summa,* I, 60. 5 ad 5, "quoddam [!] bonum commune"), St. Thomas takes care, at the outset, to keep us from over-extending the simile and, thus, getting on the wrong track. To confuse examples with formal teaching is quite inadmissible. Let us paraphrase the passage in question, in order to set its true significance in relief: Prerequisite to the exercise of infused virtues in the Heavenly City is the love of the highest good which is the divine good, the object of beatitude. In like manner, the love of the earthly city's highest good, i.e. its common good, is prerequisite to the exercise of natural virtues. In a certain sense, the divine good might also be called a common good (*quoddam bonum commune*). But the object of charity is, of course, not a common good; rather it is the divine good ("Bonum commune non est obiectum caritatis, sed summum bonum," *Q.D. De Caritate,* 5 ad 4). Considered as a common good, the highest good of the Heavenly City would be, indeed, the object of supernatural general justice, not of charity. Charity and justice must not be confused.—It is very significant that St. Thomas chooses to say *bonum commune toti societati (caelesti)* instead of *bonum commune totius societatis,* as he usually does when speaking in terms of political philosophy.

May the patient reader excuse the length to which this exegetical problem has obliged us to go. Let us now turn back to the substance of Prof. De K.'s teaching.

Is it not the most fundamental and absolutely unshakeable cornerstone of Christian ethics that the term of our ordination to God is God as He is in Himself, i.e. the Good by His essence and the essence of goodness (*bonum universale in essendo*)? Is it not the very first care of a Christian ethician to make sure that the conclusion of his very first argument directly reaches this *bonum universale in essendo?* This, at least, is the content and intention of that great argument which opens the *pars moralis* of the *Summa* (I-II, 1. 1–2. 8) and whose conclusion is: "Ex quo patet quod nihil potest quietare voluntatem hominis nisi bonum universale, quod non invenitur in aliquo creato, sed solum in Deo, quia omnis creatura habet bonitatem participatam" (ibid., 2. 8.). St. Thomas has here completely forgotten to speak about Professor De K.'s "common good" by which man's ordination to God is *très formellement* determined. I am afraid that on pages 26–27 of this book a suspicion [195] which the expert reader has felt all the way along, from the first page on, becomes definite, namely that the author has pushed the "primacy of the common good" very far, so far indeed that, if the consequences of his position are made explicit, we must in our Christian ethics re-do our work from the beginning. In setting up a "principle of the New Order" Professor De K. has done a work which is—shall we say—surprisingly radical and daring: he has at the same time taken in his stride a new foundation of Christian ethics and moral theology.

Professor De K. has confused *bonum universale in essendo* and *bonum universale in causando*. "The creature," St. Thomas says (*Summa*, I, 103. 4), "is assimilated to God in two respects: first, with regard to this that God *is good;* and thus the creature becomes like Him by *being good;* and, secondly, with regard to this that God is the *cause of goodness in others;* and thus the creature becomes like God by *causing others to be good.*"—The common good, and every common good, is formally *bonum universale in causando:* it is not, formally, *bonum universale in essendo.*

The very first and essential element of our ordination to God is not the fact that God is the first *bonum universale in causando,* the fountain of all communications, but that He is the *bonum universale in essendo.*

From this it follows that our own (personal) good is a participated good. Through this participation a "certain common good" ("quoddam bonum commune") emerges, i.e. a good which, in a certain way, is common to God and the creature. Considering the supernaturally elevated creature, this common good is constitutive for a community or "society" between God and the supernaturally elevated creature, a society which is called, by St. Thomas, *societas suae (i.e. Dei) fruitionis.*[19] It is the divine friendship to whose *essential* constitution no multitude of creatures is required.[20] The fact that there is such a multitude of creatures does not yet formally come into consideration.

This fact becomes only now, i.e. in the third place, relevant. For *if*[21] there are several creatures sharing in the same participated good they will have something in common. Here, then, there will be a common good properly speaking, i.e. a good pertaining to a multitude of beings in such manner that each and every one communicates in it. God is, as St. Thomas says, the last common good among men, [197] i.e. that good in which they finally must or should unite: "Homines non uniuntur inter se nisi in eo quod est commune inter eos. Et hoc est maxime Deus."[22]

Professor De K. has, throughout his treatise, neglected these fundamental considerations. On the very first page of the treatise proper (p. 8) he has omitted to pay due attention to St. Thomas' words: "Dicitur autem hoc [scilicet bonum commune] esse 'divinius' eo quod magis pertinet ad similitudinem Dei, qui est ultima causa omnium bonorum."[23] Obviously the words "qui est ultima causa omnium bonorum" are, in St. Thomas' mind, restrictive; and if the famous principle, "Sanctus Thomas formalissime loquitur" ever finds its application, it surely does so here. Let us paraphrase: Aristotle gives to a common good the attribute "divine," because this good, being the cause of the particular goods contained in its order and sphere, is *in this respect* more like God insofar as God is the cause of any and every good. There is, however, another respect to which the above text gives no consideration. This is the likeness to God *in linea essendi*. And in this respect the speculative intellect being, in the beatific vision, informed by God and most intimately united with Him, is by far superior to anything which is like God *in ordine causandi*. St. Thomas explictly states:

> Similitudo intellectus practici ad Deum est secundum proportionalitatem, quia scilicet se habet ad suum cognitum [the highest object of the practical intellect is a common good—II-II, 47. 12], sicut Deus ad suum. Sed assimilatio intellectus speculativi ad Deum est secundum unionem vel informationem: QUAE EST MULTO MAIOR ASSIMILATIO.

These last words are the most concise and the most explicit statement of what we now call Personalism. For, is not this act and good of the speculative intellect a personal good?

Professor De K. has constantly bypassed this most essential thesis of Thomistic ethics and, indeed, of Thomism as a whole.

IV. Professor De Koninck on Beatitude

Ever since the days when Plato stated the problem of the philosophers and kings, every occidental theory of society has ultimately proved its truth and its value by the regard it has paid to, and the place it has left open for, that which is not society nor action, viz. solitude and contemplation. The modern problem which we are now accustomed to state in terms of Person and Society is nothing but the continuation of the age-old discussion of Philosophers and Kings.

[198] Professor De Koninck will already have surprised the attentive reader by the statement quoted above, that our beatitude is a common good ("le bien commun qu'est la béatitude," p. 26). Let us have a closer look into this statement.[24]

On page 61 the author composes (one might be tempted to say concocts) the following "objection" against his thesis of the absolute primacy of "the" common good

> L'ordre pratique est tout entier ordonné à l'ordre spéculatif. Or, le bonheur parfait consiste dans la vie spéculative. Mais, la vie spéculative est solitaire. Donc, le bonheur pratique de la société est ordonné au bonheur spéculatif de la personne singulière.

Professor De K.'s answer to this "objection" is as follows:

> Nous répondons que le bonheur pratique de la communauté n'est pas, par soi, ordonné au bonheur spéculatif de la personne singulière, mais au bonheur spéculatif de la personne en tant que membre de la communauté. [Here is quoted Petrus de Alvernia, *In VII Pol*, lect. 2.] Il serait, en effet, contradictoire qu'un bien commun fût, de soi, ordonné à la personne singulière comme telle. Il est très vrai que la vie spéculative est solitaire, mais il reste vrai aussi que, même la béatitude souveraine qui consiste dans la vision de Dieu, est essentiellement bien commun. Cette apparente opposition entre la vie solitaire et le bien commun qui est l'objet de cette vie s'explique du fait que cette félicité peut être considérée, soit de la part de ceux qui en jouissent, soit de la part de l'objet même de cette félicité. Or, cet objet est, de soi, communicable à plusieurs. Sous

> ce rapport, il est le bien spéculatif de la communauté. Le bien commun pratique doit être ordonné à ce bien spéculatif qui s'étend comme bien commun aux personnes. L'indépendance des personnes les unes des autres dans la vision même n'exclut pas de l'objet cette universalité qui veut dire, pour toute intelligence créée, essentielle communicabilité à plusieurs. Loin de l'exclure, ou d'en faire abstraction, l'indépendance présuppose cette communicabilité.

Is this somehow "magisterial" *Nous répondons* in conformity with Master Thomas' famous *Respondeo. Dicendum?*

The "Thomistic" basis for the author's answer is not St. Thomas but Peter of Auvergne. The quotation from this continuator of St. Thomas' *Commentary on the Politics* is here all the more surprising since for the point in question a rich and authentically Thomistic documentation was at hand. It is, indeed, a fact as un-understandable to any serious Thomistic scholar as it is characteristic of Professor De K.'s scientific methods that at a juncture where the most proper and important point of the whole discussion is under debate—*hic Rhodus,* [199] *hic salta!*—the author completely forgets about St. Thomas. The reader is avid to get good Thomistic bread, but he must content himself with *Ersatz.*

Peter of Auvergne, as is well known, is a secular priest, a member, in the last decades of the thirteenth century, of the Faculty of Arts at the University of Paris and, at one time, a disciple of St. Thomas, whose lectures he attended in Paris, somewhere between 1269 and 1273. Although, because of his general doctrinal outlook, there is no doubt that he must be counted among the representatives of the oldest Thomist school, nevertheless, in every question of detail the quality of his Thomism is a matter, not of assumption, but of examination. For it is not impossible that the Averroistic atmosphere of the Parisian Artists might somehow have colored his doctrine, as it happened, not infrequently in those times, for instance and especially, in the case of another Parisian Artist, John Quidort, O.P. As long as the notion of a doctrinal *source* retains any proper and intelligible meaning, it is surely impossible to use Peter of Auvergne unqualifiedly as a Thomistic source; and, let it be noted, the same applies, of course, to Cajetan, John of St. Thomas, etc., commentators whom Professor De K. puts, without any distinction, on equal footing with St. Thomas himself.

In fact, the author's misfortune is that Peter of Auvergne's statement appears to be questionable in the light of St. Thomas' authentic and explicit doctrine.

In order to explain, in his own words, the suppositions of a certain passage of the Aristotelian *Politics*,[25] Peter combines the following four notions: (1) felicitas speculativa secundum unum hominem (operatio hominis secundum virtutem perfectam contemplativam quae est sapientia); (2) felicitas speculativa totius civitatis (speculatio totius civitatis); (3) felicitas practica secundum unum hominem (operatio hominis secundum perfectam virtutem hominis practicam); (4) felicitas practica totius civitatis (operatio prudentiae totius civitatis). These four notions, then, are severally combined and examined under the point of view of their respective value.

The clumsiness both of the notions themselves and of the whole procedure of combining and comparing them, is at once striking. For, what is this *operatio prudentiae totius civitatis?* And if, in spite of the manifest clumsiness of the terminology, an intelligible meaning might finally be discovered in this notion—what in the world can *speculatio totius civitatis* be? It is exactly this notion which, most unfortunately, Professor De K. has picked out to be the cornerstone of his answer.

St. Thomas speaks quite a different language:

> Sicut bonum unius consistit in actione et contemplatione, ita et bonum [200] multitudinis, secundum quod contingit multitudinem contemplationi vacare. Hoc est verum, quod . . . assecutio finis quem intellectus practicus intendit, potest esse propria et communis, inquantum per intellectum practicum aliquis [!] se et alios dirigit in finem, ut patet in rectore multitudinis [!] Sed aliquis ex hoc, quod speculatur, ipse solus dirigitur in speculationis finem. Ipse autem finis intellectus speculativi tantum praeeminet bono intellectus practici, quantum singularis assecutio eius excedit communen assecutionem boni intellectus practici. Et ideo perfectissima beatitudo in intellectu speculative consistit.[26]

How conscientious, how realistic a thinker is young St. Thomas who wrote these passages already in or about 1255 to 1256! He, indeed, never indulges in combining his notions merely for the sake of obtaining some neat scheme, but he examines them with regard to their inner possibility and truth. In the first passage it seems to be evident that St. Thomas somehow inclines towards something like Peter of Auvergne's *speculatio totius civitatis.* Yet Aquinas at once checks himself by adding, with remarkable finesse: *secundum quod contingit multitudinem contemplationi vacare.* Is contemplation, as a genuine social or common act, possible at all? In the second text to the *assecutio communis finis intellectus practici* the right, personal subject is assigned, namely the *rector multitudinis* (cf. II-II, 47. 12). And St. Thomas now vigorously sets in relief the inner impossibility of an *assecutio communis* of the end of the speculative intellect. The words IPSE SOLUS DIRIGITUR IN SPECULATIONIS FINEM and the subsequent statement of the absolute pre-eminence of the SINGULARIS ASSECUTIO of the speculative good—deserve to be written as a motto at the head of a treatise of Thomistic social philosophy. And be it noted that this whole statement is the Thomistic answer to the following *argumentum in contrarium* which most exactly states the problem of the pretended absolute pre-eminence of the common good:

> Videtur quod beatitudo magis consistat in actu intellectus practici quam speculativi. Quanto enim aliquod bonum est communius, tanto est divinius, ut patet in I Eth. Sed bonum intellectus speculativi est singulariter eius qui speculatur. Bonum autem intellectus practici potest esse commune multorum. Ergo magis consistit beatitudo in intellectu practico quam speculativo.[27]

[201] A complete collection of the Thomistic texts regarding the *dictum authenticum* of the relative primacy of a common good[28] very impressively brings to light the fact that the main preoccupation of St. Thomas, in discussing this "authority," was to explain it in such a way that the superiority of Christian contemplation and solitude, i.e. of the highest personal good, remained uncontestedly safe.

To come back to Professor De K.'s *Respondeo. Dicendum*: In the light of St. Thomas' explicit teaching, the view (*a*) that "le bonheur pratique de la communauté . . . [est ordonné] au bonheur spéculatif de la personne en tant que membre de la communauté" must be rejected. For, speaking of human communities, and of what else do we speak in mentioning "practical felicity"?—the very notion of the "speculative felicity of the person qua member of the community" is contradictory. In fact, to be a member of the community means to be imperfect, perfectible, and *in via;* whereas to have reached speculative felicity means to be perfect and *in termino.* St. Thomas says: "Sicut ergo id quod iam perfectum est praeeminet ei quod ad perfectionem exercetur, ita vita solitariorum [i.e. contemplatium] . . . praeeminet vitae sociali" (II-II, 188. 8). It is impossible to develop here this marvelous article of the *Secunda Secundae* which, in my opinion, contains the essence and the last word of St. Thomas' social theory, a theory which is, through and through, of a "personalist" stamp because it is based upon the Christian notion of contemplation. I am sorry to be obliged to state, for the sake of Thomistic truth, that Professor De K. has succeeded, in the above-quoted proposition, in disfiguring all the fundamental notions, all the essential lines and innermost intentions of this Thomistic theory.

Speaking, not of "speculative felicity"—for this denotes a final status—but of contemplation or the contemplative life as it may be lived on this earth, there is of course, a sound and intelligible meaning in saying, with St. Thomas, that "the works of the active life must be derived from the plenitude of contemplation": "Et hoc praefertur simplici contemplationi. Sicut enim maius est illuminare quam lucere solum, ita maius est contemplata aliis tradere quam solum contemplari."[29] If—to speak again and always St. Thomas' language—the Pope decides to call a man away from the "garden of contemplation" [202] in which he enjoys the sweetness of conversing with God, and to set him on the dusty roads of the active life, at the head of a diocese, for the sake of the common good, this man will obey. If—to hint at a recent splendid example—the head of a state appoints a philosopher, i.e. a lover of the contemplative life, to be ambassador to the Holy See, for the sake of the common good which, doubtlessly, is admirably served by such an appointment, made in conformity with the age-old demands of the Greek thinkers as well as the Christian Fathers, this man again will obey, although he realizes that almost everything he has fondly cherished throughout a long life will have to be abandoned. But why do both men obey? Is it because the common good is, simply and absolutely speaking, higher and more valuable than their (personal) good of contemplation? By no means! Is it because they have been enjoying the *dulcedo contemplationis* as parts of the community, and thus, already, in subordination to its interests and laws? By no means! Their obedience is, according to the clear and precise *littera Sancti Thomae*[30] intrinsically motivated by the fact that sometimes, in some circumstances (*in casu*) the common good and its necessities are more urgent. The common good has a relative and limited pre-eminence *in via utilitatis,* because it is essentially a *bonum utile*, the highest *bonum utile*, but nothing more. It has no absolute pre-eminence, i.e. no primacy *in ratione dignitatis*. And, for the sake of one of the most essential truths of Thomism, any attempt, by whomsover and in whatever way, to disfigure these elementary lines of the Thomistic social system must most energetically be rejected.

Furthermore, and again in the light of the *littera Sancti Thomae*, we refuse to accept Professor De K.'s statement, viz. (*b*) that "la béatitude souveraine qui consiste dans la vision de Dieu est essentiellement bien commun."

Objectively, i.e. viewed from the part of its uncreated object, the vision is not a common good; it is not even God as Common Good (to speak of common good in a proper and adequate language) but it is God Himself, the *Bonum universale in essendo*, as has been shown above.

Formally, i.e. viewed as a created act and good, the vision is that supreme, personal good by which a created intellect, elevated by the light of glory, is most intimately united with, and most perfectly likened to, God.

With these two elements the *essence* of the vision and of final [203] beatitude is fully circumscribed. No further element needs to be added. No further element pertains to the *intrinsic nature* of final beatitude.

Extrinsically, however, i.e. in virtue of the fact that there is a multitude of the Blessed sharing, as it were, in the same good, the vision might be called a certain common good which, then, is the constitutive of a certain "society," a society which St. Augustine has called *societas fruendi Deo et invicem in Deo*.[31] With regard to this "society" all that St. Thomas has to say is that it *quasi concomitanter se habet . . . ad perfectam beatitudinem*[32] because, speaking of the essence of things, every single "member" of it has his full sufficiency in God and in God alone.

Any serious Thomistic consideration of the Problem of Person and Society must needs lead to, and terminate in, the mystery—tremendous and consoling at the same time, as every mystery is—of the soul, and every soul, in the face of God, and God alone. St. Thomas has given this mystery yet another very illuminating but also, at first glance, disquieting formula. In the *Opusculum De Perfectione Vitae Spiritualis*, chapter 13, he says (Let us note that this work was written against the pragmatism of Gerald of Abbéville whose main mistake was to have turned the relative primacy of the common good into an absolute one!): "Proximus autem noster non est universale bonum supra nos existens, sed particulare bonum infra nos constitutum."

Will Professor De K. be able to give us a fitting explanation of this "infra nos" of St. Thomas? I cannot help but think that he will not. According to the suppositions of his system he will protest (in fact, he does so, on similar occasions) that this is the "base abomination of egoism." We have no reason to recede even one iota from the clear and precise *littera Sancti Thomae.* That it contains no egoism at all is clear to everyone who, with St. Thomas, knows how to distinguish between *amor sui ordinatus* and *amor sui inordinatus.*

Professor De K.'s root mistake, in his whole treatise on the primacy of the common good, is that he rashly assumed an absolute identification between God and "the" common good. It was inevitable that this initial error should lead to the distortion of that which is foremost in Thomism, namely, the primacy of the spiritual, which, in its turn, is all there is in the primacy of the personal. That is why I think that Professor De K.'s book will have to be re-written.

A certain danger of misrepresentation is almost inevitable in any monographic account of our problem. If you insist on the personal good above the common good, it will be very difficult, if not altogether [204] impossible, to avoid the impression that in a certain way you minimize the common good. The extant literature on "personalism"—it is large, and perhaps even a bit too large, and the authors, sometimes, do not seem to have grasped the correct synthesis in a question which is one of the most subtle and illusive of philosophy—gives ample proof of this fact. A writer cannot say everything on every occasion. Let us by no means forget that St. Thomas is among all medieval authors the one who has most extensively and most vigorously emphasized the primacy, within its order (i.e. the practical, "political" human order) of the common good over everything which falls within this same order of which the common good is the immediately last end and the supreme rule and measure. If Professor De K. meant only to protest against the manifest and, as I have said, hardly avoidable minimization in modern Thomistic literature, of the common good, his book would have been, in principle, unobjectionable. But he did infinitely (in the true sense of the word!) more than this in that, of a relative primacy he makes an absolute and absolutely all-embracing primacy. According to Professor De K.'s principles man is forever entangled in the net of common goods, without any hope of ever tearing up these chains (for chains they are, in a definite, i.e. metaphysical, not moral, sense) and of freeing himself for God and for God alone Who is first and primarily God—*Ego sum qui sum*—the divine Good, the object of our personal beatitude (*bonum universale in essendo*), rather than being, first and primarily, the creator of all things and therefore the supreme common good in which all beings are finally united (*bonum universale in causando*). According to the same principles, the common good is infinitely more than that supreme good which the *practical* intellect or reason might constitute. It (i.e. the univocally same good) is also the supreme object which can ever be offered to the *speculative* intellect. With this position, metaphysics (and theology), and the first, decisive part of ethics (and moral theology) are in ruins. This being the effect of Professor De K.'s thesis, I do not in the least hesitate to say, that from the point of view of the *littera Sancti Thomae* this book is a danger to every reader who has neither the time nor the sufficient training to discover for himself, in a problem of extreme subtlety, the genuine Thomistic truth.

V. Jacques Maritain and St. Thomas Aquinas

Our "Defense of Jacques Maritain" has been, so far, rather a defense of St. Thomas. Let us, therefore, cast a brief glance upon the specific doctrine of Jacques Maritain. Let us ask exactly how [205] it stands in relation to the *littera Sancti Thomae* and whether, or not, a *fruitful* discussion may be opened with regard to its main thesis.

It seems to me—*salvo meliore iudicio*—that the bare essence of this doctrine might be summed up in the following *enthymema:* St. Thomas says: *Ad rationem personae exigitur quod sit totum completum;* or again: *Ratio partis contrariatur personae.*[33] Hence, Jacques Maritain concludes, the person, qua person, is not a part of society: and if a person is such a part, this "being part" will not be based upon the metaphysical formality and precision of "being person."

The *antecedens* pertains to the *littera Sancti Thomae.* The conclusion is not to be found there in such explicit words, and thus, if it is correctly drawn, it will be part and parcel of that "greater Thomism" whose task it is to develop the Thomistic principles and, eventually, make them an actually living truth.

Maritain's conclusion is evident. Its necessity and intelligibility are exactly the same as the necessity and intelligibility of the following inferences:

Act as such means pure and limitless perfection. Hence, if there is a limited or participated act, this limitation will not pertain to this act, qua act, but qua mixed with potency. Or again: The intellect as such is not capable of error. Hence, if there is an intellectual being which errs, this will not happen to it, insofar as it is an intellect but insofar as it is something else.

All these inferences, the one on the person not excluded, bear a certain similarity each with the other, insofar as the perfection is always said to be perfection.

There is not the slightest doubt that Maritain, at what I assume to be the fundamental point of his doctrine, is right. The only question which can ever be raised with regard to this position is this: Why is it that St. Thomas did not draw this conclusion? Why did St. Thomas, in his ethical or social doctrine, never turn to that chapter, if I may say so, of the metaphysics of the person in which its absolute and formal essence is defined? This is a question not of living but of historic Thomism. The eminent historian of medieval philosophy, Etienne Gilson, speaking, in *The Spirit of Mediaeval Philosophy*,[34] of the age-old and undeniable fact of a Christian personalism, wonders why there is

> not a word throughout the whole of mediaeval moral philosophy on what the mediaevals themselves held to be the highest in man and therefore in all nature. How shall we account for the fact that in the very moment of a discovery [by medieval thinkers] of such immense importance, Christian thought seems to stop suddenly short and renounce all effort to exploit its success?

[206] To the historian and lover of the *littera Sancti Thomae*, in all its concrete conditions, it is, indeed, a problem of more urgent interest to know why St. Thomas, in a given case, did not envisage a certain problem than to know how he would have answered it, had he envisaged it. Any discussion of so-called Thomistic personalism will, in the first place, have to say what is meant by that term and where, in St. Thomas' works, to go looking for it.

The attentive reader of the present article will have observed the fact—and perhaps wondered at it—that in the foregoing sections our Thomistic documentation has preferably been chosen not from that chapter of metaphysics in which the absolute and formal definition of Person is stated, but from the other chapter in which metaphysics elaborates, mainly by recurring to the final cause, the relative definition of the created person, thereby, fulfilling her "royal" function which is to assign, to the particular sciences—in our case, to ethics—their subject. St. Thomas has written this chapter in the great metaphysical *Prooemium* to the *pars moralis* of the *Summa,* I-II, qq. 1–5. The chief "personalist" text is the one we have quoted above, namely, q. 3, art. 5, especially ad 1: *quae est multo maior assimilatio.*

May I submit, not that Maritain's metaphysical foundation of the ethics (and social philosophy) of the person be replaced—I will say later that and why no such replacing is needed—but that it be made more evident and gain its proper place as well as its full weight, when, first and foremost in our discussion of the problem of personalism, we insist on such a relative definition of the created person as St. Thomas has taught us.

Let me try to enumerate some of the advantages of such a procedure.

First. A closer contact with the *littera Sancti Thomae* would be obtained. It is, indeed, a fact, easily verifiable, that a very considerable and striking part of St. Thomas' discussion of the *dictum authenticum* (a common good is better and more divine than a particular good) centers around the comparison between the common and the personal good. St. Thomas, it seems to me, has, in the later years of his academic and literary career, especially after his discussion with the Geraldine pragmatism (1269), insisted with greater and greater energy on the superiority of the personal good of contemplation and divine charity over the common utility.[35] If a Thomist wishes to treat the problem of Person and Society in immediate contact with the writings and judgments of St. Thomas, he can do so only by following, step by step, the numerous efforts of the Angelic Doctor to master the problem of contemplation and action, love of [207] God and love of neighbor. To be endowed with a speculative intellect and ordained to the beatific vision is, most exactly, that relative definition, *ex parte causae finalis,* of the created and human person to which I referred above.

Second. Through this closer contact with St. Thomas a more direct approach to our problem would be possible. For is not this problem, properly speaking, a problem of ethics and social theory? Where else, therefore, should its immediate metaphysical foundation be sought for than in that part of metaphysics where the subject of ethics is determined? It will also be easier, through the same procedure, to see the problem more clearly in its historical connections. I have already pointed to the fact that anyone, who in the context of *Philosophia Perennis*, speaks of this problem, is in truth continuing the old Greek controversy between philosophers and kings, and must with the Christian Fathers, especially St. Augustine and St. Gregory, insist on the higher merit of contemplative life.

Third. If the *littera Sancti Thomae* is closely followed, it will be clear at once, and without even the slightest possibility of mistake, that our Christian personalism has nothing to do with the secularized personalism of nineteenth-century philosophy. The entire interest of Christian and Thomistic personalism is, indeed, taken up by that spiritual and personal order whose ultimate end and supreme "rule and measure" is the beatific vision. What this personalism wishes to emphasize is that universal Christian vocation to contemplation which St. Thomas liked to find expressed in the words of Psalm 45, *Vacate et videte, quoniam ego sum Dominus.*[36] To use (and extend) the language of an old and venerable papal document of the eleventh century, the so-called *Canon Urbani*—a document which has played an important role in the medieval canonist and theological discussions of our problem[37]—it is not the personalist contention that nobody dare resist the caprices of any given individual person, of Tom, Dick, and Harry, but that nobody dare resist the Holy Ghost (Act. 7:51). The resistance, in the extant anti-personalist literature, obviously has its origin in the fear that personalism is nothing but individualism and egoism. This fear is unjustified, of course, especially as far as Jacques Maritain's doctrine is concerned. The very starting point of Christian personalism should, once and forever, caution any adversary against the quixotic venture to rise and gird himself for battle against an imaginary enemy.

After all this is said and done, it will be easy to pass over to the absolute and formal definition of the person qua person. This will [208] even be necessary since it is in this chapter of metaphysics that all our knowledges find their final resolution and firmness: *Oportet quod cuicumque apprehenso per intellectum, intellectus attribuat hoc quod est ens.*[38]

Properly speaking, the principle of the primacy of the common good is valid only within the "practical," "moral," "political" human order. Outside this order the notions *common good, community,* etc. lose their significance unless they be taken analogically. The whole human order whose highest good is the common good is subordinated to things divine, among the first of which after God, is the created intellect, which is *capax summi boni.* This capacity is a personal good. With reference to our present problem, this is, it seems to me, the very quintessence of St. Thomas' doctrine, immediately and explicitly verifiable in the *littera.* If the term "personalism" (in itself, no doubt, a bad one) is purged of the connotations it has through its sources in modern philosophy, there is no objection to calling this Thomistic doctrine personalist.

NOTES

[Page numbers from the original have been inserted to aid in following De Koninck's rebuttal to this essay.]

1. Québec: Editions de l'Université Laval, Montréal: Editions Fides, 1943.

2. See, for instance, Yves Simon's recension in the *Review of Politics* 6 (October, 1944): 530–33.

3. The fact that Prof. De K. has known and intends to criticize Jacques Maritain is susceptible of a strict *proof* by the usual means of tracing an author's sources through the words and notions he employs. See for instance p. 74: "A l'individualisme [les personnalistes] opposent et recommandent la générosité de la personne"; p. 75: "la personne-tout"; and other examples elsewhere. If such words and notions were not actually and entirely Maritainesque provenance, they are, today and among Thomists, of a clearly Maritainesque stamp, and that is enough for the proof.

4. Vol. 22 (Jan., 1945): 59–75.

5. Op. cit., p. 73. Every Thomist is surely authorized to go, in this matter, just as far as St. Thomas himself has gone: "Quando homo per seipsum agit propter finem, cognoscit finem. Sed quando ab alio agitur vel ducitur, puta cum agit ad imperium alterius, vel cum movetur altero impellente, non est necessarium quod cognoscat finem. Et ita est in creaturis irrationabilibus" (*ST,* I-II, 1. 2 ad 1).

6. Op. cit., p. 74. On Father Baisnée's prospective blacklist there is also Very Reverend M. S. Gillet, O.P., Master General of the Dominican Order (cf. art. cit., p. 65). Perhaps the reader will allow a vigorous protest from a Dominican.

7. *Acta Apostolicae Sedis,* 1927, p. 79.

8. Ibid., 1943, pp. 221 ff.

9. May I be permitted to insert here, before beginning my critique, a note of a personal character. I have the privilege to regard both Jacques Maritain and Charles De Koninck as dear friends. The job, therefore, of examining and determining the truth of their respective positions is very painful to me. Yves Simon, who, if I am not

mistaken, is in the same situation, has chosen a way out of the difficulty which I am unable to follow. Yves Simon seems to be ready to admit the substance of De K.'s book, and proposes to envisage its implicit criticism of, and opposition to, Jacques Maritain as a regrettable mistake, to be sure, but, after all a merely personal affair which the reading public might more or less easily forget.—I feel obliged totally and categorically to reject De K.'s thesis. I would never have come out with this judgment, had not Father Baisnée written his article; and I protest that I do not in the least enjoy the task which the circumstances have made an urgent necessity. The question with which I am faced is not to choose between friendship and friendship, but between manifest truth and manifest error, the criterion being the *littera Sancti Thomae*.—Let it moreover be noted that this "Defense of Jacques Maritian" comprises exclusively the metaphysical doctrine of the person. None of the numerous factual or historical appreciations of Jacques Maritain is the object of the present study.

10. Pp. 28–30.

11. *Metaph.*, xii, 1075a 11.

12. On p. 35 Professor De K. writes: "Les créatures raisonnables peuvent atteindre elles-mêmes de manière explicite le bien auquel toutes choses sont ordonnées; elles diffèrent par là des créatures irraisonnables, qui sont de purs instruments, qui sont utiles seulement et qui n'atteignant pas elles-mêmes de manière explicite le bien universel auquel elles sont ordonnées." Very well! But how does this statement stand to the other one: ". . . les parties principales constituant matériellement l'univers . . ."?—Would it not be desirable that an author who uses traditional philosophical notions knew exactly what they mean? In a recent work, *Saint Joseph, Père vierge de Jésus* (Montréal, 1944), Msgr. G. Breynat, a venerable missionary Bishop, in all seriousness and against the protest of a large group of theologians, defends the following definition of St. Joseph's paternity: It is "une causalité effective, négative, par abstention" of the child Jesus (pp. 84, 117 ff.). Professor De K.'s notion of a principal part materially constituting the universe is of the same caliber.

13. *ST*, I, 93. 2.

14. *In Meta.*, XII, 12; *De Verit.*, 5. 5; *De Spirit. Creaturis*, art. 8.

15. "Inter omnes partes universi excellunt sancti Dei, ad quorum quemlibet [!] pertinet quod dicitur Matth. 24: 'Super omnia bona sua constituet eum.' Et ideo quidquid accidit vel circa ipsos vel alias res, totum in bonum eorum cedit" (*Expos. in Ep. ad Rom.*, c. 8, lect. 6).

16. *Eth.* i. 1; 1094b 9; *Metaph.* xii. 1075a 11.

17. P. 26. The Latin text is extracted from *Q.D. De Caritate*, art. 2.

18. On p. 21 De K. quotes the same text of the *Q.D. De Caritate* in its entirety and in a correct French translation. Do the readers, meeting on p. 27 a piece of the same text, now given in Latin, remember its original and authentic setting and meaning?

19. "[Deus] non tantum diligit creaturam sicut artifex opus, sed etiam quadam amicabili societate, sicut amicus amicum, inquantum trahit eos in societatem suae fruitionis, ut in hoc eorum sit gloria et beatitudo, quo Deus beatus est" (2 *Sent.*, d. 26, 1. 1 ad 2).

20. "Si esset una sola anima fruens Deo, beata esset, non habens proximum quem diligeret" (*ST*, I-II, 4. 8 ad 3).

21. "Supposito proximo" (loco cit.).

22. *In II Thess.*, c. 3, lect. 2.

23. *In Eth.*, I, 2, ed. Pirotta, n. 30. The next quotation in the article is from the same place.

24. Speaking of the Aristotelian *eudaimonia*, St. Thomas sometimes calls the felicity a common good: "Felicitas autem est finis humanae speciei, cum omnes homines ipsam naturaliter desiderent. Felicitas igitur est quoddam commune bonum possibile provenire omnibus hominibus, nisi accidat aliquibus impedimentum quo sint 'orbati' " (Arist. *Eth.* i. 1099a).—Of course this is not what Professor De K. means by *le bien commun qu'est la béatitude*. The Thomistic notion of common good is an analogical and very elusive notion. See below note 25.

25. Bk. vii, chap. 3; 1325b 14–23.

26. 3 *Sent.*, d. 35, 1. 4, sol. 3 ad 2; 4 *Sent.*, d. 49, 1. 1, sol. 3 ad. 1.

27. 4 *Sent.*, d. 49, 1. 1, sol. 3 ad 1. The first part of St. Thomas' answer to this argument (the second part has been quoted above) is as follows: "Bonum cui intellectus speculativus coniungitur per cognitionem, est communius bono, cui coniungitur intellectus practicus, inquantum intellectus speculativus magis separatur a particulari quam intellectus practicus, cuius cognitio in operatione perficitur, quae in singularibus consistit."—To understand this and similar texts (one of which is quoted by De K., p. 172, note 13) it must be noted, first, that the notion of common good is an analogical notion which St. Thomas has not always used in the same nor in its proper sense; and, secondly, that the Thomistic discussion of the primacy of the common good is frequently not, in the first place, a discussion of a doctrine, but of an "authority." A *dictum authenticum*, to a medieval writer, is always true. The only thing, therefore, that can be done about it, is to sustain it and to interpret it. A student of the Thomistic primacy of the common good must first of all know the characteristic medieval techniques of how to deal with a *dictum authenticum*. Cf. M. D. Chenu, O.P., " 'Authentica' et 'Magistralia'. Deux lieux théologiques au XII–XIII siecles." *Divus Thomas Placent.* 28 (1925): 257–285. Etienne Gilson has very aptly remarked that the results of this brief study are more valuable than a lot of certain other big volumes of "Thomism"; *Revue d'histoire franciscaine* 3 (1926), 128.

28. See *Mediaeval Studies* 5 (1943): 142 ff.

29. *ST*, II-II, 188. 6.

30. *ST*, II-II, 182. 1; 185. 2 ad 1, etc.—All these sayings might and must seem hard to us, inveterate and deep pragmatists that we are. But—there they are! For the "garden of contemplation" and the "dusty roads of the active life," see 4 *Sent.*, d. 38, 1. 4, sol. 1 ad 3 (*hortus contemplations*); 3 *Sent.*, d. 35, 1. 4, sol. 2 (*pulvis terrenorum*); 4 *Sent.*, d. 29, 4 ad 4; *ST*, II-II, 185. 2; et alibi.

31. *De Civ. Dei*, XIX, 13.

32. *ST*, I-II, 4. 8 ad 3.

33. 3 *Sent*, d. 5, III, 2.

34. Trans. A.H.C. Downes (New York: Scribners, 1936), 205.

35. See *Mediaeval Studies* 6 (1944): 62 ff.

36. 3 *Sent.*, d. 36, 3 ad 5.

37. *Corpus Iuris Canonici*, C. 19, Q. 2, c. 1: *Ed. Lips. sec.* (Friedberg), vol. I, col. 839 ff. Cf. *Mediaeval Studies* 6 (1944): 100 ff.

38. *De Ver.*, 21. 4 ad 3. I must confess that it has taken me a long time to understand Maritain's metaphysics of Person and Society. See *Bulletin thomiste* 4 (1936): 714 ff., 517 ff. I seem now to realize what was the reason for this skepticism. One cannot be, indeed, too much of a metaphysician, just as one cannot love God too much. But is it not possible to be a metaphysician too exclusively, just as, according to St. Thomas, *ST*, II-II, 27. 8, there is some possibility of loving God too exclusively? Properly speaking, the problem of Person and Society is a problem of ethics and social theory: How do Christians stand to society and, especially, to the state? It seems to me that much of the extant opposition to Maritain is due to the fact that a very sublime and absolute metaphysical theory is—if I may say so—sprung on the ethicians and takes them unawares while they are discussing, not a different problem, but the same problem on quite a different level. In truth, Maritain's elucubrations are much richer, much more varied than it might seem from the description of their "bare essence" given above. But is it not likewise true that in the literature following in his wake, not infrequently, there is no question of anything else than of that metaphysics? I cannot bring myself to admit that, in the problem of person and society, everything is said and everything is done, once an absolute, ontological comparison between person and society is instituted. After the Encyclical of Pius XII and after renewed studies I grant that thereby the very last or, if you like, the very first thing—and, in this sense, all—is said and done.

In Defence of St. Thomas

A Reply to Father Eschmann's Attack on *The Primacy of the Common Good*

1945

Table of Contents

. . . *in eligendis opinionibus vel repudiandis, non debet duci homo amore vel odio introducentis opinionem, sed magis ex certitudine veritatis.*
—St. Thomas[1]

I. On "Convenient Anonymity"

Professor Yves Simon[2] seems to agree with the doctrine contained in my brief essay on the primacy of the common good:[3]

> . . . De Koninck has outlined, with unusual profundity and accuracy, the main aspects of a theory of the common good. It would be unfair to blame such a brief treatment for what we do not find in it. We do find in it a most valuable contribution to the definition of the common good and to the vindication of its primacy.[4]

The doctrine I outlined

> calls for many specifications and further developments, but it constitutes a very sound foundation for any further development of the theory of the common good.[5] . . . Insofar as De Koninck's essay vindicates the primacy of the common good and carries out the criticism of definite positions, it is entirely praiseworthy.[6]

The positions and their necessary consequences which I consider representative of personalism and which I attack, he rightly qualifies as "vicious stupidities"[7] and "monstrosities."[8] When it comes to determining who are the personalists, Professor Simon has some understanding words to say:

> Turning to the polemical side of the essay, we realize at once that the writer was confronted by a great difficulty. De Koninck's purpose is to vindicate the primacy of the common good *against the personalists.* It is a hard job, for the obvious reason that the term personalism covers a great variety of ill-defined doctrines and attitudes.[9]

While admitting there is some difficulty in identifying the personalists, Professor Simon is yet dissatisfied that my book should have named only those whose position was well defined. And here is the reason for his dissatisfaction:

> On account of the very important part played by the concept of person in the work of Maritain, there is no reason why he [the reader] should not believe that the expression "the personalists" stands for Jacques Maritain.[10]

Now it is glaringly obvious to Professor Simon that the ideas I describe as personalist are, with few exceptions and perhaps without any exception, just as odious to M. Maritain as they are to myself; that what I maintain concerning the primacy of the common good is just as dear to the latter as it is to myself. Hence, he does not hesitate to declare that, insofar as the reader might be left to believe that Jacques Maritain would disagree with any of the fundamental positions involved,

> [t]he net effect of the essay, insofar as Maritain is concerned, resembles that which could have been brought about—perhaps not so successfully—by plain calumny.[11]

Yves Simon is indeed a friend. He does not mince words. As one of M. Maritain's most esteemed and faithful disciples, he is sure the doctrines I condemn are not those of M. Maritain; and accordingly he gives me plainly to understand, that if I intended my readers to believe otherwise, I would be committing a simple calumny.

A second critic of my little work takes an astonishingly different view. For Father I. Th. Eschmann[12] it is just as obvious that the most fundamental position of the personalism I attack is beyond a doubt that of M. Maritain. As for my own position, it is "manifest error." He does not

> in the least hesitate to say, that from the point of view of the *littera Sancti Thomae* this book is a danger to every reader who has neither the time nor the sufficient training to discover for himself, in a problem of extreme subtlety, the genuine Thomistic truth. (*DM*, 204)

If that were true, my case would be sad enough. But there is much worse than that.

> If they [Professor De K.'s doctrine and arguments] were true, then the personalists, and with them all the Christian Fathers and theologians and philosophers, should close their shops, go home and do penance, *in cinere et cilicio,* for having grossly erred and misled the Christian world throughout almost two thousand years. (*DM*, 189)

Let the reader be reminded of the sixth and seventh *loci theologici* to realize the predicament Father Eschmann has placed me in. And if such is indeed the case, the unshakable assurance and uninhibited violence of his article[13] are quite understandable. Indeed one might even understand its sneering and irony if I actually used the facile device and the absurd or dishonest methods which Father Eschmann lays to my charge:

> Will it be granted that it is inadmissible to read St. Thomas with scissors and paste, by cutting the texts out of their literary and historical context and just quoting what, in a particular instance, seems to be suitable? Will it be granted that, if St. Thomas has explicitly stated and solved a given problem, a Thomist worthy of that name is obliged to take account of this fact and can not afford to refer to some other texts which either have nothing to do with the problem or, at best, refer to it in a distant and mediate fashion? (*DM*, 189–90)

My Opponent is not just making rash statements. The criterion he uses to defend the manifest truth of the position I attack is, as he frequently repeats, the *littera Sancti Thomae.*

On page 187, my Opponent has inserted a note of a personal character, which should add to the weight of his denunciation.

> . . . I have the privilege to regard both Jacques Maritain and Charles De Koninck as dear friends. The job, therefore, of examining and determining the truth of their respective positions is very painful to me. (n. 9)

Since in spite of this protestation he discharges himself of his obligation with unconcealable gusto, it must be that Father Eschmann—who was for several years a professor in the Collegio Angelico, Rome, who taught at Laval University, Quebec, who is engaged at the Institut d'Etudes médiévales Albert-le-Grand of the University of Montreal, and is now a member of the staff of the Pontifical Institute of Mediaeval Studies, Toronto—is entirely confident the position I attack is very definitely that of Jacques Maritain, that this position is true, and that my own is a very dangerous one indeed.

Why did I not name M. Maritain? Father Eschmann has a very simple explanation. "The" personalists is

> an all too convenient anonymity which permits every attack, and leaves every avenue of retreat wide open. . . . (*DM*, 184)

Compare now Professor Simon's judgment of my essay with that of Father Eschmann. Presumably both my critics are especially qualified to judge whether or not my own position concurs with that of Maritain. Professor Simon holds that my doctrine is true, that the personalist positions I attack are vicious stupidities and monstrosities and that the net effect of letting the reader believe my essay is aimed at Maritain resembles that which could have been brought about by plain calumny. Father Eschmann feels "obliged totally and categorically to reject De K.'s thesis" (*DM*, 187, n. 9) which, at one point, he claims is opposed to *all the Christian Fathers, theologians, and philosophers;* he emphatically maintains that the doctrine I attack and he defends is that of Maritain; that "the personalists" is but a cowardly device "which permits every attack, and leaves every avenue of retreat wide open."

How is it then, that of these two critics, both especially qualified and presumably well acquainted with the writings of Jacques Maritain, the one can feel utterly confident that the latter is, while the other can feel quite as confident he definitely is not the true adversary at whom was directed *La primauté du bien commun contre les personnalistes?* Who is to blame for these contradictory judgments?

Has it occurred to anyone that I may have foreseen this very situation including the criticisms that would be heaped upon me? Or has it occurred to anyone that if

> [t]he problem of Person and Society in the philosophy of St. Thomas, for many years past a favorite topic among European Thomists, has recently become an acute question on the continent of North America, owing, in no small measure, to the publication by the Dean of the Faculty of Philosophy at Laval University, Quebec, Professor Charles De Koninck, of a book entitled *De la primauté du bien commun contre les personnalistes. Le principe de l'ordre nouveau.* (*DM*, 183);

and if Jacques Maritain is so very obviously implicated in this debate, that Jacques Maritain is still among the living and may be presumed able to speak for himself?

But let us suppose that Jacques Maritain has spoken clearly and consistently on this subject (a supposition hardly reconcilable with the contradictory judgments of Father Eschmann and of Professor Simon), that he has treated it in philosophical fashion, and that he really is the main target of my essay against the personalists. Could I have no justifiable reason for that failure to name my adversary which Father Eschmann calls "anonymity"? My Opponent cannot imagine any but this: *"The" personalists* is "an all too convenient anonymity which permits every attack, and leaves every avenue of retreat wide open" and this notwithstanding that in the same moment he finds the personalism I attack so very plainly and inescapably that "represented most prominently by Jacques Maritain" as to deprive my guilty anonymity of any sensible motive whatever (*DM*, 184).

The reader is acquainted with certain polemical *Opuscula,* such as the *De Aeternitate mundi contra murmurantes,* or the *De Unitate intellectus contra averroistas parisienses.* Of these works we may surely say that they too

> comprise more than their objective, abstract content, more than the mere words in which they are written. They embrace all the circumstances of time, place, and occasion with which their publication is surrounded. (*DM,* 184)

Yet who are these anonymous *Murmurantes* who lay claim to such subtlety in perceiving contradictions, "as if they alone were men and wisdom born with them?"[14] Did St. Thomas resort to "the" Parisian Averroists as to a convenient anonymity which permits every attack, and leaves every avenue of retreat wide open? Who speaks "*in angulis* and before young people who cannot judge of such difficult matters"?[15] That he intended to attack Siger of Brabant is susceptible of strict proof.

Indeed the circumstances of writing and publication are contingent. More than that, they are the very own circumstances of the writer himself, the *contingentia, variabilia, inenarrabilia* of human actions. That is why they should be left to the inalienable prudential judgment of the individual person. Has it occurred to my Opponent that there are circumstances, even of my public action, which he does not know and which are most certainly none of his concern? Can he conceive of no circumstances in which he might be right in attacking an anonymous adversary, or in which he might even do so anonymously? If, in some given instance, Father Eschmann might say what *he* thinks I should have done, he cannot tyrannically impose his judgment of what I should do as the ultimate norm of my own. But it has pleased him to grant me only one motive. *Qualis unusquisque est, talis ei finis videtur.*

We have all heard the story of the thief who in order to distract the attention of the people about him, cried *Thief! Thief!* Everyone looked the other way, and so forth. But there is also the saying that "you can't fool all the people all the time." It will soon be clear that the *Thief! Thief!* device, quite unconsciously, I believe, is the keynote to my Opponent's whole article *In Defense of Maritain.*

Who would suspect Father Eschmann of himself exemplifying that very subterfuge of "convenient anonymity" which he lays to my charge, and in the very section of his article in which he brands anonymity as permitting every attack and leaving every avenue of retreat wide open? Is it possible that the person he names is at the same time made the target for an adversary unnamed? That he also has in mind a person other than myself is indeed susceptible of the type of strict proof my Opponent avails himself of in such matters. Who, in connection with personalism, warns us against "a revival of the polycephalous monster of Pelagianism?" (*DM,* 185). Whom will the reader of my booklet, the Preface not excluded, have in mind when my Opponent refers to "a work which pretends to exhibit the pure wisdom of St. Thomas Aquinas?" (*DM,* 189). (And in this connection, why did Father Eschmann add to the original "pure wisdom" his own words: "of St. Thomas Aquinas"?) Is his reader, unacquainted with my text, meant to believe that *I* claim to "exhibit the pure wisdom of St. Thomas Aquinas"? If it is not deliberately planned, why does he condone with the ignominious ambiguity of jestingly referring to "a work which pretends to exhibit the pure wisdom of St. Thomas Aquinas"? And yet while indulging in practices as offensive as this, he ventures to appeal to the charity of the reader.

> Let us be charitable and forget that such a statement ("Les parties principales constituant matériellement l'univers . . .") has ever been made in a work which pretends to exhibit the pure wisdom of St. Thomas Aquinas. (*DM,* 189)

And at this very point my Opponent's kindly sentiments overflow into that note 12 of page 189, that unmistakable insinuation aimed at an anonymous target, again at the expense of his "dear friend."[16] But enough of this sort of thing.

II. St. Thomas on Part and Whole[17]

It is of no concern to us why Father Eschmann completely overlooks what I had to say on the nature of the common good. However, this omission does allow him to convey to the reader unacquainted with my text, the impression that I share my Opponent's own conception of the good and of the common good. What *he* means by a common good is already clear from the way he quotes against me a passage from the Encyclical *Divini Redemptoris* (Father Eschmann does not mention that I faced an objection construed from that very text, *BC*, 67–70) and another from the Encyclical *Mystici Corporis Christi.* The notion of common good which he has in mind throughout his attack is very distinctly the one I had emphatically and repeatedly denounced as totalitarian. To argue from his own notion most certainly leads to a "contradictory and unintelligible position" (*DM*, 184–85). But we shall return to section I of Father Eschmann's article in due course. Let us start from where he expressly claims to begin his "critique" (*DM*, 187, n. 9).

> On page 27 Professor De Koninck states that even the personalists will not have great difficulty in admitting, with him, that individual persons are subordinated to that ultimate separate and extrinsic good of the universe which is God, nor that this subordination is formally motivated by the fact that God is the common good. But this will not suffice. It must be stressed, indeed—such is the author's thesis—that persons are subordinated to the intrinsic common good of the universe, i.e. its order. And they are thus subordinated because they are material parts materially composing and materially constituting that order and common good. For, is not the ultimate reason why God has created the intellectual beings or persons none other than exactly the order and common good of the universe? (*DM*, 187)

Then Father Eschmann quotes the passage in question:

> Si l'on concède que les personnes singulières sont ordonnées au bien ultime séparé en tant que celui-ci a raison de bien commun, on ne concédera pas si volontiers que, dans l'univers même, les personnes ne sont voulues que pour le bien de l'ordre de l'univers, bien commun intrinsèque meilleur que les personnes singulières qui le constituent matériellement. (*BC*, 27)

The complete omission of what I had to say on the very nature of the common good already ensured Father Eschmann a great deal of freedom. The passage quoted above would be "revolting" indeed if we were to interpret it in the light of the notion of common good he would have the reader believe to be mine, just as revolting as would be statements such as: *Quaelibet autem persona singularis comparatur ad totam communitatem, sicut pars ad totum;*[18] or: . . . *Ipse totus homo ordinatur, ut ad finem, ad totam communitatem cujus est pars.*[19] If such statements were to be read in the light of the totalitarian notion of common good Father Eschmann would force upon me, how could we possibly attribute them to St. Thomas? But my Opponent was not satisfied with the freedom he derived from mere omission. He saw fit to paraphrase my text before quoting it. The critical reader must have observed that "les personnes singulières qui le constituent matériellement" was introduced by my Opponent's: "they are material parts materially composing and materially constituting that order. . . ." Why does he add the word "material"? Is there no difference between "parts materially composing" and "material parts materially composing?" Lest there remain any doubt in the mind of the reader, let us see how he uses this difference.

Now that, thanks to his paraphrase, the persons have become *material* parts materially constituting the order of the universe, Father Eschmann proceeds to arrest the ambiguity of the word he himself has added, by substituting for my "order of the universe" the term "cosmos." Obviously, no one could possibly object to this substitution, since everyone should know that cosmos means "order of the universe"! But, at the same time, we also know that cosmos now definitely means the order of corporeal beings—the subject of what is called cosmology. Hence, how could anyone have the effrontery to object to Father Eschmann's inferring, from his own distorted paraphrase of my text, a position so coarse and unmistakably heretical that any Catholic will be shocked?

> For, being material parts of the cosmos and subordinated, as material parts, to the stars and the spheres, they [the personalists, and with them all the Christian Fathers and theologians and philosophers] will have just as much responsibility, just as much choice, as the pistons in a steam engine. (*DM*, 189)

No wonder "Even Professor De K. somehow seems to feel that his is a 'revolting' statement (cf. p. 30)" (*DM*, 187).[20]

I must again call attention to Father Eschmann's opening paragraph of section II, which we have already quoted: ("On page 27 Professor De Koninck states . . ."). In the first part of this paragraph he allows that I distinguish between the "ultimate separate and extrinsic good of the universe which is God" and the "intrinsic common good of the universe, i.e. its order." From this it should be clear, even to the reader unacquainted with my full text that, in my view, absolutely speaking, the former alone can be the ultimate reason why God has created the intellectual beings or persons. Nevertheless, in the last sentence of his paragraph, when my Opponent ironically states: "For, is not the ultimate reason why God has created the intellectual beings or persons none other than exactly the order and common good of the universe?" he gives the reader to understand that, in my view, the "ultimate reason intrinsic to the universe" must stand for the "ultimate reason" taken absolutely.

Having bridged the gulf between persons and "the pistons in a steam engine" by means of the "material parts of the cosmos," Father Eschmann immediately adds:

> Let it be said, at once, that we simply refuse even to discuss this, Professor De K.'s own, private doctrine and thesis which is most patently erroneous. Let us be charitable and forget that such a statement ("Les parties principales constituant matériellement l'univers . . .") has ever been made in a work which pretends to exhibit the pure wisdom of St. Thomas Aquinas. (*DM*, 189)

Father Eschmann makes much of the phrase "principal parts materially constituting the universe." Indeed he will use it to deal a blow from which its author is never to recover. Let us see how he will go about this.

> Would it not be desirable that an author who uses traditional philosophical notions knew exactly what they mean? In a recent work, *Saint Joseph, Père vierge de Jésus* (Montréal, 1944), Msgr. G. Breynat, a venerable missionary Bishop, in all seriousness and against the protest of a large group of theologians, defends the following definition of St. Joseph's paternity: It is "une causalité effective, négative, par abstention" of the child Jesus (pp. 84, 117 ff.). Professor De K.'s notion of a principal part materially constituting the universe is of the same caliber. (*DM*, 189, n. 12)

In other words, to maintain that the principal or formal parts of a whole may be viewed in the line of material causality, as materially constitutive of that whole, is a grave misdemeanor deserving only ridicule. However, does not a part as part, whether principal or secondary, material or formal, corporeal or spiritual, belong to the genus of material cause? Is not any and every part "id *ex* quo"? Let us turn to St. Thomas' *Comm. in II Physicorum* (ed. Leon.), lectio 5, n. 9. Aristotle's chapter 3 raises a doubt

> de hoc quod dicit, quod partes sunt causae materiales totius, cum supra partes definitionis reduxerit ad causam formalem. Et potest dici quod supra locutus est de partibus speciei, quae cadunt in definitione totius: hic autem loquitur de partibus materiae, in quarum definitione cadit totum, sicut circulus cadit in definitione semicirculi. Sed melius dicendum est quod licet partes speciei quae ponuntur in definitione, comparentur ad suppositum naturae per modum causae formalis, tamen ad ipsam naturam cujus sunt partes comparantur ut materia: *nam omnes partes comparantur ad totum ut imperfectum ad perfectum, quae quidem est comparatio materiae ad formam.*

St. Thomas leaves no doubt as to the universality of this principle:

> Manifestum est autem ex iis quae dicta sunt in secundo (*Physic.*), quod totum habet rationem formae, *partes autem rationem materiae.*[21]

Now, whatever belongs to the very nature of a thing is inseparable from it. Will it be granted that a principal or formal part is still a *part?* If it does not have the *ratio materiae,* why call it a part?

In that same footnote 12, page 189, Father Eschmann quotes with approval the following text taken from my essay, page 35:

> Les créatures raisonnables peuvent atteindre elles-mêmes de manière explicite le bien auquel toutes choses sont ordonnées; elles diffèrent par là des créatures irraisonnables, qui sont de purs instruments, qui sont utiles seulement et qui n'atteignent pas elles-mêmes de manière explicite le bien universel auquel elles sont ordonnées.

My Opponent then exclaims:

> Very well! But how does this statement stand to the other one: ". . . les parties principales constituant matériellement l'univers . . ."?

Is it so utterly preposterous to consider the intellectual creatures as principal parts yet materially constituting the universe? Here is the *littera Sancti Thomae.*[22]

> Considerandum est quod ex omnibus creaturis constituitur totum universum sicut totum ex partibus. Si autem alicujus totius et partium ejus velimus finem assignare, inveniemus

primo quidem, quod singulae partes sunt propter suos actus; sicut oculus ad videndum.	(A)	Sic igitur et in partibus universi, unaquaeque creatura est propter suum proprium actum et perfectionem.
Secundo vero, quod pars ignobilior est propter nobiliorem; sicut sensus propter intellectum, et pulmo propter cor.	(B)	Secundo autem, creaturae ignobiliores sunt propter nobiliores, sicut creaturae quae sunt infra hominem sunt propter hominem.
Tertio vero, omnes partes sunt propter perfectionem totius, sicut et materia propter formam: *partes enim sunt quasi materia totius.*	(C)	Ulterius autem, singulae creaturae sunt propter perfectionem totius universi.

Ulterius autem, totus homo est propter aliquem finem extrinsecum, puta ut fruatur Deo.	(D)	Ulterius autem, totum universum, cum singulis suis partibus, ordinatur in Deum sicut in finem, inquantum in eis per quandam imitationem divina bonitas repraesentatur ad gloriam Dei:
	(E)	*quamvis creaturae rationales speciali quodam modo supra hoc habeant finem Deum, quem attingere possunt sua operatione, cognoscendo et amando.*[23]

Et sic patet quod divina bonitas est finis omnium corporalium.

The "singulae creaturae [quae] sunt propter perfectionem totius universi" (C) comprise the "creaturae nobiliores" as well as the "creaturae ignobiliores" (B), and all of them are "quasi materia totius" (C). In this perspective (C), the "creaturae nobiliores" are not the "perfectio totius" which is as the form. For, the "forma . . . universi consistit in distinctione et ordine partium ejus;"[24] the good of this order is "formale respectu singularium sicut perfectio totius respectu partium."[25] It is in the previous perspective (B)—"creaturae ignobiliores sunt propter nobiliores"—that the intellectual creatures may be compared to the form. But with respect to the whole universe, the "creaturae nobiliores" are still "quasi materia totius": "des parties principales constituant matériellement l'univers," for, to the order of the universe, "*quaelibet* creatura ordinatur, sicut pars ad formam totius."[26] And now my Opponent might well ask how the latter statement of Ia, q. 65, a. 2, c.: "quamvis creaturae rationales speciali quodam modo supra hoc habeant finem Deum, quem attingere possunt sua operatione, cognoscendo et amando," stands to the four preceding divisions.

While the head of the body is the principal part of the body, it is still a member, a part, of the body, and in this respect, it is "materially constitutive." Obviously this involves an imperfection. But is it an imperfection incompatible with the "partes nobiliores" of the universe? Is not Christ, Who is the Head of the Church, a *member* and a *part* according to His humanity? It is according to His divinity that He cannot be a part of the universe. And why not? Because, in this respect, He is the common good of the whole universe.

> Estis membra dependentia de Christo membro, quod quidem dicitur *membrum secundum humanitatem,* secundum quam praecipue dicitur Ecclesiae caput. Nam secundum divinitatem non habet rationem membri aut partis, *cum sit commune bonum totius universi.*[27]

Was it Father Eschmann who asked: "Would it not be desirable that an author who uses traditional philosophical notions knew exactly what they mean?" Would this be another instance of my Opponent's *Thief! Thief!* method?

III. A Thomistic Proof of a "Revolting" Statement

Let us now examine Father Eschmann's exposure of my "Thomistic proof" for a statement which, he says, even its author seems to find "revolting" (*DM*, 187). My Opponent is wholly unaware that what I had said already on the nature of the good and of the common good is essential to the problem at issue. The good, as I take it throughout my essay, is not the perfection of being that is formally identical with being, but the perfection of being as having the nature of an end (*BC*, 7). For,

> [i]n quantum . . . unum ens est secundum esse suum perfectivum alterius et conservativum, habet rationem finis respectu illius quod ab eo perficitur; et inde est quod omnes recte definientes bonum *ponunt in ratione ejus* aliquid quod pertineat ad habitudinem finis; unde Philosophus dicit in I *Ethic.* (in princip.), quod *bonum optime definiunt dicentes, quod bonum est quod omnia appetunt.*[28]

My Adversary might have been warned, too, by the adage: "the good is diffusive of itself" (*BC*, 7). And the good is diffusive of itself *inquantum hujusmodi, secundum sui ipsius rationem.* May we recall what this diffusion stands for in connection with the good taken formally?

> . . . *Diffundere,* licet secundum proprietatem vocabuli videatur importare operationem causae efficientis, tamen largo modo potest importare habitudinem cujuscumque causae, sicut influere et facere, et alia hujusmodi. Cum autem dicitur quod *bonum est diffusivum secundum sui rationem,*

> non est intelligenda effusio secundum quod importat operationem causae efficientis, sed secundum quod importat habitudinem causae finalis; et talis diffusio non est mediante aliqua virtute superaddita. Dicit autem bonum diffusionem causae finalis, et non causae agentis: tum quia efficiens, in quantum hujusmodi, non est rei mensura et perfectio, sed magis initium; tum quia effectus participat causam efficientem secundum assimilationem formae tantum; sed finem consequitur res secundum totum esse suum, et in hoc consistebat ratio boni.[29]

And now we raise the question: Is it in the very being of the individual persons taken separately that we find most perfectly realized the good which God produces, that is, *the good that is in the universe itself?* Or is it rather the total order of the universe which most perfectly represents and is closer to, the ultimate separated and extrinsic good which is God? It should be recalled that where this question is proposed in my book it is in face of the contention that the greatest perfection *within* the universe consists first and absolutely in the individual persons taken separately, whereas the perfection of the whole order of the universe would be secondary. Immediately after the "revolting" statement, I said:

> On voudrait plutôt que l'ordre de l'univers ne fût qu'une superstructure de personnes que Dieu veut, non pas comme parties, mais comme touts radicalement indépendants; et ce ne serait qu'en second que ces touts seraient des parties. En effet, les créatures raisonnables ne diffèrent-elles pas des créatures irraisonnables en ce qu'elles sont voulues et gouvernées pour elles-mêmes, non seulement quant à l'espèce, mais aussi quant à l'individu? "Les actes . . . de la créature raisonnable sont dirigés par la divine providence, non seulement en raison de leur appartenance à l'espèce, mais aussi en tant qu'ils sont des actes personnels." Donc, conclurait-on, les personnes individuelles sont elles-mêmes des biens voulus d'abord pour soi et en soi supérieurs au bien du tout accidentel qu'elles constituent par voie de conséquence et de complément. (*BC*, 27)

So the problem is not whether the universe is some kind of super-individual to whom God wills the enjoyment of all the things that He makes and governs, but whether the good that is the universe is the most perfect final cause that God has made. Now, if such is the case, it follows, in this perspective, that any particular good, any part of the universe, whether it is a person or not, will be ordered to this good of the universe, insofar as "singulae creaturae sunt propter perfectionem totius universi."

Nor can we broach my Opponent's confusions without recalling, at this juncture, what I had earlier said about the common good. Since he has quoted against me a passage from the Encyclical *Mystici Corporis Christi* (*DM*, 183, 186) with particular emphasis on the words "utpote personae sunt," it must be that, in his mind—and his reader is apparently to be left with the same understanding—the common good whose primacy I defend is not attained by the persons, that this common good, indeed, is as the good of a natural body which so unites the parts that each lacks its own individual subsistence, so that the different members are destined solely to their good through the whole. Let me choose another, of several possible citations, to show that this is indeed the interpretation he makes:

> The most essential and the dearest aim of Thomism is to make sure that the personal contact of all intellectual creatures with God, as well as their personal subordination to God, be in no way interrupted. Everything else—the whole universe and every social institution—must ultimately minister to this purpose; everything must foster and strengthen and protect the conversation of the soul, every soul, with God. It is characteristically Greek and pagan to interpose the universe between God and intellectual creatures. Is it necessary to remind Thomists that they should not, in any way whatever, revive the old pagan blasphemy of a divine cosmos? (*DM*, 192)

I think we have the right to presume that my Opponent has read my essay with care. How then can his understanding of the doctrine I defend be reconciled with even the very first pages of my little book?

> Dès lors, le bien commun n'est pas un bien qui ne serait pas le bien des particuliers, et qui ne serait que le bien de la collectivité envisagée comme une sorte de singulier. Dans ce cas, il serait commun par accident seulement, il serait proprement singulier, ou, si l'on veut, il différerait du bien singulier des particuliers en ce qu'il serait *nullius*. Or, quand nous

> distinguons le bien commun du bien particulier, nous n'entendons pas par là qu'il n'est pas le bien des particuliers: s'il n'était pas le bien des particuliers, il ne serait pas vraiment commun.
>
> Le bien est ce que toutes choses désirent en tant qu'elles désirent *leur perfection*. Cette perfection est pour chacune d'elles *son bien*—bonum suum—et, en ce sens, son bien est un bien propre. Mais alors, le bien propre ne s'oppose pas au bien commun. En effet, le bien propre auquel tend naturellement un être, le 'bonum suum', peut s'entendre de diverses manières, selon les divers biens dans lesquels il trouve sa perfection. (*BC*, 9–10)

In fact, the good that is proper to one person and distinguished from that of another person, is alien to the good of the other person. Likewise, the common good that is proper to one community, is alien to the common good that is proper to another community.

> . . . Bonum commune est finis singularum personarum in communitate existentium; sicut bonum totius, finis est cujuslibet partium. *Bonum autem unius personae singularis non est finis alterius.*[30]

That is why I insisted:

> C'est ignorer spéculativement le bien commun que de le considérer comme un bien étranger, comme un 'bonum alienum' opposé au 'bonum suum': on limite, alors, le 'bonum suum' au bien singulier de la personne singulière. Dans cette position, la subordination du bien privé au bien commun voudrait dire subordination du bien le plus parfait de la personne, à un bien étranger; le tout et la partie seraient étrangers l'un à l'autre: le tout de la partie ne serait pas 'son tout'. (*BC*, 30–31)

Since my Opponent opposes to my position the "utpote personae sunt" of the above-mentioned Encyclical, why does he ignore the following passages of my essay:

> Nous répondons que la communauté de ce bien ne doit pas s'entendre d'une communauté de prédication, mais d'une communauté

de causalité. Le bien commun n'est pas commun comme 'animal' par rapport à 'homme' et 'brute', mais comme le moyen universel de connaître, qui dans son unité atteint les connus dans ce qu'ils ont de plus propre. Il s'étend à plusieurs, non pas grâce à une confusion, mais à cause de sa détermination très élevée qui s'étend principalement à ce qu'il y a de plus élevé dans les inférieurs: "une cause plus élevée a un effet propre plus élevé." Il s'étend à Pierre, non pas d'abord en tant que Pierre est animal, ni même en tant qu'il est nature raisonnable seulement, mais en tant qu'il est 'cette' nature raisonnable: il est le bien de Pierre envisagé dans sa personnalité la plus propre. C'est pourquoi le bien commun est aussi le lien le plus intime des personnes entre elles et le plus noble. (*BC*, 54–55)

L'indépendance des personnes les unes des autres dans la vision (béatifique) même n'exclut pas de l'objet cette universalité qui veut dire, pour toute intelligence créée, essentielle communicabilité à plusieurs. Loin de l'exclure, ou d'en faire abstraction, l'indépendance présuppose cette communicabilité. (*BC*, 62)[31]

Comme les précédentes, cette objection suppose admise l'interprétation que les collectivistes font de notre conception de la société. Or, la société n'est pas une entité séparable de ses membres: elle est constituée de personnes qui sont à l'image de Dieu. Et c'est cette société, non pas une entité quasi abstraite, mais constituée de personnes, qui est de l'intention principale de Dieu. (*BC*, 64)

La cité n'est pas, ou ne peut pas être, un 'pour soi' figé et refermé sur soi, opposé comme un singulier à d'autres singuliers; son bien doit être identiquement le bien de ses membres. Si le bien commun était le bien de la cité en tant que celle-ci est, sous un rapport accidentel, une sorte d'individu, il serait du coup bien particulier et proprement étranger aux membres de la société. Il faudrait même accorder à l'organisation ainsi ravie à ses membres, intelligence et volonté. La cité serait alors comme un tyran anonyme qui s'assujettit l'homme. L'homme serait pour la cité. Ce bien ne serait ni commun ni bien de natures raisonnables. L'homme serait soumis à un bien étranger. (*BC*, 68–69)

> En fait, le personnalisme fait sienne la notion totalitaire de l'Etat. Sous les régimes totalitaires, le bien commun s'est singularisé, et il s'oppose en singulier plus puissant à des singuliers purement et simplement assujettis. Le bien commun a perdu sa note distinctive, il devient bien étranger. Il a été subordonné à ce monstre d'invention moderne qu'on appelle l'Etat, non pas l'état pris comme synonyme de société civile ou de cité, mais l'Etat qui signifie une cité érigée en une sorte de personne physique. (*BC*, 75)

I am not aware that Father Eschmann has anywhere said that my notion of the common good is false, although he violently attacks its application to God as the object of created beatitude. But I think that, from the above quotations, it is unmistakably clear that his notion of the common good as such is not mine; that he hopelessly distorts my notion; that the doctrine he attributes to me is, in fact, his own distortion and that the texts just quoted from my essay are definitely opposed to his own notion of the common good. These citations make it plain, in short, that I must energetically reject all possibility of a subordination of the person to Father Eschmann's common good, or to anything like the common good as he understands it. Hence, when he says that

> [t]here is a proper and profound Thomistic doctrine of the relative superiority, within definite orders, of *their* respective common goods over the particular goods contained in those orders (*DM*, 184),

we may be certain that, even within definite orders, my Opponent's totalitarian common good could not possibly be accepted, by any Thomist, as superior in any sense over the particular good of persons.

No Thomist could accept Father Eschmann's unfortunate notion of part and whole. That it is not even applicable to the moral whole and part is obvious not only from what we have already quoted, but from what he calls the *antecedens* of the proof of personalism.

> It seems to me—*salvo meliore iudicio*—that the bare essence of this doctrine might be summed up in the following *enthymema:* St. Thomas says: *Ad rationem personae exigitur quod sit totum completum;* or again:

> *Ratio partis contrariatur personae.* Hence, Jacques Maritain concludes, the person, qua person, is not a part of society: and if a person is such a part, this "being part" will not be based upon the metaphysical formality and precision of "being person."
>
> The *antecedens* pertains to the *littera Sancti Thomae.* (*DM,* 205)

Since Father Eschmann has asserted that I "read St. Thomas with scissors and paste, by cutting texts out of their literary and historical context and just quoting what, in a particular instance, seems to be suitable" (*DM,* 189), the reader will hardly be inclined to suspect him of doing just that with every single quotation from St. Thomas he brandishes against me. Let us examine the two phrases here brought to our attention. He refers the reader to "3 *Sent.,* d. 5, III, 2." The article in question considers: *Utrum anima separata sit persona.* The immediate context of the first line quoted by my Opponent is:

> *Ad tertium* dicendum quod anima rationalis dicitur *hoc aliquid* per modum quo esse subsistens est hoc aliquid, etiam si habeat naturam partis; sed ad rationem personae exigitur ulterius quod sit totum completum.

The context of the four words which form his second quotation is:

> Sed haec opinio (Platonis) non potest stare; quia sic corpus animae accidentaliter adveniret. Unde hoc nomen *homo* de cujus intellectu est anima et corpus, non significaret *unum per se,* sed per accidens; et ita non esset in genere substantiae.
>
> Alia opinio est Aristotelis, . . . quam omnes moderni sequuntur, quod *anima unitur corpori sicut forma materiae.* Unde anima est pars humanae naturae, et non natura quaedam per se. Et quia ratio partis contrariatur rationi personae, ut dictum est, ideo anima separata non potest dici persona; quia quamvis separata non sit pars actu, tamen habet naturam ut sit pars.

No person could be part of a substantial "unum per se." But the human soul is but a part of man. Therefore the soul alone is not the person. Now why should Father Eschmann confront me with these texts, in which the term 'part' is used exclusively of the soul as part of the human person, unless for him 'to be a part' means to be a part of such a whole as is implied in these phrases, namely, an "unum per se"? If we are to understand that his notion of part has a wider range than this, of what worth is his "*enthymema*"?

The whole of any society or of the universe is but an accidental unity (*BC*, 56). When St. Thomas calls the intellectual creature a part of society,[32] a part of the universe, or a part when compared with the divine good,[33] he is obviously not using the term 'part' in the sense in which it is understood in the article referred to by Father Eschmann, i.e. as part of an "unum per se." Yet my Opponent allows the person to be a part in this latter, strictly totalitarian sense which contradicts the very nature of any person no matter how imperfect and limited. The reader will recall his argument:

> *Ratio partis contrariatur personae.* Hence, Jacques Maritain concludes, the person, qua person, is not a part of society: and if a person is such a part, this "being part" will not be based upon the metaphysical formality and precision of "being person." (*DM*, 205)

Since the argument calls for a consistent meaning of the term "part," and since the "part" of the *antecedens* means "part of an *unum per se*," "to be a part of society" must mean "to be a part of an *unum per se*." When my Opponent attempts to show just how obvious is his conclusion, he merely makes more clear his own error:

> Maritain's conclusion is evident. Its necessity and intelligibility are exactly the same as the necessity and intelligibility of the following inferences: Act as such means pure and limitless perfection. Hence, if there is a limited or participated act, this limitation will not pertain to this act, qua act, but qua mixed with potency. Or again: The intellect as such is not capable of error. Hence, if there is an intellectual being which errs, this will not happen to it, insofar as it is an intellect but insofar as it is something else. (*DM*, 205)

A person, then, may be rendered capable of being "such a part of society" by reason of some limitation. This is to say that a person, by some limitation, can be that which is contrary to the very nature of person; that is, a person, while person, can also be non-person. Does my Opponent realize that what is contrary to the very nature of a thing can in no case belong to it? It is for that very reason we hold no person, however imperfect, can possibly be a part in the sense in which St. Thomas uses the term in the passage cited by Father Eschmann.

But let us suppose for a moment that my Opponent is taking the notion of part in all its amplitude—which he decidedly could not do without destroying his own argument or distorting the meaning of the *littera Sancti Thomae.* Even then, it would be very true that no person could be a part because of his being a person, for, if "to be a part" were of the very nature of person, every person would necessarily be a part, including the Divine Persons. But granted no person is a part merely because a person, it surely does not follow that the created person, who is essentially and inalienably a finite person, cannot be a part *secundum hoc ipsum quod est.* What my Opponent overlooks is that the concept of person is an analogical concept, just as much as the concepts of act and of intellect. If his argument is to be at all conclusive, he must maintain that *we* created, finite persons do indeed possess the pure and limitless perfection of the person who is not a part; that insofar as we are in act, *we* possess the pure and limitless perfection of pure actuality. There is not the slightest doubt that this is what Father Eschmann must hold if his *antecedens* is to lead to his conclusion. How else could the pure and limitless perfection of personality, which precludes being a part in any real sense, and which is proper to the Divine Persons,[34] have anything to do in this connection with the persons that *we* are? Logically, he has no alternative. He must conceive the potency and limitation which make us finite beings and finite persons as adventitious to pure actuality and pure personality. In this respect, that which was pure actuality should now become subject; and since it would have to be the subject of a being substantially one, it would have to be pure subject, that is, pure potentiality. In other words, if he carried through the inescapable implications of his argumentation, my Opponent would be faced with something like the position of David of Dinant "qui stultissime posuit Deum esse materiam primam."[35]

And now we shall return to Father Eschmann's criticism of my "Thomistic proof" of a "'revolting' statement."

> Let us here for the moment consider the second part of this thesis, viz. the statement regarding the intrinsic common good of the universe

> and its relation to the intellectual beings or persons. Even Professor De K. somehow seems to feel that his is a "revolting" statement (cf. p. 30). He, therefore, makes every effort to be very careful in establishing a Thomistic proof of it. In fact, he asks, is not the same statement repeatedly implied in St. Thomas' discussions of the question: What is the end God has proposed to Himself in the production of all things? Four texts are cited by the author. Let us here reproduce, in Latin, the first two, taken from *Contra Gentiles*, III, 64; they will sufficiently show in what specific set of Thomistic texts Professor De K. has found a proof, satisfying to his mind, of his assertion. The italicized sentences are not held worthy of quotation, by the author. (*DM*, 187–88)

My Opponent then quotes the two paragraphs in question. The italicized sentence completing the first paragraph is:

> *Ipse igitur Deus omnia suo intellectu et voluntate gubernat.*

The italicized sentence completing the second is:

> *Est igitur gubernator ipsius.*

Just what did I want to prove by these texts? It is important to note that Father Eschmann opens his criticism by assailing "the second part of this thesis." Whether he uses this procedure willfully or not, it does obscure the issue and create a convenient confusion. Do I seek to prove that the order of the universe is the most profound and absolutely ultimate good of persons? On the contrary, I had spoken in the early part of my essay of this ultimate good of persons as being the absolute separated and extrinsic good of the universe, which is God. In "the second part of this thesis," however, I consider persons as *parts of the universe*, and I enquire what is their greatest good as parts of the universe. Now this is merely to ask what is the greatest good that God produces and that most perfectly imitates His own goodness. St. Thomas' answer leaves no doubt. Here are the two paragraphs I quoted (*BC*, 28–29; 175) from *Contra gentes* III, c. 64, to which I now add the concluding sentences which Father Eschmann says I hold "not . . . worthy of quotation":

> Deus res omnes in esse produxit, non ex necessitate naturae, sed per intellectum et voluntatem. Intellectus autem et voluntatis ipsius non potest esse alius finis ultimus nisi bonitas eius, ut scilicet eam rebus communicaret, sicut ex praemissis (lib. I, capp. 75 sq.) apparet. Res autem participant divinam bonitatem per modum similitudinis, inquantum ipsae sunt bonae. Id autem quod est *maxime bonum in rebus causatis, est bonum ordinis universi, quod est maxime perfectum,* ut Philosophus dicit (XII *Metaph.*, x, 1; 1075a): cui etiam consonat Scriptura divina, *Gen.* 1, cum dicitur (vers. 31), *Vidit Deus cuncta quae fecerat, et erant valde bona,* cum de singulis operibus dixisset simpliciter quod *erant bona.* Bonum igitur ordinis rerum causatarum a Deo est id quod est *praecipue volitum et causatum* a Deo. Nihil autem aliud est gubernare aliqua quam eis ordinem imponere. Ipse igitur Deus omnia suo intellectu et voluntate gubernat.
>
> Amplius. Unumquodque intendens aliquem finem, magis curat de eo quod est propinquius fini ultimo: quia hoc etiam est finis aliorum. Ultimus autem finis divinae voluntatis est bonitas ipsius, *cui propinquissimum in rebus creatis est bonum ordinis totius universi: cum ad ipsum ordinetur, sicut ad finem, omne particulare bonum huius vel illius rei, sicut minus perfectum ordinatur ad id quod est perfectius; unde et quaelibet pars invenitur esse propter suum totum. Id igitur quod maxime curat Deus in rebus creatis, est ordo universi.* Est igitur gubernator ipsius.

How the omission of those last sentences makes me guilty of my Opponent's practice of reading "St. Thomas with scissors and paste, by cutting the texts out of their literary and historical context and just quoting what, in a particular instance, seems to be suitable" (*DM*, 189), I fail to see. I quoted that part of the text which shows that St. Thomas expressly teaches the order of the universe to be the greatest good which God produces and that it is the *praecipue volitum.* It is from this truth that St. Thomas infers: *Ipse igitur Deus omnia suo intellectu et voluntate gubernat. Est igitur gubenator ipsius* [*ordinis universi*]. In stating that these "sentences are not held worthy of quotation, by the author" (*DM*, 188) my Opponent may distract attention, but his accusation should not blind the critical reader to the fact that he is distorting the perspective by stressing those last sentences, as if the premises of St. Thomas' conclusion did not properly and immediately belong to the question I undertook to treat, or as if the truth of the premises were irrelevant to their conclusion.

Before examining Father Eschmann's interpretation of these quotations, let me recall again the latter four of the six texts I quoted to support the doctrine that, of all created goods, the good of the universe is the greatest. The third is taken from Ia, q. 47, a. l: *Utrum rerum multitudo et distinctio sit a Deo.*

> . . . Distinctio rerum et multitudo est ex intentione primi agentis, quod est Deus. Produxit enim res in esse propter suam bonitatem communicandam creaturis, et per eas repraesentandam. Et *quia per unam creaturam sufficienter repraesentari non potest, produxit multas creaturas et diversas,* ut quod deest uni ad repraesentandam divinam bonitatem, suppleatur ex alia: nam bonitas quae in Deo est simpliciter et uniformiter, in creaturis est multipliciter et divisim. Unde *perfectius participat divinam bonitatem, et repraesentat eam, totum universum, quam alia quaecumque creatura.* (*BC*, 29; 175, n. 31)

The fourth text was taken from Ia, q. 15, a. 2, where St. Thomas proves that God has an idea of the order of the whole universe, because the *bonum ordinis universi* is the *optimum in rebus existens.*

> . . . In quolibet effectu illud quod est ultimus finis, proprie est intentum a principali agente; sicut ordo exercitus a duce. Illud autem quod est *optimum in rebus existens, est bonum ordinis universi,* ut patet per Philosophum in XII *Metaph.* Ordo igitur universi est proprie a Deo intentus, et non per accidens proveniens secundum successionem agentium . . . Sed . . . ipse ordo universi est per se creatus ab eo, et intentus ab ipso. (*BC*, 29; 176, n. 32)

In the fifth text St. Thomas proves (*Contra gentes,* II, c. 39) that the order of the parts of the universe and their distinction is the end of the production of the universe.

> Id quod est bonum et optimum in effectu, est finis productionis ipsius. Sed *bonum et optimum universi consistit in ordine partium eius ad invicem,* qui sine distinctione esse non potest: per hunc enim ordinem universum in sua totalitate constituitur, quae est optimum ipsius. Ipse igitur ordo partium universi et distinctio earum est finis productionis universi. (*BC*, 30; 176, n. 33)

From this St. Thomas further concludes: "non est igitur distinctio rerum a casu." These words I did not quote because we are concerned with whether or not the good of the universe is the greatest of all created goods, and not with the various conclusions that must be drawn from this principle. However, the manifold conclusions St. Thomas does draw from this fundamental truth illustrate its importance and fecundity.

The sixth text is taken from the *Q.D. de spiritualibus creaturis*, q. un., a. 8, where St. Thomas shows (*secunda ratio*) that the separated substances, occupying the *suprema pars universi*, constitute a *per se* order, differing in species, because, being the superior parts of the universe, they must have a greater participation in the good of the universe, which is its order.

> Manifestum est enim quod duplex est bonum universi; quoddam separatum, scilicet Deus, qui est sicut dux in exercitu; et quoddam in ipsis rebus, et hoc est ordo partium universi, sicut ordo partium exercitus est bonum exercitus. Unde Apostolus dicit *Rom.* 13, 1: *Quae a Deo sunt, ordinata sunt.* Oportet autem quod superiores universi partes magis de bono universi participent, quod est ordo. Perfectius autem participant ordinem ea in quibus est ordo per se, quam ea in quibus est ordo per accidens tantum. (*BC*, 33; 176, n. 35)

My Opponent states that the group of texts involving this principle is a very large one. In view of his interpretation, we shall quote a few more.

> Id quod est optimum in rebus causatis, reducitur ut in primam causam in id quod est optimum in causis: oportet enim effectus proportionales esse causis. Optimum autem in omnibus entibus causatis est ordo universi, in quo bonum universi consistit: sicut et in rebus humanis *bonum gentis est divinius quam bonum unius* (I *Ethic.*, ii, 8; 1094b). Oportet igitur ordinem universi sicut in causam propriam reducere in Deum, quem supra (lib. I, cap. 41) ostendimus esse summum bonum. Non igitur rerum distinctio, in qua ordo consistit universi, causatur ex causis secundis, sed magis ex intentione causae primae.
>
> Adhuc. *Absurdum videtur id quod est optimum in rebus reducere sicut in causam in rerum defectum. Optimum autem in rebus causatis est distinctio et ordo ipsarum, ut ostensum est* (arg. praec. et cap. 39). Inconve-

niens igitur est dicere quod talis distinctio ex hoc causetur quod secundae causae deficiunt a simplicitate causae primae.

Item. In omnibus causis agentibus ordinatis, ubi agitur propter finem, oportet quod fines causarum secundarum sint propter finem causae primae: sicut finis militaris et equestris et frenifactricis est propter finem civilis. Processus autem entium a primo ente est per actionem ordinatam ad finem: cum sit per intellectum, ut ostensum est (cap. 24); intellectus autem omnis propter finem agit. Si igitur in productione rerum sunt aliquae causae secundae, oportet quod fines earum et actiones sint propter finem causae primae, qui est *ultimus finis in rebus causatis. Hoc autem est distinctio et ordo partium universi, qui est quasi ultima forma. Non igitur est distinctio in rebus et ordo propter actiones secundarum causarum: sed magis actiones secundarum causarum sunt propter ordinem et distinctionem in rebus constituendam.*

Adhuc. Si distinctio partium universi et ordo earum est proprius effectus causae primae, quasi *ultima, forma et optimum in universo,* oportet rerum distinctionem et ordinem esse in intellectu causae primae. . . .[36]

Quanto enim aliquid est melius in effectibus, tanto est prius in intentione agentis. *Optimum autem in rebus creatis est perfectio universi, quae consistit in ordine distinctarum rerum: in omnibus enim perfectio totius praeminet perfectioni singularium partium.* Igitur diversitas rerum ex principali intentione primi agentis provenit, non ex diversitate meritorum.[37]

Item, *cum bonum totius sit melius quam bonum partium singularium, non est optimi factoris diminuere bonum totius ut aliquarum partium augeat bonitatem:* non enim aedificator fundamento tribuit eam bonitatem quam tribuit tecto, ne domum faciat ruinosam. *Factor igitur omnium, Deus, non faceret totum universum in suo genere optimum, si faceret omnes partes aequales:* quia multi gradus bonitatis in universo deessent, et sic esset imperfectum.[38]

Bonum ordinis universi nobilius est qualibet parte universi: cum partes singulae ordinentur ad bonum ordinis qui est in toto sicut ad finem, ut per Philosophum patet, in XII *Metaphysicae* (cap. x, 1; 1075a). Si igitur

Deus cognoscit aliquam aliam naturam nobilem, *maxime cognoscet ordinem universi.* Hic autem cognosci non potest nisi cognoscantur et nobiliora et viliora, in quorum distantiis et habitudinibus ordo universi consistit. Relinquitur igitur quod Deus cognoscit non solum nobilia, sed etiam ea quae vilia reputantur.[39]

Si Deus cognoscit aliquid aliud a se, *maxime cognoscet quod est optimum. Hoc autem est ordo universi, ad quem sicut ad finem omnia particularia bona ordinantur.*[40]

Providentia divina quibusdam rebus necessitatem imponit: non autem omnibus, ut quidam crediderunt. Ad providentiam enim pertinet ordinare res in finem. *Post bonitatem autem divinam, quae est finis a rebus separatus, principale bonum in ipsis rebus existens, est perfectio universi:* quae quidem non esset, si non omnes gradus essendi invenirentur in rebus. Unde ad divinam providentiam pertinet omnes gradus entium producere. Et ideo quibusdam effectibus praeparavit causas necessarias, ut necessario evenirent; quibusdam vero causas contingentes, ut evenirent contingenter, secundum conditionem proximarum causarum.[41]

. . . Malum quod in corruptione rerum aliquarum consistit, reducitur in Deum sicut in causam. Et hoc patet tam in naturalibus quam in voluntariis. Dictum est enim quod aliquod agens, inquantum sua virtute producit aliquam formam ad quam sequitur corruptio et defectus, causat sua virtute illam corruptionem et defectum. *Manifestum est autem quod forma quam principaliter Deus intendit in rebus creatis, est bonum ordinis universi.* Ordo autem universi requirit, ut supra dictum est, quod quaedam sint quae deficere possint, et interdum deficiant. Et sic Deus, in rebus causando bonum ordinis universi, ex consequenti, et quasi per accidens, causat corruptiones rerum; secundum illud quod dicitur I *Reg.* 2, *Dominus mortificat et vivificat.* Sed quod dicitur *Sap.* 1, quod *Deus mortem non fecit,* intelligitur *quasi per se intentam. Ad ordinem autem universi pertinet etiam ordo justitiae,* qui requirit ut peccatoribus poena inferatur. Et secundum hoc, Deus est auctor mali quod est poena: non autem mali quod est culpa, ratione supra dicta.[42]

. . . Providentia Dei, qua res gubernat, est similis, ut dictum est, art. praeced., providentiae, qua paterfamilias gubernat domum, et rex civitatem aut regnum: in quibus gubernationibus hoc est commune, quod bonum commune est eminentius quam bonum singulare; sicut bonum gentis est eminentius quam civitatis vel familiae vel personae, ut habetur, in principio *Ethic.* (cap. ii, in fin.). Unde quilibet provisor plus attendit quid communitati conveniat, si sapienter gubernat, quam quid conveniat uni tantum.

Hoc autem quidam non attendentes, considerantes in rebus corruptibilibus aliqua quae possent meliora esse secundum seipsa considerata, non attendentes ordinem universi, secundum quem optime collocatur unumquodque in ordine suo, dixerunt ista corruptibilia non gubernari a Deo sed sola incorruptibilia; ex quorum persona dicitur *Job* 22, 14: *Nubes latibulum ejus, scilicet Deus, neque nostra considerat; sed circa cardines caeli perambulat.* Haec autem corruptibilia posuerunt vel omnino absque gubernatore esse et agi, vel a contrario principio gubernari. Quam opinionem Philosophus in XII *Metaphysic.* (com. lii et seq.) reprobat per similitudinem exercitus, in quo invenimus duplicem ordinem: unum quo exercitus partes ordinantur ad invicem, alium quo ordinantur ad bonum exterius, scilicet ad bonum ducis; et ordo ille quo ordinantur partes exercitus ad invicem, est propter illum ordinem quo totus exercitus ordinatur ad ducem; unde si non esset ordo ad ducem, non esset ordo partium exercitus ad invicem. Quantumcumque ergo multitudinem invenimus ordinatam ad invicem, oportet eam ordinari ad exterius principium. Partes autem universi, corruptibiles et incorruptibiles, sunt ad invicem ordinatae, non per accidens, sed per se: videmus enim ex corporibus caelestibus utilitates provenire in corporibus corruptibilibus vel semper vel in majori parte secundum eumdem modum; unde oportet omnia, corruptibilia et incorruptibilia, esse in uno ordine providentiae principii exterioris, quod est extra universum. Unde Philosophus concludit, quod necesse est ponere in universo unum dominatum et non plures.

Sciendum tamen, quod aliquid provideri dicitur dupliciter: uno modo propter se, alio modo propter alia; sicut in domo propter se providentur ea in quibus essentialiter consistit bonum domus, sicut filii, possessiones, et hujusmodi: alia vero providentur ad horum utilitatem, ut vasa, animalia, et hujusmodi. Et similiter in universo illa propter se

> providentur in quibus essentialiter consistit perfectio universi; et haec perpetuitatem habent, sicut et universum perpetuum est. Quae vero perpetua non sunt, non providentur nisi propter alium; et ideo substantiae spirituales et corpora caelestia, quae sunt perpetua et secundum speciem, et secundum individuum, sunt provisa propter se et in specie et in individuo; sed corruptibilia perpetuitatem non possunt habere nisi in specie; unde species ipsae sunt provisae propter se, sed individua eorum non sunt provisa nisi propter perpetuum esse speciei conservandum.[43]
>
> . . . Quamvis res corruptibilis melior esset si incorruptibilitatem haberet, melius tamen est universum quod ex corruptibilibus et incorruptibilibus constat, quam quod ex incorruptibilibus tantum constaret, quia utraque natura bona est, scilicet corruptibilis et incorruptibilis; melius autem est esse duo bona quam unum tantum. Neque multiplicatio individuorum in una natura posset aequivalere diversitati naturarum, cum bonum naturae, quod est communicabile, praeemineat bono individui, quod est singulare.[44]

These form part of the body of texts I argue from. They prove that according to sound Thomistic doctrine, *optimum in omnibus entibus creatis est ordo universi, in quo bonum universi consistit.* Now, what does my Opponent have to say about this group of texts?

> . . . Against Greco-Arabian necessitarianism St. Thomas states that there exists an intelligent and loving Creator, i.e. a personal God and a divine and all-embracing Providence. Were this not so, he argues, the universe would fall apart into so many unconnected and unconnectable bits, and it would be impossible to maintain the fact of the order of the universe on whose existence and sublime beauty both the Greeks, and especially the Christian Fathers, have so energetically insisted. In this group of texts—it is a very large one—St. Thomas frequently, and with obvious enjoyment, avails himself of two quotations from Aristotle, viz., (*a*) *bonum commune est divinius* . . . and, (*b*) *quod est optimum in rebus existens est bonum universi.*[45] By these citations no proper doctrine on the common good is taught; and still less is anything said about the relations between the common good and the proper good of the intellectual substances. Their impact is clearly to show, against a

> Greek heresy, that, even in the Greek thinkers themselves, and above all in Aristotle, who was so fondly cherished in the Arabian world, there are principles upon which one may proceed to prove the fact of divine Providence.
>
> This is the group of texts Professor De K. argues from. He should not have done so, because they do not properly and immediately belong to the question he undertook to treat. (*DM*, 191–92)

In other words, according to Father Eschmann, when St. Thomas says that God governs the order of the universe and bestows upon it His greatest care (*maxime curat*) because it is the *maxime bonum in rebus causatis,* the *praecipue volitum et causatum,* and because the good of the order of the universe is the *propinquissimum in rebus creatis* to His own goodness, *cum ad ipsum ordinetur, sicut ad finem, omne particulare bonum hujus vel illius rei, sicut minus perfectum ad id quod est perfectum,* he does not really mean the reasons he gives to be taken as the true reasons. When St. Thomas exposes these reasons, and does so in language so unmistakable that even a reader who finds his view unacceptable must grant the obvious significance of these passages, still we are not to take the Angelic Doctor as meaning what he says. What he does mean, my Opponent explains, is that if there were no all-embracing Providence,

> the universe would fall apart, into so many unconnected and unconnectable bits, and it would be impossible to maintain the fact of the order of the universe on whose existence and sublime beauty both the Greeks, and especially the Christian Fathers, have so energetically insisted. (*DM*, 191)

Hence, according to my Opponent, the reason St. Thomas actually gives in the texts concerned, namely that the order of the universe is what is best in all creation, is not a universal, metaphysical, true reason at all, nor does the quotation from Genesis 1:31 express a theological principle. The true reason is the mere

> fact of the order of the universe on whose existence and sublime beauty both the Greeks, and especially the Christian Fathers, have so energetically insisted. (*DM*, 191)

In St. Thomas' arguments it is of no importance that this order of the universe—and by *universe* is meant the whole of creation and not just the *cosmos*—is what is best in all creation. True, he does infer: "Id igitur quod *maxime curat* Deus in rebus creatis est ordo universi," but that, presumably, is merely because he had used a premise designed to achieve a greater impact against a Greek heresy. Likewise, when St. Thomas declares that the *bonum ordinis universi* is the good which is closest to the divine goodness, *quum ordinetur ad ipsum, sicut ad finem omne particulare bonum hujus vel illius rei, sicut minus perfectum ordinatur ad id quod est perfectius,* the *omne bonum particulare* here distinguished from the *bonum ordinis universi,* has no more to do with the proper doctrine on the common good than the *bonum commune* of St. Thomas' quotations from I *Ethics,* c. l, and from XII *Metaphysics,* c. 10.

> In this group of texts—it is a very large one—St. Thomas frequently, and with obvious enjoyment, avails himself of two quotations from Aristotle, viz., (*a*) *bonum commune est divinius* . . . and, (*b*) *quod est optimum in rebus existens est bonum universi.* By these citations no proper doctrine on the common good is taught; and still less is anything said about the relations between the common good and the proper good of the intellectual substances. (*DM,* 191–92)

Indeed, how could the intellectual substances be included here unless *quaelibet res* or *quaelibet creatura* actually meant *quaelibet res* or *quaelibet creatura?* The addicts of the Historical Point of View know better: St. Thomas is not concerned here with strictly doctrinal truth, but with creating an impact against a Greek heresy, even at the cost of making false or misleading statements.

> Their impact is clearly to show, against a Greek heresy, that, even in the Greek thinkers themselves, and above all in Aristotle, who was so fondly cherished in the Arabian world, there are principles upon which one may proceed to prove the fact of divine Providence. (*DM,* 192)

Is my Opponent insinuating that St. Thomas uses the wiles of sophistry? Of course not! For, is it not true that the quotations from Aristotle are actually in Aristotle? One cannot contradict historical fact. St. Thomas is merely using the true facts of history to get results. Whether or not what Aristotle actually held is also true is another matter.

Father Eschmann says:

> By these citations no proper doctrine on the common good is taught; and still less is anything said about the relations between the common good and the proper good of the intellectual substances. (*DM*, 191–92)

And why not? No justification is necessary, for it is only too obvious that the term *bonum* could not mean *bonum*, i.e. *perfectivum alterius;* and it is just as obvious that the good to which *ordinatur, sicut ad finem* (and therefore as to what perfects), *omne particulare bonus hujus vel illius rei, sicut minus perfectum ad id quod est perfectum,* is a good which belongs to one creature to the exclusion of the other, and by no means to the one and the other as a good which is more perfect than their exclusive proper good. Why? Because Father Eschmann says so. It is for the same unquestionable reason that the following text (quoted *BC*, 15–16; 172, nn.13, 14) has nothing to do with the common good, nor with the relations between the common good and the proper good of the intellectual substances:

> Cum affectio sequatur cognitionem; quanto cognitio est universalior, tanto affectio eam sequens magis respicit commune bonum; et quanto cognitio est magis particularis, tanto affectio ipsam sequens magis respicit privatum bonum; unde et in nobis privata dilectio ex cognitione sensitiva exoritur; dilectio vero communis et absoluti boni ex cognitione intellectiva. Quia ergo angeli quanto sunt altiores, tanto habent scientiam magis universalem . . . ideo eorum dilectio maxime respicit commune bonum. Magis ergo diligunt se invicem, si specie differunt, quod magis pertinet ad perfectionem universi . . . quam si specie convenirent, quod pertineret ad bonum privatum unius speciei.[46]

IV. Why Did God Make Things Many?

The texts I quoted in this connection were to prove that "dans l'univers même," the greatest perfection of the created persons is the good of the universe. The question of the relation of the intellectual creature to God as He is in Himself apart from the universe had already been treated in substance. The reader will recall Father Eschmann's complaint:

> Will it be granted that it is inadmissible to read St. Thomas with scissors and paste, by cutting the texts out of their literary and historical context and just quoting what, in a particular instance, seems to be suitable? Will it be granted that, if St. Thomas has explicitly stated and solved a given problem, a Thomist worthy of that name is obliged to take account of this fact and can not afford to refer to some other texts which either have nothing to do with the problem or, at best, refer to it in a distant and mediate fashion? (*DM*, 189–90)

Faithful to his normal practice, my Opponent again proceeds to do just that. Not only does he cloud the distinction between the two questions (that of the relation of the person to the ultimate good and that of his relation to the intrinsic good of the universe) but he neglects to inform the reader that I had formulated and answered the very objection he levels against me (*BC*, 63–64). Just what is the problem we are to have in mind when Father Eschmann says, "Here is the problem as stated by St. Thomas" (*DM*, 190), is conveniently undetermined, but let us allow him to quote it:

> Videtur quod imago Dei inveniatur in irrationabilibus creaturis . . . [for, and this is the third *argumentum in contrarium*] quanto aliquid est magis perfectum in bonitate, tanto magis est Deo simile. Sed totum universum est perfectius in bonitate quam homo, quia etsi bona sint singula, tamen simul omnia dicuntur "valde bona," *Gen.* 1 (St. Augustine). Ergo totum universum est ad imaginem Dei et non solum homo.

The objection is taken from Ia, q. 93, a. 2. The answer to this objection is:

> Universum est perfectius in bonitate quam intellectualis creatura: extensive et diffusive. Sed intensive et collective similitudo divinae perfectionis magis invenitur in intellectuali creatura, quae est capax summi boni.—Vel dicendum, quod pars non dividitur contra totum, sed contra aliam partem. Unde cum dicitur quod sola natura intellectualis est

> ad imaginem Dei, non excluditur quin universum secundum aliquam sui partem sit ad imaginem Dei; sed excluduntur aliae partes universi. (*DM*, 190)

If this quotation is directed against me, it must mean that the greatest created good of the person—of the person viewed within the order of the universe—is not that which was said to be closest to the divine good, namely, the order of the universe itself whose principal parts are the intellectual substances in all their manifold and variety; rather the greatest good of the person is held to be each individual person himself, taken separately, so that each and every one of them is, absolutely speaking, a greater good than any or all of the other persons. Each person, then, because he is in the image of God, is a better created good than any and all of the other created persons who are also in the image of God. If this is not what Father Eschmann means, where is his objection?

Immediately following the quotation of St. Thomas' answer, Father Eschmann adds:

> St. Thomas' solution of the problem is so clear, so complete, and so perfectly balanced that it needs no explanation. (*DM*, 190)

Let us see, then, just how simple this matter is.

Why did God make things many and varied? Let us consider a few texts on this subject.

> Cum enim omne agens intendat suam similitudinem in effectum inducere secundum quod effectus capere potest, tanto hoc agit perfectius quanto agens perfectius est: patet enim quod quanto aliquid est calidius, tanto facit magis calidum; et quanto est aliquis melior artifex, formam artis perfectius inducit in materiam. Deus autem est perfectissimum agens. Suam igitur similitudinem in rebus creatis ad Deum pertinebat inducere perfectissime, quantum naturae creatae convenit. Sed *perfectam Dei similitudinem non possunt consequi res creatae secundum unam solam speciem creaturae: quia, cum causa excedat effectum, quod est in causa simpliciter et unite, in effectu invenitur composite et multipliciter, nisi effectus pertingat ad speciem causae; quod in proposito dici non potest,*

non enim creatura posset esse Deo aequalis. Oportuit igitur esse multiplicitatem et varietatem in rebus creatis, ad hoc quod inveniretur in eis *Dei similitudo perfecta secundum modum suum.*[47]

Item. *Plura bona uno bono finito sunt meliora:* habent enim hoc et adhuc amplius. *Omnis autem creaturae bonitas finita est:* est enim deficiens ab infinita Dei bonitate. *Perfectius est igitur universum creaturarum si sunt plures,* quam si esset unus tantum gradus rerum. Summo autem bono competit facere quod melius est. Ergo conveniens ei fuit ut plures faceret creaturarum gradus.[48]

Amplius. Operi a summe bono artifice facto non debuit deesse *summa perfectio. Sed bonum ordinis diversorum est melius quolibet illorum ordinatorum per se sumpto: est enim formale respectu singularium, sicut perfectio totius respectu partium.* Non debuit ergo bonum ordinis operi Dei deesse. Hoc autem bonum esse non posset si diversitas et inaequalitas creaturarum non fuisset. Est igitur diversitas et inaequalitas in rebus creatis non a casu (cap. 39); non ex materiae diversitate (cap. 40); non propter interventum aliquarum causarum (capp. 41–43), vel meritorum (cap. 44); sed ex propria Dei intentione perfectionem creaturae dare volentis qualem possibile erat eam habere. Hinc est quod dicitur *Gen.* 1, 31: *Vidit Deus cuncta quae fecerat, et erant valde bona:* cum de singulis dixisset quod sunt *bona.* Quia singula quidem sunt in suis naturis bona: simul autem omnia, *valde bona,* propter ordinem universi, *quae est ultima et nobilissima perfectio in rebus.*[49]

Ostensum enim est quod Deus per suam providentiam omnia ordinat in divinam bonitatem sicut in finem (cap. 64): non autem hoc modo quod suae bonitati aliquid per ea quae fiunt accrescat, sed ut similitudo suae bonitatis, quantum possibile est, imprimatur in rebus (capp. 18 sq.). Quia vero omnem creatam substantiam a perfectione divinae bonitatis deficere necesse est, ut perfectius divinae bonitatis similitudo rebus communicaretur, *oportuit esse diversitatem in rebus, ut quod perfecte ab uno aliquo repraesentari non potest, per diversa diversimode perfectiori modo repraesentaretur:* nam et homo, cum mentis conceptum uno vocali verbo videt sufficienter exprimi non posse, verba diversimode multiplicat ad exprimendam per diversa suae mentis conceptionem. Et in hoc etiam divinae perfectionis eminentia considerari potest, quod *perfecta bonitas, quae in Deo est unite et simpliciter, in creaturis esse non potest nisi secundum modum diversum et per plura.* Res autem per hoc diversae sunt, quod formas habent diversas, a quibus speciem sortiuntur. Sic igitur ex fine sumitur ratio diversitatis formarum in rebus.[50]

In praedicto autem ordine, secundum quem ratio divinae providentiae attenditur, primum esse diximus divinam bonitatem, quasi ultimum finem, qui est primum principium in agendis; dehinc vero rerum numerositatem; ad quam constituendam necesse est gradus diversos in formis et materiis, et agentibus et patientibus, et actionibus et accidentibus esse. *Sicut ergo prima ratio divinae providentiae simpliciter est divina bonitas, ita prima ratio in creaturis est earum numerositas, ad cuius institutionem et conservationem omnia alia ordinari videntur.* Et secundum hoc rationabiliter videtur esse a Boetio dictum, in principio suae *Arithmeticae* (lib. I, cap. ii), *quod omnia quaecumque a primaeva rerum natura constituta sunt, ex numerorum videntur ratione esse formata.*[51]

Uterque enim error [scil. Manichaeorum et Origenis] *ordinem universi praeterire videtur in sua consideratione, considerando tantummodo singulas partes ejus.* Ex ipso enim ordine universi potuisset ejus ratio apparere, quod ab uno principio, nulla meritorum differentia praecedente, oportuit diversos gradus creaturarum institui, ad hoc quod universum esset completum (repraesentante universo per multiplices et varios modos creaturarum quod in divina bonitate simpliciter et indistincte praexistit) sicut et ipsa perfectio domus et humani corporis diversitatem partium requirit. Neutrum autem eorum esset completum si omnes partes unius conditionis existerent; sicut si omnes partes humani corporis essent oculus, aliarum enim partium deessent officia. Et similiter si omnes partes domus essent tectum, domus complementum et finem suum non consequeretur, ut scilicet ab imbribus et caumatibus defendere posset. Sic igitur dicendum est, quod ab uno primo multitudo et diversitas creaturarum processit, non propter materiae necessitatem, nec propter potentiae limitationem, nec propter bonitatem, nec propter bonitatis obligationem; *sed ex ordine sapientiae,* ut in diversitate creaturarum perfectio consisteret universi.[52]

Hence, a single creature, unless it were equal to God, could never sufficiently express that which exists in God *simpliciter et unite.* If, then, according to God's actual design, the fullness of divine perfection is to be more profoundly represented by His work, divine wisdom must bring this about "perfectissime, quantum naturae creatae convenit" through multiplicity and variety. Therefore, it is what is realized in creation *composite et multipliciter* which imitates most perfectly what is in God *simpliciter et unite.* Hence, to deem secondary the perfection which in creation is accomplished by way of composition and multiplicity is to deny value to that which most perfectly imitates what is in God *simpliciter et unite.*

In the context of this general problem, it would be true to say that, *intensive,* any single creature represents more perfectly the uniqueness of anything it has in common with God. *Intensive,* any single created intelligible species represents more perfectly than a multiplicity of species the unique intelligible species which is God's essence. However, the superabundance of whatever exists in God *simpliciter et uniformiter,* is more perfectly expressed by what exists in creation *multipliciter et divisim.* The inexhaustible richness of the divine intelligible species is, absolutely speaking, more perfectly represented by the multiplicity of created species. Thus, if the texts already quoted have any meaning, the single creature's imitation of God by intension could not possibly be more perfect absolutely than that realized by the manifold to which it belongs unless a creature could be equal to God in perfection. This St. Thomas brings out clearly in his answers to the objections of *De potentia,* q. 3, a. 16, from which our last quotation was taken.

> 1. Sextodecimo quaeritur utrum ab uno primo possit procedere multitudo. Et videtur quod non. Sicuti enim Deus est per se bonum, et per consequens summum bonum; ita est per se et summe unum. Sed ab eo in quantum est bonum, non potest procedere nisi bonum. Ergo nec ab eo procedere potest nisi unum.
>
> Ad primum ergo dicendum, quod *sicut Deus est unus, ita et unum produxit, non solum quia unumquodque in se est unum, sed etiam quia omnia quodammodo sunt unum perfectum, quae quidem unitas diversitatem partium requirit,* ut ostensum est.
>
> 2. Praeterea, sicut bonum convertitur cum ente, ita et unum. Sed in his quae sunt entia, oportet attendi assimilationem creaturae ad Deum,

ut supra, art. praeced., dictum est. Ergo sicut in bonitate, ita et in unitate oportet Deo creaturam assimilari, ut scilicet sit una ab uno.

Ad secundum dicendum, quod *creatura assimilatur Deo in unitate, in quantum unaquaeque in se una est, et in quantum omnes unum sunt unitate ordinis,* ut dictum est.

5. Praeterea, uniuscujusque effectus est aliquam propriam causam accipere. Sed impossibile est unum esse proprium multorum. Ergo impossibile est quod unum sit causa multitudinis.

Ad quintum dicendum, quod appropriatio causae ad effectum attenditur secundum assimilationem effectus ad causam. Assimilatio autem creaturae ad Deum attenditur secundum hoc quod creatura implet id quod de ipsa est in intellectu et voluntate Dei; sicut artificiata similantur artifici in quantum in eis exprimitur forma artis, et ostenditur voluntas artificis de eorum constitutione. Nam sicut res naturalis agit per formam suam, ita artifex per suum intellectum et voluntatem. Sic igitur Deus propria causa est uniuscujusque creaturae, in quantum intelligit et vult unamquamque creaturam esse. *Quod autem dicitur idem non posse esse plurium proprium, intelligendum est quando fit propriatio per adaequationem;* quod in proposito non contingit.

7. Praeterea, oportet esse conformitatem inter causam et effectum. Sed Deus est omnino unus et simplex. Ergo in creatura, quae est ejus effectus, nec multitudo nec compositio debet inveniri.

Ad septimum dicendum, quod *licet sit quaedam similitudo creaturae ad Deum, non tamen adaequatio; unde non oportet, si unitas Dei caret omni multitudine et compositione, quod propter hoc oporteat talem esse creaturae unitatem.*

10. Sed dices, quod universitas creaturarum est quodammodo unum secundum ordinem.—Sed contra, effectum oportet assimilari causae. Sed unitas Dei non est unitas ordinis, quia in Deo non est prius nec posterius, nec superius et inferius. Ergo non sufficit unitas ordinis ad hoc quod ab uno Deo plura possint educi.

Ad decimum dicendum, quod *non oportet,* sicut dictum est, *uniusmodi unitatem esse in creatura et in Deo; licet creatura Deum in unitate imitetur.*

12. Praeterea, creatura procedit a Deo, non solum sicut effectus a causa efficiente, sed etiam sicut exemplatum ab exemplari. Sed unius exemplati est unum exemplar proprium. Ergo a Deo non potest procedere nisi una creatura.

Ad decimumsecundum dicendum quod *quando exemplatum perfecte repraesentat exemplar, ab uno exemplari non est nisi unum exemplatum, nisi per accidens, in quantum exemplata materialiter distinguuntur.* Creaturae vero non perfecte imitantur suum exemplar. Unde diversimode possunt ipsum imitari, et sic esse diversa exemplata. *Perfectus autem modus imitandi est unus tantum: et propter hoc Filius, qui perfecte imitatur Patrem, non potest esse nisi unus.*

13. Praeterea, Deus est causa rerum per intellectum. Agens autem per intellectum agit per formam sui intellectus. Cum igitur in divino intellectu non sit nisi una forma, videtur quod ab eo non possit procedere nisi una creatura.

Ad decimumtertium dicendum, quod licet forma intellectus divini sit una tantum secundum rem, est tamen multiplex ratione secundum diversos respectus ad creaturam, *prout scilicet intelliguntur creaturae diversimode formam divini intellectus imitari.*

18. Praeterea, unusquisque agens propter finem, facit effectum suum propinquiorem fini quantum potest. Sed Deus producendo creaturam ordinat eam in finem. Ergo facit eam propinquissimam fini quantum potest. Sed hoc non potest nisi uno modo fieri. Ergo Deus non producit nisi unam creaturam.

Ad decimumoctavum dicendum, quod *ratio illa tenet quando id quod est ad finem, potest totaliter et perfecte consequi finem per modum adaeqationis:* quod in proposito non contingit.

22. Praeterea, quidquid Deus facit, est unum. Ergo ab eo non est nisi unum; et ita ipse non erit causa multitudinis.

Ad vicesimumsecundum dicendum, quod *licet quidquid Deus facit, in se sit unum, tamen haec unitas,* ut dictum est, *non removet omnem multitudinem, sed manet illa cujus unum est pars.*

My Opponent's simplistic understanding of the terms *intensive* and *extensive* shows itself to be a defence of the doctrines St. Thomas consistently attacks, namely, that the ordered manifold of creation is, at best, only secondarily intended by God. Of course, *intensive,* any indivisible part of a creature is, as to the formality "indivisible," a better imitation of divine simplicity than any created whole; in this respect, even the *per se* unity of any single created being is inferior to that of any of its parts. However, absolutely speaking,

> apud nos composita sunt meliora simplicibus, quia perfectio bonitatis creaturae non invenitur in uno simplici, sed in multis. Sed perfectio divinae bonitatis invenitur in uno simplici. . . .[53]

The imperfection of intensive imitation is compensated by extension, by the manifold. By manifold, we do not mean the mere homogeneous multiplicity of predicamental quantity;[54] nor do we mean that the manifold of creation is an end insofar as it is a manifold.

> . . . Nullum agens intendit pluralitatem materialem ut finem: quia materialis multitudo non habet certum terminum, sed de se tendit in infinitum; infinitum autem repugnat rationi finis.[55]

Material multiplicity is for the sake of formal multiplicity.[56]

As an intensive imitation of divine perfection, any single term of any manifold is admittedly more perfect than the manifold itself. Yet we cannot afford to take this facile observation as an adequate solution to our problem save at the cost of being led into the trap into which Father Eschmann has fallen. For it must be noted that, whereas any higher term of the formal manifold of creation is a more perfect intensive imitation of divine perfection than an inferior one, the lower groups of the ordered manifold nevertheless approach intensive imitation more perfectly than the higher in that they have fewer members.[57] In other words, the more numerous the terms of the manifold, the less perfect it is from the viewpoint of intensive imitation. Hence, with respect to *what* is in God *simpliciter et indivisim,* if intensive imitation by the creature were absolutely better than that which is achieved through extension, the universe could not possibly be the *praecipue intentum;* and since in the higher regions of the universe, the spiritual creatures are more numerous than in the lower, those higher regions would be, absolutely speaking, less perfect than the lower. How would this compare with the doctrine of the texts already quoted? Or with that of the following?

> . . . *Cum perfectio universi sit illud quod praecipue Deus intendit in creatione rerum, quanto aliqua sunt magis perfecta, tanto in maiori excessu sunt creata a Deo.* Sicut autem in corporibus attenditur excessus secundum magnitudinem, ita in rebus incorporeis potest attendi excessus secundum multitudinem. Videmus autem quod corpora incorruptibilia, quae sunt perfectiora inter corpora, excedunt quasi incomparabiliter secundum magnitudinem corpora corruptibilia: nam tota sphaera activorum et passivorum est aliquid modicum respectu corporum caelestium. Unde rationabile est quod substantivae immateriales excedant secundum multitudinem substantias materiales, quasi incomparabiliter.[58]

Father Eschmann's understanding of the distinction between *intensive* and *extensive* destroys the Thomistic doctrine concerning the reason why God made things many and varied. Like the Manicheans and Origen, "ordinem universi praeterire videtur in sua consideratione, considerando tantummodo singulas partes ejus." In truth, the extensive perfection of the universe is not just a purely quantitative addition; extension is not intended for the mere sake of numerosity. The varied manifold of creation, its unity of order, is intended *per se* as the only manner in which what is in God *simpliciter et unite* can be more fully represented in His work. The divine "simpliciter et unite" is the principle and term of the created "composite et multipliciter." In comparison with the fullness of what is in God *simpliciter et indivisim,* the manifold of creation as a whole is more profoundly one, than any single part. ". . . Sicut Deus est unus, ita et unum produxit, non solum quia unumquodque est unum, sed etiam quia omnia quodammodo sunt *unum perfectum,* quae quidem unitas diversitatem partium requirit."

The errors concerning the procession of the Many from the One, which St. Thomas attacks, follow from considering the Many as something absolute, which could not properly proceed from the One. But a deeper grasp of the problem reveals that, ultimately, the Many is but an imitation of the One. The ultimate principle of the unity of the manifold of creation is the identity of the superabundant unity of the divine essence.[59] Only through the unity of the manifold can there be in creation that *Dei similitudo perfecta secundum modum suum.*[60]

If the manifold of the intellectual creatures were but the result of intending *this* person to be, and *that* person to be, and so on, God would be primarily the *propria ratio* of One and of the other, and only secondarily the *ratio communis* of the many. Indeed, God would be reduced to the status of a univocal cause, and the created person elevated to that of a perfect *exemplatum*, "quod non multiplicatur nisi materialiter."

> . . . Ratio illa teneret de exemplato quod perfecte repraesentat exemplar, quod non multiplicatur nisi materialiter. Unde imago increata, quae est perfecta, est una tantum. Sed nulla creatura repraesentat perfecte exemplar primum, quod est divina essentia.[61]

> . . . Deus cognoscit omnia uno, quod est ratio plurium, scilicet essentia sua, quae est similitudo rerum omnium; et quia essentia sua est propria ratio uniuscujusque rei, ideo de unoquoque propriam cognitionem habet. Qualiter autem unum possit esse multorum ratio propria et communis, sic considerari potest. Essentia enim divina secundum hoc est ratio alicujus rei, quod res illa divinam essentiam imitatur. *Nulla autem res imitatur divinam essentiam ad plenum;* sic enim non posset esse nisi una imitatio ipsius; nec sua essentia esset per modum istum nisi unius propria ratio, sicut una sola est imago Patris perfecte eum imitans, scilicet Filius. Sed *quia res creata imperfecte imitatur divinam essentiam, contingit esse diversas res diversimode imitantes;* in quarum nulla est aliquid quod non deducatur a similitudine divinae essentiae; et ideo illud quod est proprium unicuique rei, habet in divina essentia quod imitetur; et secundum hoc divina essentia est similitudo rei quantum ad proprium ipsius rei, et sic est propria ipsius ratio: et eadem ratione est propria alterius, et omnium aliorum. Est igitur communis omnium ratio, in quantum est res ipsa una, quam omnia imitantur: sed est propria hujus ratio vel illius, secundum quod res eam diversimode imitantur: et sic propriam cognitionem divina essentia facit de unaquaque re, in quantum est propria ratio uniuscujusque.[62]

Indeed, we would have to reverse the doctrine of the following passage from Ia, q. 15, a. 2: *Utrum sint plures ideae,* which we have already quoted in part.

> Respondeo dicendum quod necesse est ponere plures ideas. Ad cuius evidentiam, considerandum est quod *in quolibet effectu illud quod est*

> *ultimus finis, proprie est intentum a principali agente;* sicut ordo exercitus a duce. Illud autem quod est *optimum in rebus existens, est bonum ordinis universi,* ut patet per Philosophum in XII *Metaphys.* Ordo igitur universi est proprie a Deo intentus, et non per accidens proveniens secundum successionem agentium: prout quidam dixerunt quod Deus creavit primum creatum tantum, quod creatum creavit secundum creatum, et sic inde quousque producta est tanta rerum multitudo: secundum quam opinionem, Deus non haberet nisi ideam primi creati. Sed, si ipse ordo universi est per se creatus ab eo, et intentus ab ipso, *necesse est quod habeat ideam ordinis universi. Ratio autem alicuius totius haberi non potest, nisi habeantur propriae rationes eorum ex quibus totum constituitur:* sicut aedificator speciem domus concipere non posset, nisi apud ipsum esset propria ratio cuiuslibet partium eius. Sic igitur oportet quod in mente divina sint propriae rationes omnium rerum. Unde dicit Augustinus, in libro *Octoginta trium Quest.*, quod *singula propriis rationibus a Deo creata sunt.* Unde sequitur quod in mente divina sint plures ideae.

We may now apply this general doctrine to the more restricted problem of why God made the intellectual creatures, who are properly in His image, many and varied. Since their manifold *qua* manifold cannot constitute a single image; since to be in His image is proper to each intellectual creature taken separately; since, singly, any one of them is *intensive* a better expression of the unique uncreated original, what then can it be that is added by the varied manifold of images? The answer is that no single created image is a perfect image of God; to achieve a fuller created representation of the uncreated original, divine wisdom has made the created images many and varied. Absolutely speaking, this manifold is more expressive of the fullness of the original than any single created image. The manifold was conceived by divine wisdom for that very purpose, and it remains the greatest perfection that God produced in all spiritual creation.

To demand that, in order to be better absolutely than any of its parts, the whole possess intensively the perfection of its parts, is to misunderstand the nature and purpose of the whole.

> . . . Optimi agentis est producere totum effectum suum optimum: non tamen quod quamlibet partem totius faciat optimam simpliciter, sed optimam secundum proportionem ad totum: tolleretur enim bonitas

> animalis, si quaelibet pars eius oculi haberet dignitatem. Sic igitur et Deus totum universum constituit optimum, secundum modum creaturae: non autem singulas creaturas, sed unam alia meliorem. Et ideo de singulis creaturis dicitur *Gen.* 1: *Vidit Deus lucem quod esset bona,* et similiter de singulis: sed de omnibus simul dicitur: *Vidit Deus cuncta quae fecerat, et erant valde bona.*[63]

If the animal could not be better absolutely than its eye except by being better intensively, then, in order to be superior to this single organ, the entire animal would have to be an eye. Likewise, the universe itself would have to have an intellect and will; it would have to be a proper image of God. And since "nec per se de toto potest dici, et primo, quod non convenit sibi ratione omnium partium," even the irrational part of the universe would have to be in the image of God.

To this simplistic reasoning, St. Thomas answers:

> Universum est perfectius in bonitate quam intellectualis creatura: extensive et diffusive. Sed intensive et collective similitudo divinae perfectionis magis invenitur in intellectuali creatura, quae est capax summi boni.

Let us now read Father Eschmann's paraphrase of this text.

> St. Thomas' solution of the problem is so clear, so complete, and so perfectly balanced that it needs no explanation. Let us however try to paraphrase: Which is more like God, i.e. more to the image of God, the whole universe, or one single intellectual creature? The whole universe is more like God "extensively and diffusively." That is, if you consider God as the cause and fountain-head of the whole universe and of every creature pertaining to it, you will judge that there is quantitatively more likeness in the whole than in the parts. But before you consider God as cause, you must first look at Him as He is in Himself the supreme good by His essence. In this way a single intellectual creature is more perfectly likened to Him, because only the intellectual substance (every single intellectual substance) is capable of being, by knowledge and love, united with God as God is in Himself. "Intensively," thus, and "collectively," i.e. considering the fact that the essentially most perfect likeness is gathered together in one single point, a single intellectual substance by far surpasses everything that

> might, in a certain sense, be said to be like God. The intellectual substance is, indeed, the only proper image of God. (*DM*, 190–91)

Just what does my Opponent mean by: "there is quantitatively more likeness in the whole than in its parts"? Does he mean that whether God makes one image of Himself, or many, the difference is merely quantitative? That, *absolutely* speaking, there is no better expression of Himself when He produces images many and varied than when He produces a single one? By his superficial understanding of the term "extensive"[64] Father Eschmann destroys the Thomistic doctrine of the reason why God made the intellectual creatures many and varied.

When we consider God "as He is in Himself the supreme good by His essence" and the intellectual creature as "capable of being, by knowledge and love, united with God as God is in Himself," the good in question is beyond that universe to which the intellectual creature is compared as a part to a whole. In this respect, the intellectual creature is not to be considered formally as a part of the universe at all. Father Eschmann had promised to

> consider the second part of this thesis, viz. the statement regarding the intrinsic common good of the universe and its relation to the intellectual beings or persons. (*DM*, 187)

Now he suddenly shifts to the first part of the thesis and speaks as if I had maintained that the intrinsic common good of the universe is to be identified with the absolutely ultimate good of the intellectual creatures. Why does my Opponent do these things? He might have quoted IaIIae, q. 2, a. 8, ad 2:

> . . . Si totum aliquod non sit ultimus finis, sed ordinetur ad finem ulteriorem, ultimus finis partis non est ipsum totum, sed aliquid aliud. Universitas autem creaturarum, ad quam comparatur homo ut pars ad totum, non est ultimus finis, sed ordinatur in Deum sicut in ultimum finem. Unde bonum universi non est ultimus finis hominis, sed ipse Deus.

But then it might be too obvious that, when stressing this aspect of the intellectual creature "capax summi boni," he is not confining himself to "the second part of this thesis."

Because the comparison between the perfection of the universe as a likeness to God and the perfection of the intellectual creature as "capax summi boni," called for by the objection, is not an adequate comparison, St. Thomas adds:

> Vel dicendum, quod pars non dividitur contra totum, sed contra aliam partem. Unde cum dicitur quod sola natura intellectualis est ad imaginem Dei, non excluditur quin universum secundum aliquam sui partem sit ad imaginem Dei; sed excluduntur aliae partes universi.

The universe may be said to be in the image of God, but only because of the rational natures.

> . . . Similitudo divinae bonitatis, quantum ad nobilissimas participationes ipsius, non resultat in universo nisi ratione nobilissimarum partium ejus, quae sunt intellectuales naturae: nec per se de toto potest dici, et primo, quod non convenit sibi ratione omnium partium, ut in VI *Physic.* dicitur frequenter: et ideo universum non potest dici imago Dei, sed intellectualis natura.[65]

But here is Father Eschmann's paraphrase of the second part of St. Thomas' answer:

> The Angelic Doctor then continues, not by proposing another solution, but by stressing a certain aspect of the same solution which in the foregoing has been left aside. Are not the intellectual substances parts, i.e. of course, principal, formal, constitutive, primary, parts of the universe? Are they not, as it were, the sons of that great family or economy of the universe of which God is the *paterfamilias?*[66] Are they not, just as sons are, very deeply interested in the vicissitudes of that which is their possession and heredity—and the possession and heredity of each one of them, according to Holy Scripture (Matth. 24:47): "Super omnia bona sua constituet eum"?[67] The statement, therefore, that the intellectual substance alone is *ad imaginem Dei,* might be expanded by saying that the universe in one of its parts, and precisely in its first and foremost constitutive parts, is *ad imaginem Dei.* In this way a solution of the

> problem is obtained which is most properly "Thomistic" in that it takes account of every possible aspect of the problem. (*DM,* 191)

We shall not try to unravel in what sense this passage may be considered as a paraphrase of St. Thomas' text. But whatever it may be as a paraphrase, insofar as my Opponent turns it against the second part of my thesis, he is again exploiting his own confusion as I have already pointed out: he confuses the good of the persons that is the universe with the good that is the persons; he confuses the persons as contributing to the essential perfection of the universe (which perfection is, within this order, their *finis cujus gratia*) with the persons considered as "for whom" (*finis cui*) is the perfection of the universe. Why did he overlook this distinction?

> La substance intellectuelle étant "comprehensiva totius entis," étant une partie de l'univers dans laquelle peut exister, selon la connaissance, la perfection de l'univers tout entier, son bien le plus propre en tant qu'elle est une substance intellectuelle sera le bien de l'univers, bien essentiellement commun. La substance intellectuelle n'est pas ce bien comme elle est l'univers selon la connaissance. En effet, il convient de marquer ici la différence radicale entre la connaissance et l'appétit: "le connu est dans le connaissant, le bien est dans les choses." Si, comme le connu, le bien était dans l'aimant, nous serions à nous-mêmes le bien de l'univers. (*BC,* 13)

> C'est donc tout autre chose de dire que les créatures raisonnables sont gouvernées et ordonnées pour elles-mêmes, et de dire qu'elles le sont à elles-mêmes et pour leur bien singulier: elles sont ordonnées pour elles-mêmes au bien commun. Le bien commun est pour elles, mais il est pour elles comme bien commun. Les créatures raisonnables peuvent atteindre elles-mêmes de manière explicite le bien auquel toutes choses sont ordonnées; elles diffèrent par là des créatures irraisonnables, qui sont de purs instruments, qui sont utiles seulement et qui n'atteignent pas elles-mêmes de manière explicite le bien universel auquel elles sont ordonnées. Et c'est en cela que consiste la dignité de la nature raisonnable. (*BC,* 35)

> . . . La créature raisonnable, en tant qu'elle peut elle-même atteindre à la fin de la manifestation de Dieu au dehors, existe pour elle-même. Les

> créatures irraisonnables n'existent que pour cet être qui pourra lui-même atteindre à cette fin qui ne fut qu'implicitement la leur. L'homme est la dignité qui est leur fin. Mais, cela ne veut pas dire que la créature raisonnable existe pour la dignité de son être propre et qu'elle est elle-même la dignité pour laquelle elle existe. Elle tire sa dignité de la fin à laquelle elle peut et doit atteindre; sa dignité consiste en ce qu'*elle* peut atteindre à la fin de l'univers, la fin de l'univers étant, sous ce rapport, pour les créatures raisonnables, à savoir, pour chacune d'elles. Cependant, le bien de l'univers n'est pas pour elles comme si celles-ci étaient la fin pour laquelle il est. Il est le bien de chacune d'elles en tant qu'il est *leur* bien comme bien commun. (*BC*, 38)

Since the good of the universe is the same "pour chacune d'elles," since it is a good which does not belong to one person to the exclusion of the other person, it is strictly a common good. In support of this position I might have quoted the very text my Opponent levels at me (*DM*, 191, n. 15):

> Inter omnes . . . partes universi excellunt sancti Dei, ad quorum quemlibet pertinet quod dicitur Matth. xxiv: *Super omnia bona sua constituet eum.* Et ideo quicquid accidit, vel circa ipsos vel alias res, totum in bonum eorum cedit. . . . [68]

Let us now turn to section III of Father Eschmann's article, which he has seen fit to entitled: "Professor De Koninck's Notion of God."

V. *Quis Ut Deus?*

Let us be certain that we grasp clearly the distinction between a common good and a proper good. The proper good of one person is never the proper good of another person; the proper good of the one is never the proper good of another; if the good aimed at by one person is a proper good, it is impossible for it to be the proper good of another, for the good in these two cases differs by a numerical distinction. A proper good may indeed be spoken of as common to many persons, but we are then using the term "common" in the sense of "common according to predication" (*BC*, 54–55). The following objection and answer, taken from *In IV Sententiarum*, d. 49, q. l, a. l, qa. 1, obj. 3, and sol. l, ad 3, are to the point:

> Praeterea, quanto aliquod bonum est communius, tanto divinius, ut patet in I *Ethic.*, cap. l. Sed bonum corporale communius est quam spirituale: quia corporale ad plantas et animalia bruta extendit, non autem spirituale. Ergo corporale bonum spirituali praeeminet; et ita in corporalibus bonis magis est beatitudo quaerenda.
>
> Ad tertium dicendum, quod dupliciter aliquid dicitur esse commune. Uno modo per praedicationem; hujusmodi autem commune non est idem numero in diversis repertum; et hoc modo habet bonum corporis communitatem. Alio modo est aliquid commune secundum participationem unius et ejusdem rei secundum numerum; et haec communitas maxime potest in his quae ad animam pertinent, inveniri; quia per ipsam attingitur ad *id quod est commune bonum omnibus rebus, scilicet Deum;* et ideo ratio non procedit.

When St. Thomas asserts that God is a common good, he means a good which is numerically one, yet which can be the end of many.

> Bonum particulare ordinatur in bonum commune sicut in finem: esse enim partis est propter esse totius; unde et *bonum gentis est divinius quam bonum unius hominis.* Bonum autem summum, quod est Deus, est bonum commune, cum ex eo universorum bonum dependeat: bonum autem quo quaelibet res bona est, est bonum particulare ipsius et aliorum quae ab ipso dependent. Omnes igitur res ordinantur sicut in finem in unum bonum, quod est Deus.[69]

The most striking text my Opponent quotes (*DM*, 195) against the first part of my thesis, that namely which maintains that with respect to any created person God is most properly a common good, consists undoubtedly in the nine words he has extracted from the *Q.D. de caritate*, a. 5, ad 4: *Bonum commune non est objectum caritatis, sed summum bonum.*

Does the expression *bonum commune* stand for a *bonum commune in praedicando* or for a *bonum commune in causando?* That we can hardly know without taking a look at the context. The complete text, the objection and the answer, will settle this difficulty. The problem St. Thomas is discussing is: *Utrum caritas sit virtus specialis.*

> 4. Praeterea, bonum est objectum generale omnium virtutum: nam virtus est quae bonum facit habentem, et opus ejus bonum reddit. Sed bonum est objectum caritatis. Ergo caritas habet objectum generale; et ita est generalis virtus.

Hence, according to this objection, the object of the theological virtue of charity would be the general good sought by all the virtues, that is, the good which is *predicable* of the object of any virtue. To this St. Thomas answers:

> Ad quartum dicendum, quod bonum commune non est objectum caritatis, sed summum bonum; et ideo non sequitur quod caritas sit generalis virtus, sed quod sit summa virtutum.

Concerning this text my Opponent commits several gross errors. Those nine words cannot be lifted from their context without rendering them hopelessly ambiguous. For, the "common good" of this text is to be taken, not as the common good of persons, but as the good common to the different virtues nor is it a *commune in causando,* but *in praedicando* and *in essendo.* If the "bonum commune" of this text were to be understood as a *commune secundum virtutem* or *in causando* (the objection shows that it is not), we should then conclude that charity is a general virtue.

A glance at the reply which St. Thomas elsewhere offers to a similar question will suffice to make this last point clear. Discussing the nature of general justice (IIaIIae, q. 58, a. 6, c.) he asks: *Utrum justitia, secundum quod est generalis, sit idem per essentiam cum omni virtute.* The first two objections of this article had referred to two statements from V *Ethics.,* c. 1: "virtus et justitia legalis *est eadem omni virtuti, esse autem non est idem,*" and "justitia praedicta . . . *non est pars virtutis, sed tota virtus.*"

> Respondeo dicendum quod generale dicitur aliquid dupliciter. Uno modo, per praedicationem: sicut *animal* est generale ad hominem et equum et ad alia hujusmodi. Et hoc modo generale oportet quod sit idem essentialiter cum his ad quae est generale: quia genus pertinet ad

essentiam speciei et cadit in definitione eius.—Alio modo dicitur aliquid generale secundum virtutem: sicut causa universalis est generalis ad omnes effectus, ut sol ad omnia corpora, quae illuminantur vel immutantur per virtutem ipsius. Et hoc modo generale non oportet quod sit idem in essentia cum his ad quae est generale: quia non est eadem essentia causae et effectus.

Hoc autem modo, secundum praedicta, justitia legalis dicitur esse virtus generalis: inquantum scilicet ordinat actus aliarum virtutum ad suum finem, quod est movere per imperium omnes alias virtutes. *Sicut enim caritas potest dici virtus generalis inquantum ordinat actus omnium virtutum ad bonum divinum,* ita etiam justitia legalis inquantum ordinat actus omnium virtutum ad bonum commune. Sicut ergo caritas, quae respicit bonum divinum ut proprium objectum, est quaedam specialis virtus secundum suam essentiam; ita etiam justitia legalis est specialis virtus secundum suam essentiam, secundum quod respicit commune bonum ut proprium objectum. Et sic est in principe principaliter, et quasi architectonice; in subditis autem secundario et quasi ministrative.

Potest tamen quaelibet virtus, secundum quod a praedicta virtute, speciali quidem in essentia, generali autem secundum virtutem, ordinatur ad bonum commune, dici justitia legalis. Et hoc modo loquendi justitia legalis est idem in essentia cum omni virtute, differt autem ratione. Et hoc modo loquitur Philosophus.

It should be noted that, even when we call the theological virtue of charity a general virtue in this sense, we do so, not because it has as its object a good communicable to many persons, but because charity "ordinat actus omnium virtutum ad bonum divinum." Hence Father Eschmann's quotation, if rightly understood, is not even concerned with the position he attacks.

Nor is this all, for it can be readily shown how his misinterpretation of the words "bonum commune" in this phrase of which he makes so much leads him into impossible difficulties. That it may be quite clear the adversary is being done no injustice, let me make one or two preliminary remarks to establish beyond doubt that he does understand this term in the sense we are attributing to him. I will ask the reader to recall that throughout my own essay I most unambiguously use the expression "common good" for a *bonum commune in causando;* let us note, moreover, that all my quotations from

St. Thomas concern this good and that I maintain God is most formally good in this sense. Now this is, of course, precisely the position Father Eschmann attacks: so that when he finally brings forth this text from the *Q.D. de caritate,* a. 5, ad 4, as a climactic *littera Sancti Thomae,* it is impossible that, in the phrase *bonum commune non est objectum caritatis, sed summum bonum,* he can be taking *bonum commune* to mean anything other than *bonum commune in causando.* Consequently, when St. Thomas elsewhere expressly says that every creature naturally loves God more than itself because He is their common good, and that this also holds true for love according to the theological virtue of charity, my Opponent is compelled to maintain that in such passages, God is understood to be a common good only "in a certain sense" (*DM,* 195).

Then to prove this interpretation by the *littera Sancti Thomae,* Father Eschmann quotes three entire words from Ia, q. 60, a. 5, ad 5: "quoddam (!) bonum commune."[70] Anyone sufficiently acquainted with Latin will know that when those three words alone are taken, it is impossible to determine the meaning of "quoddam." It may be intended as an indefinite pronoun, "a certain one" or simply "a"—in which case we would translate "a common good"—or it may be taken as an adjective meaning "as it were," "so to speak," or "in a certain sense."[71] The first is the principal meaning of "quoddam," and St. Thomas most frequently uses it in this sense, as anyone familiar with his text must know.[72] But let us turn to the context of this extremely succinct quotation which Father Eschmann has taken from the article: *Utrum Angelus naturali dilectione diligat Deum plus quam seipsum.*

> Ad quintum dicendum quod, cum *in Deo sit unum et idem ejus substantia et bonum commune,* omnes qui vident ipsam Dei essentiam, eodem motu dilectionis *moventur in ipsam Dei essentiam* prout est ab aliis distincta, et *secundum quod est quoddam bonum commune.* Et quia *inquantum est bonum commune, naturaliter amatur ab omnibus;* quicumque videt eum per essentiam, impossibile est quin diligat ipsum. Sed illi qui non vident essentiam ejus, cognoscunt eum per aliquos particulares effectus, qui interdum eorum voluntati contrariantur. Et sic hoc modo dicuntur odio habere Deum: cum tamen, inquantum est *bonum commune omnium,* unumquodque naturaliter diligat plus Deum quam seipsum.

My Opponent does not realize that, besides making the gratuitous assumption that "quoddam" must mean "in a certain sense," he is implicitly accusing St. Thomas of constructing a syllogism with four terms.[73] For unless "bonum universale" is a "bonum commune" in the strict sense ("cum in Deo sit unum et idem ejus substantia et bonum commune"), the whole proof of this article 5 is sophistical. The same holds for the following texts:

> . . . Diligere Deum super omnia plus quam seipsum, est naturale non solum angelo et homini, sed etiam cuilibet creaturae, secundum quod potest amare aut sensibiliter aut naturaliter. Inclinationes enim naturales maxime cognosci possunt in his quae naturaliter aguntur absque rationis deliberatione; sic enim agit unumquodque in natura, sicut aptum natum est agi. Videmus autem quod unaquaeque pars naturali quadam inclinatione operatur ad bonum totius, etiam cum periculo aut detrimento proprio: ut patet cum aliquis manum exponit gladio ad defensionem capitis, ex quo dependet salus totius corporis. Unde *naturale est ut quaelibet pars suo modo plus amet totum quam seipsam.* Unde et secundum hanc naturalem inclinationem, et secundum politicam virtutem, bonus civis mortis periculo se exponit pro bono communi. Manifestum est autem quod *Deus est bonum commune totius universi et omnium partium ejus; unde quaelibet creatura suo modo naturaliter plus amat Deum quam seipsam;* insensibilia quidem naturaliter, bruta vero animalia sensitive, creatura vero rationalis per intellectivum amorem, quae dilectio dicitur.[74]

> . . . Diligere autem Deum super omnia est quidem connaturale homini; et etiam cuilibet creaturae non solum rationali, sed irrationali et etiam inanimatae, secundum modum amoris qui unicuique creaturae competere potest. Cujus ratio est quia unicuique naturale est quod appetat et amet aliquid, secundum quod aptum natum est esse: sic enim *agit unumquodque, prout aptum natum est,* ut dicitur in 2 *Physic. Manifestum est autem quod bonum partis est propter bonum totius.* Unde etiam naturali appetitu vel amore *unaquaeque res particularis amat bonum suum proprium propter bonum commune totius universi, quod est Deus.* Unde et Dionysius dicit, in lib. *de Divin. Nomin.* quod *Deus convertit omnia ad amorem sui ipsius.* Unde homo in statu naturae integrae dilectionem sui ipsius referebat ad amorem Dei sicut ad finem, et similiter dilectionem

omnium aliarum rerum. Et ita Deum diligebat plus quam seipsum, et super omnia. Sed in statu naturae corruptae homo ab hoc deficit secundum appetitum voluntatis rationalis, quae propter corruptionem naturae sequitur bonum privatum, nisi sanetur per gratiam Dei.[75]

To show that, according to charity, we must love God more than ourselves, St. Thomas uses the same reason.

Respondeo dicendum quod a Deo duplex bonum accipere possumus: scilicet bonum naturae, et bonum gratiae. Super communicatione autem bonorum naturalium nobis a Deo facta fundatur amor naturalis, quo non solum homo in suae integritate naturae super omnia diligit Deum et plus quam seipsum, sed etiam quaelibet creatura suo modo, idest vel intellectuali, vel rationali vel animali, vel saltem naturali amore, sicut lapides et alia quae cognitione carent: quia *unaquaeque pars naturaliter plus amat commune bonum totius quam particulare bonum proprium.* Quod manifestatur ex opere: quaelibet enim pars habet inclinationem principalem ad actionem communem utilitati totius. Apparet etiam hoc in politicis virtutibus, secundum quas *cives pro bono communi et dispendia propriarum rerum et personarum interdum sustinent.—Unde multo magis hoc verificatur in amicitia caritatis,* quae fundatur super communicatione donorum gratiae. Et ideo *ex caritate magis debet homo diligere Deum, qui est bonum commune omnium, quam seipsum: quia beatitudo est in Deo sicut in communi et fontali omnium principio qui beatitudinem participare possunt.*

Ad primum dicendum quod Philosophus ("amicabilia quae sunt ad alterum veniunt ex amicabilibus quae sunt ad seipsum)[76] loquitur de amicabilibus quae sunt ad alterum in quo bonum quod est objectum amicitiae invenitur secundum aliquem particularem modum: non autem de amicabilibus quae sunt ad alterum in quo bonum praedictum invenitur *secundum rationem totius.*

Ad secundum dicendum quod bonum totius diligit quidem pars secundum quod est sibi conveniens: *non autem ita quod bonum totius ad se referat, sed potius ita quod seipsam refert in bonum totius.*

Ad tertium dicendum quod hoc quod aliquis velit frui Deo, pertinet ad amorem quo Deus amatur amore concupiscentiae. Magis autem

> amamus Deum amore amicitiae quam amore concupiscentiae: quia *maius est in se bonum Dei quam participare possumus fruendo ipso.* Et ideo simpliciter homo magis diligit Deum ex caritate quam seipsum.[77]

My Opponent simply does not realize that the notion of common good is an analogical notion. That is why, when we call God a common good, he will allow it to be a common good only "in a certain sense."

But we do maintain that, for any intellectual creature, God can never be aught than a common good. Nor need there be any hesitation in declaring that to prescind from the superabundant and inexhaustible communicability of divine goodness to other persons amounts to prescinding from the infinite plenitude of divine goodness. There is a solid argument for this profound truth which it is not difficult to defend against the attack which Father Eschmann makes by means of a quotation taken from IaIIae, q. 4, a. 8, ad 3 (*DM*, 196, n. 20):

> Si esset una sola anima fruens Deo, beata esset, non habens proximum quem diligeret.

My Opponent might have mentioned that I used this very objection, and might have tried to refute my answer:

> 4. La béatitude de la personne singulière ne dépend pas de la communication de cette béatitude à plusieurs. De plus, il faut aimer Dieu en premier lieu et le prochain *ex consequenti.* Donc le caractère commun de la béatitude est secondaire: celle-ci est d'abord et en premier le bien de la personne singulière.
>
> Nous répondons que si de soi la béatitude de la personne singulière ne dépend pas de la communication actuelle de cette béatitude à plusieurs, elle n'en dépend pas moins de son essentielle communicabilité à plusieurs. Et la raison en est la surabondance de ce bien qu'est la béatitude, et son incommensurabilité au bien singulier de la personne. Le péché des anges consistait à vouloir tout bien commensurable à leur bien propre. L'homme pèche quand il veut le bien de l'intelligence commensurable au bien privè. Dès lors, quand même une seule personne jouirait de la béatitude, elle aurait toujours raison de partie en face de ce bien surabondant: même si en fait elle était seule pour en jouir, jamais

la personne singulière ne pourrait considérer ce bien comme le sien singulier. (*BC*, 55–56)

Rappelons encore une fois que le bien commun est dit commun dans sa surabondance et dans son incommensurabilité au bien singulier. Or le bien proprement divin est si grand qu'il ne pourrait pas être le bien propre, même de la création tout entière: celle-ci gardera toujours en quelque façon raison de partie. Il est très vrai qu'en face du bien commun la personne singulière peut le dire 'mien', mais il n'est pas pour cela approprié à la personne comme bien singulier. Le bien qu'elle dit 'mien' n'est pas pour elle prise comme fin. S'il était tel, le bien qu'est la personne elle-même serait la fin pour laquelle il est voulu. (*BC*, 65)

Why is God so insistent that we love our neighbour? Why does our very salvation depend upon the love of our neighbour? *If any man say: I love God, and hateth his brother; he is a liar.*[78] It can surely be only because it is impossible to love God as He is in Himself without loving Him in His communicability to others. If God had created and beatified but a single intellectual creature, He would still have to be loved in His communicability to other intellectual creatures. God is the *bonum universale simpliciter.* There can never be a proportion of equality between this infinite good and the intellectual creature's capacity for beatitude. The divine good can never be other than a common good for the creature. To prescind from the inexhaustible communicability of the divine good to others, whether it is actually communicated or not, is to prescind from the *bonum universale* itself.

When St. Thomas says that God is a common good according to His substance, he does not mean that God is a common good with respect to Himself, nor that the *actual* diffusion of His goodness to others is of the very nature of God; the Angelic Doctor means that it is of the very nature of God to be a common good for any creature He freely chooses to create. A similar distinction must be made in connection with the following text from IIIa, q. l, a. l, c.: *Utrum fuerit conveniens Deum incarnari:*

Respondeo dicendum quod unicuique rei conveniens est illud quod competit sibi secundum rationem propriae naturae: sicut homini conveniens est ratiocinari quia hoc convenit sibi inquantum est rationalis secundum suam naturam. Ipsa autem natura Dei est bonitas: ut patet per Dionysium, i cap. *de Div. Nom.* Unde quidquid pertinet ad rationem boni, conveniens est Deo.

Pertinet autem ad rationem boni ut se aliis communicet: ut patet per Dionysium, iv cap. *de Div. Nom.* Unde ad rationem summi boni pertinet quod summo modo se creaturae communicet. Quod quidem maxime fit per hoc quod *naturam creatam sic sibi conjungit ut una persona fiat ex tribus, Verbo, anima et carne;* sicut dicit Augustinus, XIII *de Trin.* Unde manifestum est quod conveniens fuit Deum incarnari.[79]

Just as we say that "ad *rationem,* summi boni pertinet quod summo modo se creaturae communicet," we say also that it is of the very nature of the divine good to be a common good. *We* cannot love the *bonum universale* except as the common good, that is, the good which incommensurably surpasses anything which might be the proper good of a creature and which, because of its very infinity, is communicable to others as *bonum universale.* If God could be the proper good (proper as opposed to common) of any created person, He could not be the good of another person. *Bonum unius personae singularis non est finis alterius.*[80] If our ultimate end were a proper good, we ourselves would be our ultimate end. Why does my Opponent choose to overlook the text I quoted (*BC,* 24–25; 174–75) from *De caritate,* a. 4, ad 2? The objection was:

> . . . Philosophus dicit in IX *Ethic.* (cap. viii, parum a princ.), quod amicabilia quae sunt ad alterum, venerunt ex amicabilibus quae sunt ad seipsum. Sed id quod est principium et causa, est potissimum in unoquoque genere. Ergo homo ex caritate diligit seipsum tamquam principale objectum, et non Deum.

To this St. Thomas answers:

> . . . Quod cum amor respiciat bonum, secundum diversitatem boni est diversitas amoris. Est autem *quoddam bonum proprium* alicujus hominis in quantum est singularis persona; et *quantum ad dilectionem respicientem hoc bonum, unusquisque est sibi principale objectum dilectionis.* Est autem *quoddam bonum commune quod pertinet ad hunc vel ad illum in quantum est pars alicujus totius,* sicut ad militem, in quantum est pars exercitus, et ad civem, in quantum est pars civitatis; et quantum ad di-

> lectionem respicientem hoc bonum, principale objectum dilectionis est illud in quo principaliter illud bonum consistit, sicut bonum exercitus in duce, et bonum civitatis in rege; unde ad officium boni militis pertinet ut etiam salutem suam negligat ad conservandum bonum ducis, sicut etiam homo naturaliter ad conservandum caput, brachium exponit; *et hoc modo caritas respicit sicut principale objectum, bonum divinum, quod pertinet ad unumquemque, secundum quod esse potest particeps beatitudinis;* unde ea sola ex caritate diligimus quae nobiscum beatitudinem participare possunt, ut Augustinus dicit in lib. de *Doctrina Christiana.*

What does "et hoc modo" stand for? Is it not unmistakably opposed to the love of the "bonum proprium alicujus hominis in quantum est singularis persona"? It is difficult to understand how Father Eschmann can manage so explicitly to contradict the *littera Sancti Thomae,* and to be so unaware of destroying the very root of charity toward our neighbour, which is the divine good *prout est beatitudinis objectum.*

Let us now turn to his diatribe against my use of a text taken from *De caritate,* a. 2, c:

> To prove his assertion by a text of St. Thomas, Professor De K. extracts a few words from the *Q.D. De Caritate* (art. 2). The content of these words is so important that we ask the patient reader to excuse us for transcribing the relevant text in its entirety. We shall italicize the words to which Professor De K. draws attention.
>
>> Si [?sicut] autem homo, inquantum admittitur ad participandum bonum alicuius civitatis et efficitur civis illius civitatis: Competunt ei virtutes quaedam ad operandum ea quae sunt civium, et amandum bonum illius civitatis, ita, cum homo per divinam gratiam admittatur in participationem caelestis beatitudinis, quae in visione et fruitione Dei consistit, fit quasi civis et socius illius beatae societatis, quae vocatur Caelestis Jerusalem, secundum illud *Eph.* 2, 19: "Estis cives sanctorum et domestici Dei." Unde homini sic ad caelestia adscripto competunt quaedam virtutes gratuitae, quae sunt virtutes infusae, ad quarum debitam operationem *praeexigitur amor boni communis toti societati, quod est bonum divinum, prout est beatitudinis objectum.*

> To begin our criticism of this whole position, let us first say a few words with regard to the author's exegetical methods. The word "praeexigitur," extracted from St. Thomas' text, is at once commandeered by Professor De K. to supplement the arsenal of his own ammunition. Whereas, according to St. Thomas' text, there is something prerequisite for the exercise of the infused virtues, according to Professor De K. this something is made a prerequisite for a moral philosophy and a social metaphysics. A facile device to support one's own assertions by authority! The solemn gravity of an apparently authentic quotation, given in Latin, turns out to be an empty show. Was this quotation intended to impress the reader or is it possible that the author himself was impressed by his pseudo-discovery? Strictly speaking, the disclosure of such an inept method of dealing with a text would authorize us in taking no further account whatsoever either of this excerpt or of the teaching based upon it. (*DM*, 193–94)

Father Eschmann has not quoted the relevant text in its entirety. Strangely enough, he omits the most important section. Here is the complete text of the passages I had already quoted both in French and in Latin:

> Proprium autem bonum hominis oportet diversimode accipi, secundum quod homo diversimode accipitur. Nam proprium bonum hominis in quantum homo, est bonum rationis, eo quod homini esse est rationale esse. Bonum autem hominis secundum quod est artifex, est bonum artis; et sic etiam secundum quod est politicus, est bonum ejus bonum commune civitatis. . . . Ad hoc quod aliquis sit bonus politicus, requiritur quod amet bonum civitatis.[81] Si autem homo, in quantum admittitur ad participandum bonum alicujus civitatis, et efficitur civis illius civitatis; competunt ei virtutes quaedam ad operandum ea quae sunt civium, et ad amandum bonum civitatis; ita cum homo per divinam gratiam admittatur in participationem caelestis beatitudinis, quae in visione et fruitione Dei consistit, fit quasi civis et socius illius beatae societatis, quae vocatur caelestis Jerusalem secundum illud, *Ephes.* 2, 19: *Estis cives sanctorum et domestici Dei.* Unde homini sic ad caelestia adscripto competunt quadam virtutes gratuitae, quae sunt virtutes infusae; ad quarum debitam operationem praeexigitur amor boni commu-

> nis toti societati, quod est bonum divinum, prout est beatitudinis objectum.[82] Amare autem bonum alicujus civitatis contingit dupliciter; uno modo ut habeatur: alio modo ut conservetur. *Amare autem bonum alicujus civitatis ut habeatur et possideatur, non facit bonum politicum; quia sic etiam aliquis tyrannus amat bonum alicujus civitatis ut ei dominetur; quod est amare seipsum magis quam civitatem; sibi enim ipsi hoc bonum concupiscit, non civitati. Sed amare bonum civitatis ut conservetur et defendatur, hoc est vere amare civitatem; quod bonum politicum facit; in tantum quod aliqui propter bonum civitatis conservandum vel ampliandum, se periculis mortis exponant et negligant privatum bonum. Sic igitur amare bonum quod a beatis participatur ut habeatur vel possideatur, non facit hominem bene se habentem ad beatitudinem, quia etiam mali illud bonum concupiscunt; sed amare illud bonum secundum se, ut permaneat et diffundatur, et ut nihil contra illud bonum agatur, hoc facit hominem bene se habentem ad illam societatem beatorum; et haec est caritas, quae Deum per se diligit, et proximos qui sunt capaces beatitudinis, sicut seipsos.*

How could St. Thomas state more clearly that in order to love the "bonum divinum, prout est beatitudinis objectum," it is not enough to love it "ut habeatur et possideatur," for this is the way tyrants love the common good; we must love it "secundum se, ut permaneat et diffundatur." Even the wicked have the kind of love of the divine good which my Opponent advocates: "amare bonum quod a beatis participatur ut habeatur vel possideatur, non facit hominem bene se habentem ad beatitudinem, quia etiam mali illud bonum concupiscunt." To urge that God is to be loved as the object of beatitude, yet not loved as the divine good "secundum se, ut permaneat et diffundatur" would be to defend a most perverted form of selfishness.

And now for Father Eschmann's interpretation of as much of the "relevant text" as he quotes:

> Is it true that St. Thomas taught, as Professor De K. would have us to believe, that the object of our beatitude, the very first and essential element of our ordination to God, is the divine good, insofar as this good is a common good, constituting, first and foremost, a society ("amor boni communis toti societati, quod est bonum divinum, prout est beatitudinis

objectum")? By no means! This interpretation is false. St. Thomas' argument in the *De Caritate,* loco cit., proceeds *a simili,* i.e. by comparing two highest goods, each taken in its own order, not, properly speaking, two common goods. The highest good of the earthly city is called a common good. No description or definition of it is given in this text. St. Thomas is here not lecturing on social metaphysics or political philosophy, but on charity; and the example of the city is only used as an *argumentum ad hominem.* To the earthly city, referred to in the example, the Heavenly City corresponds as the thing exemplified; and, through the words "quasi [!] civis" (to which corresponds in the parallel text, *Summa,* I, 60, 5 ad 5, "quoddam [!] bonum commune"), St. Thomas takes care, at the outset, to keep us from over-extending the simile and, thus, getting on the wrong track. To confuse examples with formal teaching is quite inadmissible. Let us paraphrase the passage in question, in order to set its true significance in relief: Prerequisite to the exercise of infused virtues in the Heavenly City is the love of the highest good which is the divine good, the object of beatitude. In like manner, the love of the earthly city's highest good, i.e. its common good, is prerequisite to the exercise of natural virtues. In a certain sense, the divine good might also be called a common good (*quoddam bonum commune*). But the object of charity is, of course, not a common good; rather it is the divine good ("Bonum commune non est obiectum caritatis, sed summum bonum," *Q.D. de Caritate,* 5 ad 4). Considered as a common good, the highest good of the Heavenly City would be, indeed, the object of supernatural general justice, not of charity. Charity and justice must not be confused.—It is very significant that St. Thomas chooses to say *bonum commune toti societati (caelesti)* instead of *bonum commune totius societatis,* as he usually does when speaking in terms of political philosophy. (*DM,* 194–95)

Just what does my Opponent mean by a "common good, constituting, first and foremost, a society"? Is he again forcing upon me his own totalitarian notion of common good and society? From the section of the text which he does not quote, it is clear what St. Thomas means by "amor boni communis toti societati, quod est bonum divinum, prout est beatitudinis objectum." The article aims to show "quod caritas absque dubio virtus est" (ibid., circa princ). A virtue requires the love of the good for which it operates. But the supernatural divine good cannot be reached by a natural virtue and hence the necessity of the infused virtues, "ad quarum debitam operationem praeexigitur amor boni communis toti societati, quod est bonum divinum, *prout est beatitudinis objectum.*" And this is the good which is the proper object of the virtue of charity. Now, because this divine good, *prout est beatitudinis objectum,* is a common good; it is not to be loved merely "ut habeatur et possideatur," for the evil, too, desire it in this manner, and such love of the divine good is not charity. If St. Thomas understood the expression "bonum commune toti societati" in Father Eschmann's sense, his reasoning, besides using four terms, would prove that charity is a virtue by means of a secondary object, namely, the love of our neighbour. Furthermore, the obligation to love our neighbour is not the reason why the divine good is a common good; nor does it become a common good because of the actual existence of the "beata societas," but because the object of charity is a common good, that is, a good which because of its very superabundance is communicable to others, and because it is "secundum se" communicable to others we must also love all those who are capable of beatitude. The principal object of charity is thus the reason of the secondary object. And this reason why we must love our neighbour is prior to our neighbour as well as to our act of loving him. Because the love of our neighbour follows from the true love of God, the former is, for us, a test of the latter. Unless we love God "secundum se, ut permaneat et diffundatur"—and this means to love Him as a common good, we simply do not love Him by charity. We must love the universal good as a common good, otherwise we shall not truly love the universal good; we shall love it merely "ut habeatur et possideatur," that is, in the manner in which "etiam mali illud bonum concupiscunt."

Father Eschmann suggests that "the example of the city is only used as an *argumentum ad hominem.*" Setting aside the problem of the identity of the *homo* to whom it is said to be addressed, let us merely try to determine what St. Thomas is seeking to prove in this article. He is teaching that charity is a virtue: "quod caritas non solum est virtus, sed potissima virtutum" (ibid., in fine). Now he plainly must have some reason for using the example of the city. The comparison between the earthly city and the heavenly must strengthen his argument in some way. It follows that, in his mind, the two have something in common; and, in fact, they must have something in common if his proof is to be valid. In a word, what St. Thomas establishes here is that the divine good, *prout est beatitudinis objectum,* must be loved as the good citizen loves the good of the earthly city; and this means that it must be loved "ut permaneat et diffundatur," and not, like the tyrant, "ut habeatur et possideatur."

My Opponent's "quoddam bonum commune" and his "bonum commune non est objectum caritatis" have been sufficiently exposed. So, let us pass immediately to what he says regarding supernatural general justice. When he writes off-hand: "Considered as a common good, the highest good of the Heavenly City would be, indeed, the object of supernatural general justice, not of charity," he reveals a strange understanding of the nature and object of infused moral virtues, "per quas homines bene se habent in ordine ad hoc quod sint *cives sanctorum et domestici Dei.*"[83] Since infused general justice is not a theological virtue, God could not possibly be that common good which is the object of justice. God is the norm and the ultimate end of infused justice, but this does not make Him its object. In the text under discussion, St. Thomas is speaking of the proper object of the theological virtue of charity, of the "amor boni communis toti societati, quod [bonum commune] est bonum divinum, *prout est beatitudinis objectum.*" By this virtue we love the divine good "secundum se, ut permaneat et diffundatur," "et haec est caritas, quae Deum per se diligit, et proximos qui sunt capaces beatitudinis, sicut seipsos." And this has formally nothing to do with any kind of justice. Nor has charity toward our neighbour anything to do, in the present discussion, with the proper object of justice whether acquired or infused. Even in the love of our neighbour the divine good is the "ratio formalis objecti."

> ... *Caritas in diligendo proximum habet Deum ut rationem formalem objecti, et non solum ut finem ultimum,* ut ex supradictis, art. praec, patet: sed aliae virtutes habent Deum non ut rationem formalem objecti, sed ut ultimum finem; et ideo, cum dicitur quod caritas diligit proximum propter Deum, illud *propter* denotat non solum causam materialem, sed quodammodo formalem. Cum autem dicitur de aliis virtutibus quod operantur propter Deum, illud *propter* denotat causam finalem tantum.[84]

If, as my Opponent suggests, the distinction between "bonum commune toti societati" and "bonum commune totius societatis" is significant, its significance would be to bring out more clearly that we are treating of the separated common good which is the object of charity, and not of the intrinsic, created, finite common good of the heavenly city toward which we become well disposed by the infused moral virtues.

Was it the *Thief! Thief!* method which prompted my Opponent to say that,

> Strictly speaking, the disclosure of such an inept method of dealing with a text would authorize us in taking no further account whatsoever either of this excerpt or of the teaching based upon it. (*DM*, 194)

Father Eschmann is ineffable.

Indeed I recognize the distinct though unenvied polemic advantage of his faulty Latin, his shallow acquaintance with philosophy and theology when allied to such unclouded confidence. His article has produced the proper rejoicing in personalist quarters, but what is more important, it has disturbed and poisoned by anticipation the mind of many an unsuspecting reader unable to see through the sham of his legerdemain with the *littera Sancti Thomae*. After all, even if his cliché sneers are discreditable, who could still hold that the object of the virtue of charity is the divine good as a common good, now that Father Eschmann has produced the trenchant *littera Sancti Thomae*: "Bonum commune non est objectum caritatis, sed summum bonum"? However, even if there were no point in trying to refute my Opponent for his own sake, it would be unfair to let the unsuspecting reader be misled by his pretense. My persistence in demolishing his criticisms, which I can hardly do without disclosing his own method, will be thought merciless perhaps and surely dull, yet I must pursue this course, lest I should seem to be evading difficulties and leaving unanswered such accusations as:

> In setting up a "principle of the New Order" Professor De K. has done a work which is—shall we say—surprisingly radical and daring: he has at the same time taken in his stride a new foundation of Christian ethics and moral theology. (*DM*, 196)[85]

We can all agree that the accusation is not lacking in gravity, at least insofar as it concerns the truth of the doctrine I defend. Let us examine his reasons.

VI. *Bonum universale in essendo* and *bonum universale in causando*

Immediately following his hopelessly muddled exegesis of St. Thomas' text (*De caritate*, a. 2, c.) Father Eschmann goes on to say:

> May the patient reader excuse the length to which this exegetical problem has obliged us to go. Let us now turn back to the substance of Prof. De K.'s teaching.
>
> Is it not the most fundamental and absolutely unshakeable cornerstone of Christian ethics that the term of our ordination to God is God as He is in Himself, i.e. the Good by His essence and the essence of goodness (*bonum universale in essendo*)? Is it not the very first care of a Christian ethician to make sure that the conclusion of his very first argument directly reaches this *bonum universale in essendo?* This, at least, is the content and intention of that great argument which opens the *pars moralis* of the *Summa* (I-II, 1, 1–2, 8) and whose conclusion is: "Ex quo patet quod nihil potest quietare voluntatem hominis nisi bonum universale, quod non invenitur in aliquo creato, sed solum in Deo, quia omnis creatura habet bonitatem participatam" (ibid., 2, 8). St. Thomas has here completely forgotten to speak about Professor De K.'s "common good" by which man's ordination to God is *très formellement* determined. I am afraid that on pages 26–27 of this book a suspicion which the expert reader has felt all the way along, from the first page on, becomes definite, namely that the author has pushed the "primacy of the common good" very far, so far indeed that, if the consequences of his position are made explicit, we must in our Christian ethics re-do our work from the beginning. In setting up a "principle of the New Order" Professor De K. has done a work which is—shall we say—surprisingly radical and daring: he has at the same time taken in his stride a new foundation of Christian ethics and moral theology.
>
> Professor De K. has confused *bonum universale in essendo* and *bonum universale in causando.* "The creature," St. Thomas says (*Summa*, I, 103, 4), "is assimilated to God in two respects: first, with regard to this that God *is good;* and thus the creature becomes like Him by *being good;* and, secondly, with regard to this that God is the *cause of goodness in others;* and thus the creature becomes like God by *causing others to be good.*"—The

> common good, and every common good, is formally *bonum universale in causando:* it is not, formally, *bonum universale in essendo.*
>
> The very first and essential element of our ordination to God is not the fact that God is the first *bonum universale in causando,* the fountain of all communications, but that He is the *bonum universale in essendo.* (*DM,* 195–90)

When we first read that God is "the Good by His essence and the essence of goodness" and that this is the *bonum universale in essendo,* we might think we know what Father Eschmann is talking about. But when he opposes this to the *bonum universale in causando* we may well wonder whether he himself knows what he is talking about. My Opponent's argumentation is so confused that, in order to unravel it and arrest the possible meaning of his terms, we must beg leave to make several distinctions.

Let us first consider the more elementary distinction between *bonum in essendo* and *bonum in causando. Bonum in essendo* may be used to mean *bonum per essentiam.* And this in turn may be understood to mean *bonum a se* as opposed to *bonum ab alio* or *per participationem;* again it may mean *bonum per se* as opposed to *bonum secundum quid.* There is yet another meaning of *bonum in essendo,* namely, the good that a thing is insofar as it is. In this case, *bonum in essendo* is not opposed to *bonum per se,* unless we understand it to mean *in essendo tantum.* For, in the creature, the good that it is from the mere fact that it is and that it has substantial being is only *bonum secundum quid,* whereas it will be good *per se* or *simpliciter* only according to added perfections which, from the viewpoint of being, are accidental.[86]

Since Father Eschmann opposes *bonum universale in essendo* to *bonum universale in causando,* let us now consider what *bonum in causando* may mean.—In *De veritate,* q. 1, a. 1, c., St. Thomas says:

> [Modus generaliter consequens omne ens] dupliciter accipi potest: uno modo secundum quod consequitur unumquodque ens in se; alio modo secundum quod consequitur unumquodque ens in ordine ad aliud. . . . Si autem modus entis accipiatur secundo modo, scilicet secundum ordinem unius ad alterum, hoc potest esse dupliciter. Uno modo secundum divisionem unius ab altero. . . . Alio modo secundum convenientiam

unius entis ad aliud; et hoc quidem non potest esse nisi accipiatur aliquid quod natum sit convenire cum omni ente. Hoc autem est anima, quae quodammodo est omnia, sicut dicitur in III *De anima* (text. 37). In anima autem est vis cognitiva et appetitiva. Convenientiam ergo entis ad appetitum exprimit hoc nomen *bonum,* ut in principio *Ethic.* dicitur: *Bonum est quod omnia appetunt.* Convenientiam vero entis ad intellectum exprimit hoc nomen *verum.*

In q. 21, a. l, c., of the same work, St. Thomas goes further into this matter:

. . . Verum et bonum super intellectum entis [addunt] respectum *perfectivi.* In quolibet autem ente est duo considerare: scilicet ipsam rationem speciei, et esse ipsum quo aliquid aliud subsistit in specie illa; et sic aliquod ens potest esse perfectum dupliciter. *Uno modo secundum rationem speciei tantum;* et sic ab ente perficitur intellectus, qui perficitur per rationem entis; nec tamen ens est in eo secundum esse naturale; et ideo hunc modum perficiendi addit verum super ens. Verum enim est in mente, ut Philosophus dicit in VI *Metaph.*; et unumquodque ens in tantum dicitur verum, in quantum conformatum est vel conformabile intellectui; et ideo omnes recte definientes verum, ponunt in ejus definitione intellectum. *Alio modo* ens est perfectivum alterius non solum secundum rationem speciei, sed *etiam secundum esse quod habet in rerum natura:* et per hunc modum est perfectivum bonum; bonum enim in rebus est, ut Philosophus dicit in VI *Metaph.* (com. 8). In quantum autem unum ens est secundum esse suum perfectivum alterius et conservativum, habet rationem finis respectu illius quod ab eo perficitur; et inde est quod omnes recte definientes bonum ponunt in ratione ejus aliquid quod pertineat ad habitudinem finis; unde Philosophus dicit in I *Ethic.* (in princip.), quod *bonum optime definiunt dicentes, quod bonum est quod omnia appetunt.*

And in the body of the following article he adds:

. . . Cum *ratio boni* in hoc consistat quod aliquid sit *perfectivum alterius per modum finis,* omne id quod invenitur habere rationem finis, habet et rationem boni.

From this it should be clear that the most proper and profound meaning of the term "good" is: *perfectivum alterius per modum finis.* Now, if such is the very *ratio boni,* the divine good will be called good in the strict sense of the word only insofar as it is *perfectivum alterius per modum finis,* that is, because of the *convenientia ad appetitum.* As it is in itself, the divine good may be considered either with respect to the divine will, or with respect to a created will elevated by the infused virtue of charity. Obviously, when we consider the divine good with respect to the divine will, the term *finis* cannot be taken in the strict sense of final cause, since causality involves dependence. In this case, "perfectivum alterius per modum finis" or "aliquid quod pertineat ad rationem finis" merely express the proper perfection of the good which draws and attracts the will toward it as to its proper object. This "drawing toward" and "attracting" involves no potentiality on the part of the divine will. The first article of *De veritate,* q. 23: *Utrum Deo conveniat voluntatem habere,* contains the following objection and answer:

> 3. Praeterea, ei quod non habet aliquam causam, non competit aliquid quod importet respectum ad causam. Sed Deus, cum sit prima causa omnium, non habet aliquam causam. Ergo, cum voluntas importet habitudinem in causam finalem, quia voluntas est finis, secundum Philosophum in III *Ethic.* (c. ii, ante med.), videtur quod voluntas Deo non competat.
>
> Ad tertium dicendum, quod voluntas est alicujus dupliciter; uno modo principaliter, et alio modo secundario. Principaliter quidem voluntas est finis, qui est ratio volendi omnia alia; secundario autem est eorum quae sunt ad finem, quae propter finem volumus. Voluntas autem non habet habitudinem ad volitum quod est secundarium, sicut ad causam; sed tantummodo ad volitum principale, quod est finis. Sciendum est autem, quod *voluntas et volitum aliquando distinguuntur secundum rem; et tunc volitum comparatur ad voluntatem sicut realiter causa finalis. Si autem voluntas et volitum distinguuntur tantum ratione, tunc volitum non erit causa finalis voluntatis nisi secundum modum significandi.* Voluntas ergo divina comparatur, sicut ad finem, ad bonitatem suam, quae secundum rem idem est quod sua voluntas; distinguitur autem solum secundum modum significandi. Unde relinquitur quod voluntatis divinae nihil sit causa realiter, sed solum secundum modum significandi. Nec est

> inconveniens, in Deo significari aliquid per modum causae; sic enim divinitas significatur in Deo ut habens se ad Deum per modum causae formalis. Res vero creatae, quas Deus vult, non se habent ad divinam voluntatem ut fines, sed ut ordinata ad finem; propter hoc enim Deus creaturas vult esse, ut in eis sua bonitas manifestetur, et ut sua bonitas, quae per essentiam multiplicari non potest, saltem similitudinis participatione in plures effundatur.

Hence, when we consider the divine good with respect to the created will, it is an end in the strict sense of the word, a final cause attained as "ratio formalis objecti." Accordingly, to deny that the divine good is in this respect a good "sicut realiter causa finalis," is to imply that the created will is not merely on a plane with the divine will, but that it is identical with the divine good and will.

Again the good may be called *bonum in causando* in two ways: either to mean the good as a final cause, i.e. "perfectivum alterius per modum finis" or to signify the good as an efficient and exemplary cause of another good. Both meanings apply to God. As the good of His will "per modum finis," He attracts all things as their ultimate end; and as the divine good to be attained as it is in itself by the rational creature, He is the "ratio formalis objecti" of charity. In the second sense, He is a *bonum in causando* as the exemplary and efficient cause of all created goodness. Hence, God may be loved as He is in Himself, or again we may love His goodness as exemplified in the finite good of which He is final, exemplary, and efficient cause.[87]

When we oppose *bonum in essendo* to *bonum in causando,* the former can mean either of two things: the perfection of a being considered absolutely, as *formally* constitutive of that being, and this meaning prescinds from the good as "perfectivum alterius per modum finis" (for the good proper "non solum habet rationem perfecti, sed perfectivi");[88] or it may mean the good that a being is from the mere fact that it is, and this signification prescinds from the distinction between *bonum per se* and *bonum per accidens.* When *bonum in essendo* is taken in the latter sense and opposed to *bonum in causando,* it can once more have only two possible meanings: the good as a *commune in praedicando,*[89] which is not the object of appetite and hence not a good at all in the proper sense, or it may mean *bonum in essendo tantum* as opposed to that *bonum per se* which is found in creatures, and which is really *bonum per acci-*

dens of the created *ens per se.* Now this last significance of *bonum in essendo* does not permit it to be a good in the full sense of the word, either, for

> secundum esse substantiale non dicitur aliquid bonum simpliciter et absolute, nisi superaddantur perfectiones aliae debitae: et ideo ipsum esse substantiale non est absolute appetibile nisi debitis perfectionibus adjunctis.[90]

Let us now consider the expressions *bonum universale in essendo* and *bonum universale in causando.* The former may bear three distinct meanings: first, it may be taken to mean *bonum universale in praedicando* which is common to all things insofar as they are good in any way; secondly, it may mean the perfection of divine being considered in itself without formal reference to will; thirdly, it may mean *bonum universale per essentiam,* where the good is understood in the rigorous sense of "perfectivum alterius per modum finis," and this is the divine good, for God is good *simpliciter* by His very essence, "in quantum ejus essentia est suum esse."[91]

Bonum universale in causando may mean the divine good considered according to the strict formality of the good, i.e. as "perfectivum alterius per modum finis." It has already been emphasized that, when so considered with respect to the divine will, the divine good is a final cause only "secundum modum significandi," because in God, "voluntas et volitum distinguuntur tantum ratione." However, unless we use this "modus significandi," we do not express the proper formality of the good. But the divine good becomes a final cause in the strict sense of "cause," when considered with respect to a will which is not identical with the divine good: "voluntas et volitum aliquando distinguuntur secundum rem; et tunc volitum comparatur ad voluntatem sicut realiter causa finalis." In either case, however, God is called *bonum universale in causando,* and this term is opposed to the second meaning of *bonum universale in essendo.* Finally, the same expression—*bonum universale in causando*—may also be used to signify the divine good as the universal effective and exemplary cause of all created goodness.

Hence, *bonum universale in essendo* understood in its third sense and *bonum universale in causando* taken in its first sense are the same thing, the only difference being that the former expresses the identity of divine goodness and divine being; the latter brings out the proper formality of the divine good as final cause, either "per modum significandi," or "sicut realiter causa finalis." When we oppose the two and apply them to God, then *bonum universale in essendo* must be taken in the second sense, which prescinds from the proper formality of the good as "perfectivum alterius per modum finis."

And now let us examine Father Eschmann's reasoning more closely. *In forma,* it amounts to this: The term of our ordination to God is *bonum universale in essendo.* But *bonum universale in essendo* is not *bonum universale in causando.* Therefore, the term of our ordination to God is not *bonum universale in causando.*

To this we answer that if *bonum universale in essendo* means *bonum per essentiam,* and *bonum universale in causando* means *bonum universale per modum finis,* the major of the argument is true, but the minor is false. If, on the contrary, *bonum universale in essendo* is taken to mean the perfection of divine being considered absolutely, i.e. prescinding from the formality "perfectivum alterius per modum finis," the minor is true, but the major is false. In either case, the conclusion is null.—When Father Eschmann uses the expression *bonum universale in causando* to mean the divine good as exemplified in the goodness of which it is the cause, he evades the issue and is tilting at windmills.

Let us examine the text, together with its context, which my Opponent quotes in support of his distinction between *bonum universale in essendo* and *bonum universale in causando.* The problem is: *Utrum effectus gubernationis sit unus tanum, et non plures.*

> Respondeo dicendum quod effectus cujuslibet actionis ex fine ejus pensari potest: nam per operationem efficitur ut pertingatur ad finem. Finis autem gubernationis mundi est bonum essentiale, ad cujus participationem et assimilationem omnia tendunt. Effectus igitur gubernationis potest accipi tripliciter. *Uno modo,* ex parte ipsius finis: et sic est unus effectus gubernationis, scilicet assimilari summo bono.—*Alto modo* potest considerari effectus gubernationis secundum ea quibus ad Dei assimilationem creatura perducitur. Et sic in generali sunt duo effectus gubernationis. Creatura enim assimilatur Deo *quantum ad duo:* scilicet quantum ad id quod Deus bonus est, inquantum creatura est bona; et quantum ad hoc quod Deus est aliis causa bonitatis, inquantum una creatura movet aliam ad bonitatem. Unde duo sunt effectus gubernatio-

> nis: scilicet conservatio rerum in bono, et motio earum ad bonum.—*Tertio modo* potest considerari effectus gubernationis in particulari: et sic sunt nobis innumerabiles.[92]

Obviously, our problem is related to the first consideration of the effect of divine government, that is, of the effect, "ex parte finis: . . . scilicet assimilari summo bono," where the supreme good is *bonum universale in causando per modum finis.* We are not now concerned with the effect of divine government "secundum ea quibus ad Dei assimilationem creatura perducitur," that is, whereby the creature is assimilated to God insofar as the creature itself is a good, and by its own goodness is enabled to move another toward the good. The following text deals with the principle of this inherent goodness:

> . . . Omne agens invenitur sibi simile agere; unde si prima bonitas sit effectiva omnium bonorum, oportet quod similitudinem suam imprimat in rebus effectis; et sic unumquodque dicetur *bonum sicut forma inhaerente* per similitudinem summi boni sibi inditam, et ulterius per bonitatem primam, *sicut per exemplar et effectivum omnis bonitatis creatae.*[93]

But we are concerned with an assimilation of the creature to God which is not mentioned in the portion of St. Thomas' text brought forth by Father Eschmann. The following passage from *De veritate,* q. 20, a. 4, c., describes the latter assimilation (which had been briefly stated in that first portion of the text not quoted by my Opponent) while carefully distinguishing it from the former:

> . . . Cum Deus sit principium omnium rerum et finis; *duplex habitudo ipsius ad creaturas invenitur: una secundum quam omnia a Deo procedunt in esse; alia secundum quam ad eum ordinantur ut in finem;* quaedam per viam assimilationis tantum, sicut irrationales creaturae; quaedam vero tam per viam assimilationis, quam pertingendo ad ipsam divinam essentiam. Cuilibet enim creaturae procedenti a Deo inditum est ut in bonum tendat per suam operationem. In cujuslibet autem boni consecutione creatura Deo assimilatur; sed creaturae rationales super hoc habent ut ad ipsum Deum cognoscendum et amandum sua operatione pertingant; unde prae ceteris creaturis beatitudinis sunt capaces.

> In utraque autem praedictarum habitudinum invenitur creaturas Creator excedere. Quantum ad primam quidem, quod super omnia quae Deus fecit, adhuc possit alia dissimilia facere, et novas species et nova genera, et alios mundos; nec unquam id quod factum est, facientis virtutem adaequare potest. Quantum vero ad secundam, quia creatura quantumcumque fiat boni particeps, nunquam tamen pertingit ad hoc quod Dei bonitatem adaequet. *Quantumcumque etiam creatura rationalis Deum cognoscat et amet, nunquam tamen ita perfecte eum cognoscit et amat, quantum ipse cognoscibilis et diligibilis est.* Sicut autem creaturae imperfectae essent, si a Deo procederent, et ad Deum non reordinarentur; ita *imperfectus esset creaturarum a Deo exitus, nisi reditio in Deum exitum adaequaret.*

It is surely absurd to suggest that the ultimate end of the rational creature could be that similitude of divine goodness in which its own goodness consists and that highest operation in which it leads to or produces another created good. And yet, if Father Eschmann's quotation is to be relevant to the problem we are treating, he must interpret the quotation to mean exactly that. He cannot mean the divine goodness as it is in itself, since he does not allow the *bonum universale in essendo* to be "realiter causa finalis."

My Opponent allows us to call God a *bonum universale in causando,* but by this he means the divine good as the exemplary and efficient cause of the created good. He thereby denies that the divine good is good in the strict sense of the word as "perfectivum alterius per modum finis," and that the ultimate good of the rational creature is a final cause in the strict sense. It is this denial which determines his notion of common good.

> The common good, and every common good, is formally *bonum universale in causando:* it is not, formally, *bonum universale in essendo.* (*DM,* 196)

Hence, according to Father Eschmann, a good is a common good only insofar as it produces a multiplicity of other goods, and not, formally, in that it is the end of this multiplicity, that is, a good communicable to many *per modum finis*—an exceedingly shallow understanding of the nature of the common good for a Thomist.

When my Opponent goes on to say that "God is the first *bonum universale in causando,* the fountain of all communications," what does he mean by "communication"? Clearly he must understand and restrict it to mean: "effusio secundum quod importat operationem causae efficientis." Yet St. Thomas is quite intolerant of such a narrow concept of the diffusive power of the good, as we have seen in a text already quoted:

> . . . *Diffundere,* licet secundum proprietatem vocabuli videatur importare operationem causae efficientis, tamen largo modo potest importare habitudinem cujuscumque causae sicut influere et facere, et alia hujusmodi. Cum autem dicitur quod bonum est diffusivum secundum sui rationem, *non est intelligenda effusio secundum quod importat operationem causae efficientis, sed secundum quod importat habitudinem causae finalis;* et talis diffusio non est mediante aliqua virtute superaddita. *Dicit autem bonum diffusionem, causae finalis, et non causae agentis:* tum quia efficiens, in quantum hujusmodi, non est rei mensura et perfectio, sed magis initium; tum quia effectus participat causam efficientem secundum assimilationem formae tantum; sed *finem consequitur res secundum totum esse suum, et in hoc consistebat ratio boni.*[94]

It is not, formally, because God produces the good that is the universe or the good seen in any single creature that He is creation's final cause, but because He is the *bonum universale in causando* for all the good that He produces. His own goodness is the *finis cujus gratia,* and all being of which He is the efficient and exemplary cause is for this end. If God is a common good in producing the creature, "secundum quod importat operationem causae efficientis," He is, *a fortiori,* a common good "secundum quod importat habitudinem causae finalis," since the final cause is the *causa causarum.*

Let us now return to Father Eschmann's text. Immediately following the long passage quoted above, page 47, he writes:

> From this it follows that our own (personal) good is a participated good. Through this participation a "certain common good" ("quoddam bonum commune") emerges, i.e. a good which, in a certain way, is common to God and the creature. Considering the supernaturally elevated creature, this common good is constitutive for a community or "society" between God and the supernaturally elevated creature, a society which is called, by St. Thomas, *societas suae (i.e. Dei) fruitionis.*[95] It is the divine friendship to whose *essential* constitution no multitude of creatures is

> required.[96] The fact that there is such a multitude of creatures does not yet formally come into consideration.
>
> This fact becomes only now, i.e. in the third place, relevant. For if[97] there are several creatures sharing in the same participated good they will have something in common. Here, then, there will be a common good properly speaking, i.e. a good pertaining to a multitude of beings in such manner that each and everyone communicates in it. God is, as St. Thomas says, the last common good among men, i.e. that good in which they finally must or should unite: "Homines non uniuntur inter se nisi in eo quod est commune inter eos. Et hoc est maxime Deus."[98]
>
> Professor De K. has, throughout his treatise, neglected these fundamental considerations. (*DM*, 196–97)

Just what follows and how "it follows that our own (personal) good is a participated good" is not quite clear. What does my Opponent mean by "participated good"? If he uses the expression "our own (personal) good" in the strict sense, that is, for a good which belongs to the person as a personal, proper good and therefore to no other, then the "participated good" is necessarily a created good—created beatitude, the formal, essential beatitude of the created person as distinct from objective beatitude which is God Himself. This formal beatitude is indeed a good which belongs to the person as a purely personal good, in the strict sense, since it consists in the very operation of the intellect by which the divine essence is attained. If this were what Father Eschmann meant by "our own (personal) good is a participated good," then, when he says: "Through this participation a 'certain common good' ('quoddam bonum commune') emerges, i.e. a good which, in a certain way, is common to God and the creature," this "certain common good" could only be a *bonum commune in essendo et in praedicando*. Now, compared to the common good in the strict sense, that is, *bonum commune in causando*, the former is indeed a common good only in a certain sense. But such a good is not a good at all in the proper sense and it most certainly is not the good of the *societas suae (i.e. Dei) fruitionis*.

If, on the contrary, the phrase "our own (personal) good is a participated good" is used by the author to mean something other than what these words should mean (namely, the imperfect participation "ex parte ipsius participantis, qui quidem ad ipsum objectum beatitudinis secundum seipsum attingit, scilicet Deum, sed imperfecte, per respectum ad modum quo

Deus seipso fruitur,"[99] that is, if the objective beatitude is called "participated" (by purely extrinsic denomination) insofar as it is the object communicated to the created intellect, but imperfectly, though intuitively, attained by that intellect, it is *formalissime* a common good. The *Summum Bonum*, God, the objective beatitude of the supernaturally elevated creature, can never be anything else than a common good because our intellect and will are not identical with the divine intellect and will. What we must never lose sight of is that our formal beatitude is created and intrinsically participated and that "majus est in se bonum Dei quam participare possumus fruendo ipso." Objective beatitude, in short, is the proper good of God alone. It could not possibly be the common good of God and of the creature unless we used the expression to mean *bonum commune in praedicando,* which is indeed common good only "in a certain way" since its foundation lies in the identity of the object of the divine *beatitudo per essentiam* and of the created formal beatitude. Hence, whatever Father Eschmann means by "our own (personal) good," the "good which, in a certain way, is common to God and the creature" could be common only according to predication.

We now begin to understand what he means by his "quoddam bonum commune." I say "his," because the term as he uses it has absolutely nothing to do with the text from which it was lifted. The reader will recall the context:

> . . . Cum in Deo sit *unum et idem ejus substantia et bonum commune,* omnes qui vident ipsam Dei substantiam, eodem motu dilectionis moventur in ipsam Dei essentiam prout est ab aliis distincta, et secundum quod est *quoddam bonum commune.* Et quia inquantum est bonum commune, naturaliter amatur ab omnibus; quicumque videt eum per essentiam, impossibile est quin diligat ipsum.[100]

And this St Thomas states in connection with the problem: *Utrum angelus naturali dilectione diligat Deum plus quam seipsum.* The reader will also remember that he used the same principle ("unaquaeque pars naturaliter plus amat commune bonum totius quam particulare bonum proprium") to answer the question: *Utrum homo debeat ex caritate plus Deum diligere quam seipsum.*[101] Now if we understand St. Thomas' "quoddam bonum commune" in Father Eschmann's sense, the result will be that we will love God more than ourselves, not because He is the *bonum universale,* to which we are compared as a part, but because the divine good, "in a certain way, is common to God and the creature" (*DM*, 196). Surely no such reason has ever occurred to St. Thomas! Actually, it would mean that we love ourselves more than God. This alone (that we must love God more than ourselves because of such a community) should be enough to show how preposterous is my Opponent's confident understanding of the *littera Sancti Thomae*. I shall not insult the reader's intelligence by drawing out all the ridiculous consequences which would necessarily follow this contradiction, that, is, from Father Eschmann's type of "fundamental considerations."

On the basis of a text which he himself invokes and which the reader may recall as used above to expose the shallowness of this concept of the common good, still another and perhaps more fatal attack may be made on Father Eschmann's view of what really constitutes the common good. Here are the words of St. Thomas:

> . . . Perfectio caritatis est essentialis beatitudini quantum ad dilectionem Dei, non autem quantum ad dilectionem proximi. Unde si esset una sola anima fruens Deo, beata esset, non habens proximum quem diligeret. Sed supposito proximo, sequitur dilectio ejus ex perfecta dilectione Dei. Unde quasi concomitanter se habet amicitia ad beatitudinem perfectam.

In Father Eschmann's opinion, a good may be called common only when it is actually communicated to many; its being common depends upon its being actually imparted to a community. In other words, the denomination "common good" is founded, not on the superabundance and incommensurability of the divine good (which, for that very reason, can never be the proper good of any person) but on the fact of a manifold of persons who actually share in this good. According to my Opponent, God is a common good only *supposito proximo*. Now, since the existence of a neighbour and his sharing in the divine good is not essential to beatitude, it follows that, with respect to the objective beatitude of any single created person, God is a common good only *per accidens*. That God is a common good, then, merely follows from His decree to beatify, and from the fact of the existence of many persons. For any single person God is a common good only because there happen to be other created persons. The denomination is taken from the existing manifold of the Blessed; it is a purely extrinsic denomination.

This is, inescapably, Father Eschmann's position. It is an opinion wholly in keeping with what he calls "a common good properly speaking" (*DM*, 196). And it is equally inescapable that if, *per impossibile*, God were common good in such a sense, i.e. as the proper good of this person and the proper good of that person and so forth, He would be merely *bonum commune in praedicando*. Now, in this sense, He could not be loved by anyone as common good, since *bonum commune in praedicando* cannot be the object of love. Furthermore, when, in loving our neighbour, we want him to share in the divine good, in this respect God would be loved by us as a proper good for our neighbour. But the love of a proper good (which is always a particular good as opposed, not to a good common according to predication, but to a common good in the full sense) for our neighbour proceeds *ex amicabilibus ad seipsum*, and not from the common good. And this in turn implies that in loving our neighbour we would love ourselves more than we love God. The following objection and answer taken from IIaIIae, q. 26, a. 3, bear this out:

> . . . Videtur quod homo non debeat ex caritate plus Deum diligere quam seipsum. Dicit enirn Philosophus, in IX *Ethic.*, quod *amicabilia quae sunt ad alterum veniunt ex amicabilibus quae sunt ad seipsum.* Sed causa est potior effectu. Ergo maior est amicitia hominis ad seipsum quam ad quemcumque alium. Ergo magis se debet diligere quam Deum.
>
> Ad primum dicendum quod *Philosophus loquitur de amicabilibus quae sunt ad alterum, in quo bonum quod est objectum amicitiae invenitur secundum aliquem particularem modum; non autem de amicabilibus quae sunt ad alterum in quo bonum praedictum invenitur secundum rationem totius.*

St. Thomas is speaking of the theological virtue of charity whose object, even in the love of our neighbour, is God "ut ratio formalis objecti." The paper shortage notwithstanding, I shall again quote the answers to the other two objections from the same article:

> Ad secundum dicendum quod bonum totius diligit quidem pars secundum quod est sibi conveniens: *non autem ita quod bonum totius ad se referat, sed potius ita quod seipsam refert in bonum totius.*

Ad tertium dicendum quod hoc quod aliquis velit frui Deo, pertinet ad amorem quo Deus amatur amore concupiscentiae. *Magis autem amamus Deum amore amicitiae quam amore concupiscentiae: quia maius est in se bonum Dei quam participare possumus fruendo ipso.* Et ideo simpliciter homo magis diligit Deum ex caritate quam seipsum.

Supposito proximo, we love him by charity because we already love God as a common good. This love of neighbour presupposes the common good as common good. Obviously, our neighbour is not the formal reason why we love the common good as common good. This principle is true of any love toward our fellow man which has its root in a common good. If the common good is to be loved more than the purely personal good,

> [c]ela ne veut pas dire que les autres sont la raison de l'amabilité propre du bien commun; au contraire, sous ce rapport formel, les autres sont aimables en tant qu'ils peuvent participer à ce bien. (*BC,* 9)

The love of a good which presupposes our neighbour and which radically and formally proceeds from this presupposition alone, is not a love of our neighbour, for the sake of God, but for the sake of our neighbour.[102] This love may be generous, but the good which properly depends upon this presupposition alone, "invenitur secundum aliquem particularem modum," it is not the "bonum commune totius"; when thus isolated it has formally nothing to do with the divine common good *prout est beatitudinis objectum.*

If the divine good is to be loved as a common good only *supposito proximo,* why is it that we must love that good more than our proper good, and yet, at the same time, love ourselves more than our neighbour? The following objection and answer are taken from IIaIIae, q. 26, a. 4:

> 3. Praeterea, I *ad Cor.* 13 dicitur quod caritas *non quaerit quae sua sunt.* Sed illud maxime amamus cujus bonum maxime quaerimus. Ergo per caritatem aliquis non amat seipsum magis quam proximum.
>
> Ad tertium dicendum quod, sicut Augustinus dicit, in *Regula, quod dicitur, Caritas non quaerit quae sua sunt, sic intelligitur quia communia propriis anteponit.* Semper autem commune bonum est magis amabile unicuique quam proprium bonum; sicut etiam ipsi parti est magis amabile bonum totius quam bonum partiale sui ipsius, ut dictum est.

Ludwig Feuerbach (1804–1872), in *Das Wesen des Christenthums*, produces the very text Father Eschmann quotes against me from IaIIae, q. 4, a. 8 (*Si esset una sola anima fruens Deo . . .*) in proof of the opinion held by my Opponent. I quoted this text in my book: Appendice IV, "Ludwig Feuerbach interprète saint Thomas." Its complete agreement with Father Eschmann's interpretation is so striking that it is worth quoting again.

> Aristoteles sagt bekanntlich ausdrücklich in seiner Politik, dass der Einzelne, weil er für sich selbst nicht sich genüge, sich gerade so zum Staate verhalte, wie der Theil zum Ganzen, dass daher der Staat der Natur nach früher sei als die Familie und das Individuum, denn das Ganze sei nothwendig früher als der Theil.—Die Christen "opferten" wohlauch "das Individuum," d. h. hier den Einzelnen als Theil dem Ganzen, der Gattung, dem Gemeinwesen auf. Der Theil, sagt der heilige Thomas Aquino, einer der grössten christlichen Denker und Theologen, opfert sich selbst aus natürlichem Instinkt zur Erhaltung des Ganzen auf. "Jeder Theil liebt von Natur mehr das Ganze als sich selbst. Und jedes Einzelne liebt von Natur mehr das Gut seiner Gattung, als sein einzelnes Gut oder Wohl. Jedes Wesen liebt daher auf seine Weise naturgemäss Gott, als das allgemeine Gut, mehr, als sich selbst." (Summae P. I. Qu. 60. Art. V.) Die Christen denken daher in dieser Beziehung wie die Alten. Thomas A. preist (de Regim. Princip. 1. III. c. 4) die Römer, das sie ihr Vaterland uber alles setzten, seinem Wohl ihr Wohl aufopferten. Aber alle diese Gedanken und Gesinnungen gelten im Christenthum nur auf der Erde, nicht im Himmel, in der Moral, nicht in der Dogmatik, in der Anthropologie, nicht in der Theologie. Als Gegenstand der Theologie ist das Individuum, der Einzelne übernatürliches, unsterbliches, selbstgenüges, absolutes, göttliches Wesen. Der heidnische Denker Aristoteles erklärt die Freundschaft (Ethik 9, B. 9. K.) für nothwendig zur Glückseligkeit, der christliche Denker Thomas A. aber nicht. "Nicht gehört nothwendig, sagt er, Gesellschaft von Freunden zur Seligkeit, weil der Mensch die ganze Fülle seiner Vollkommenheit in Gott hat." "Wenn daher auch eine Seele allein für sich im Genusse Gottes wäre, so wäre sie doch selig, wenn sie gleich keinen Nächsten hätte, den sie liebte." (Prima Secundae. Qu. 4. 8.) Der Heide weiss sich also auch in der Glückseligkeit als Einzelnen, als Individuum und desswegen als bedürftig eines andern Wesens seines Gleichen, seiner

> Gattung, der Christ aber bedarf keines andern Ich, weil er als Individuum zugleich nicht Individuum, sondern Gattung, allgemeines Wesen ist, weil er "die ganze Fülle seiner Vollkommenheit in Gott" d. h. in sich selbst hat.[103]

Feuerbach, too, believes that when St. Thomas speaks of the ordination of the supernaturally elevated creature to the highest good, he means to deny the primacy of the common good which applies only in the natural order. Feuerbach seems never to have reached the IIaIIae.

We have the right to presume that Father Eschmann read my book. Yet he has completely ignored those passages, some of which I have already quoted, in which I explained why God is a common good in the strict sense and why the created person can never be referred to this good except as a part. I repeated this explanation in connection with Feuerbach's interpretation of St. Thomas:

> Feuerbach est obligé de recourir à cette distinction parce qu'il ne voit pas que c'est tout autre chose d'être sous la dépendance du tout et de ses parties pour atteindre le bien du tout, et d'atteindre le bien du tout. La raison fondamentale pour laquelle nous appelons toute personne créée partie, c'est que son plus grand bien est incommensurable au bien de la personne singulière prise comme telle; c'est bien plutôt comme individu que la personne humaine est un tout. Aucune personne créée n'est une nature proportionnée ni proportionnable au bien purement et simplement universel comme à son bien propre en tant que personne singulière. Autrement toute personne serait Dieu. Aussi, pour Feuerbach, l'homme est-il Dieu. (*BC*, 140)

In drawing this conclusion, the father of Marxist materialism was logical. He had confused *bonum universale in causando* with *bonum universale in praedicando* (*BC*, 142–43): he was unacquainted with the proper formality of the good. In all fairness we must add that he did not claim to be a Thomist.

In this connection, Gabriel Vasquez' (c. 1551–1604) interpretation of St. Thomas may be of even greater interest. The object of his criticism is the following text from IIaIIae, q. 25, a. l, ad 2:

> . . . Amor respicit bonum in communi, sed honor respicit proprium bonum honorati: defertur enim alicui in testimonium propriae virtutis. Et *ideo amor non diversificatur specie propter diversam quantitatem bonitatis diversorum, dummodo referuntur ad aliquod unum bonum commune:* sed honor diversificatur secundum propria bona singulorum. Unde *eodem amore caritatis diligimus omnes proximos, inquantum referuntur ad unum bonum commune, quod est Deus:* sed diversos honores diversis deferimus, secundum propriam virtutem singulorum. Et similiter Deo singularem honorem latriae exhibemus, propter ejus singularem virtutem.

Here is Vasquez' paraphrase of this text:

> . . . S. Thomas in eo a. l. ad 2. assignare nititur discrimen inter charitatem et religionem, ut defendat, etiamsi diligatur homo eadem charitate propter Deum, numquam coli eadem religione propter ipsum. Asserit ergo, amorem versari circa bonum in universum: honorem vero circa bonum proprium ejus, quem colimus: ideo amorem non esse speciei diversum, propter diversas bonitates particulares, dummodo omnes illae referantur ad aliquod unum commune bonum. Quocirca, inquit, charitate eadem diligimus, Deum, et proximum, etiamsi respiciamus in proximo diversam bonitatem particularem. Quoniam bonum proximi, quod ei volumus, ad commune bonum, quod est Deus, refertur. Honor vero distinguitur ex diversis bonis particularibus eorum, quos honoramus; quia in bonum particulare semper tendit; ac proinde, cum latria respiciat bonum Dei, dulia autem bonum hominis particulare: fit, ut cultus Dei, et hominis, ad diversas virtutes, non ad unam religionem pertineat.

The relevant section of Vasquez' criticism immediately follows the above paraphrase:

> Ego sane fateor meam ingenii tarditatem. Vix enim discrimen hoc mente concipere possum, nedum de illo judicium ferre, in ea tamen doctrina, *primum illud mihi est difficile quod asserit, amorem tantum ferri in bonum universe, cum revera etiam particulate ipsius dilecti respicere possit.*

> *Deinde non satis apparet quo pacto bonum particulare, quod volumus proximo, cum ipsum diligimus, referatur in bonum universum, quod est Deus: sed cultus particularis sancti in cultum et honorem ipsius Dei non referatur;* cum verum sit, eum, qui martyrem adorat, ipsum quoque Deum et dominum martyris quodammodo adorare. . . .[104]

Vasquez does not seem to realize that when the expressions *bonum commune* or *bonum in communi* are used to mean *bonum commune in praedicando,* St. Thomas expressly denies that God is a common good. Father Eschmann attacked me with the text in which St. Thomas uses *bonum commune* in this sense (*in praedicando*): "Bonum commune non est objectum caritatis, sed summun bonum" (*DM,* 195). As I have already pointed out, my Opponent in using this text to deny that the object of charity is *bonum commune in causando* must suppose that St. Thomas is taking the expression in the latter sense. By this interpretation he not only distorts the meaning of the *littera Sancti Thomae,* but at the same time denies that "eodem amore caritatis diligimus omnes proximos, inquantum referuntur ad unum bonum commune, quod est Deus." When he does allow that God is *bonum commune in causando*—not however as the object of charity—he actually reduces even this common good to a *bonum commune in praedicando,* for, in his opinion, God is a common good only insofar as He is the proper good of this person and of that person.

John of St. Thomas' answer to Vasquez applies to my Opponent as well:

> . . . Respondetur non dixisse S. Thomam, quod amor fertur in bonum universe, hoc est, in bonum universale in essendo, et praedicando (crassa est haec intelligentia) sed fertur amor in bonum particulare singulare, imo ad personam cum qua habetur amicitia. Dicitur autem ferri in bonum in communi communitate causalitatis, non praedicationis, quatenus scilicet bonum est diffusivum sui, et potest esse ratio formalis objectiva, non solum diligendi se, sed etiam aliud per respectum ad se.[105]

Vasquez equally fails to distinguish the divine common good as the effective principle of the created goods, from the common good as that to which the created goods are referred as to their final cause. John of St. Thomas' answer is again to the point:

. . . Divinae bonitatis communicatio dupliciter consideratur, et effective, et respective: *effective* quatenus se, vel dona sua creata communicat hominibus, et sic ponit in eis bonitatem intrinsecam, ratione cujus homo est diligibilis ex propria perfectione: *respective* quatenus homo praecise consideratur ut aliquid Dei, et quasi ab extrinseco diligibilis redditur; et hoc modo Deus non consideratur ut bonum concupitum homini, vel donatum illi, sed consideratur ut principale objectum diligibile, et per respectum ad illum diligitur proximus, sicut medicina diligitur ratione salutis, quatenus non attenditur ibi alia ratio diligendi, quam bonitas principalis objecti, et finis non communicata intrinsece mediis, sed respecta a mediis.[106]

It remains for us to consider the last lines of Father Eschmann's section III ("Professor De Koninck's Notion of God"). They are the continuation of the text quoted above:

Professor De K. has, throughout his treatise, neglected these fundamental considerations. On the very first page of the treatise proper (p. 8) he has omitted to pay due attention to St. Thomas' words: "Dicitur autem hoc [scilicet bonum commune] esse 'divinius' eo quod magis pertinet ad similitudinem Dei, qui est ultima causa omnium bonorum."[107] Obviously the words "qui est ultima causa omnium bonorum" are, in St. Thomas' mind, restrictive; and if the famous principle, "Sanctus Thomas formalissime loquitur" ever finds its application, it surely does so here. Let us paraphrase: Aristotle gives to a common good the attribute "divine," because this good, being the cause of the particular goods contained in its order and sphere, is *in this respect* more like God insofar as God is the cause of any and every good. There is, however, another respect to which the above text gives no consideration. This is the likeness to God *in linea essendi*. And in this respect the speculative intellect being, in the beatific vision, informed by God and most intimately united with Him, is by far superior to anything which is like God *in ordine causandi*. St. Thomas explicitly states:

Similitudo intellectus practici ad Deum est secundum proportionalitatem, quia scilicet se habet ad suum cognitum [the highest object of the practical intellect is a common good—II-II, 47. 12], sicut

> Deus ad suum. Sed assimilatio intellectus speculativi ad Deum est secundum unionem vel informationem: QUAE EST MULTO MAIOR ASSIMILATIO.
>
> These last words are the most concise and the most explicit statement of what we now call Personalism. For, is not this act and good of the speculative intellect a personal good?
>
> Professor De K. has constantly bypassed this most essential thesis of Thomistic ethics and, indeed, of Thomism as a whole. (*DM*, 197)

Father Eschmann cannot have read carefully "the very first page of the treatise proper," and has perhaps neglected to read in its entirety the text of St. Thomas to which he refers and which I also quoted. The very first lines of what my Opponent calls "the treatise proper" are:

> Le bien est ce que toutes choses désirent en tant qu'elles désirent leur perfection. Donc, le bien a raison de cause finale. Donc, il est la première des causes, et par conséquent, diffusif de soi. (*BC*, 7)

And the first lines of St. Thomas' text are:

> Manifestum est enim, quod unaquaeque causa tanto prior est et potior quanto ad plura se extendit. Unde et bonum, *quod habet rationem causae finalis,* tanto potius est quanto ad plura se extendit.[108]

To overlook these texts is to neglect the fundamental consideration. Neither Aristotle nor St. Thomas is here speaking of the common good "secundum quod importat operationem causae efficientis," but of the common good as a final cause, therefore, "secundum quod importat habitudinem causae finalis." The "ultima causa omnium bonorum" cannot be taken as anything but the final cause of all created good.[109] The created common good is "more divine" than a proper good of the same order for the very reason that it is a more perfect imitation of the ultimate final cause which draws all things to itself.[110] To suit Father Eschmann's purpose, "the words 'qui est ultima causa omnium bonorum' are, in St. Thomas' mind, restrictive; and if the famous principle, 'Sanctus Thomas formalissime loquitur' ever finds its application, it surely does so here" (*DM*, 197). In other words, when St. Thomas, in this very text, speaks of the good "quod habet rationem causae finalis," he actually means (just for the sake of speaking *formalissime*) something quite different, namely the good as an "effective" cause of other goods!

Nor does the next development of Father Eschmann's thought seem to represent much of an improvement. The ambiguity of the passage beginning: "There is, however, another respect . . ." makes it difficult to discuss, but whatever interpretation we put upon it, no acceptable doctrine emerges. If he means that formal beatitude, which is a purely personal proper good, is greater than any created good considered as the cause of another good, we emphatically agree: "beatitudo . . . quantum ad actum, in creaturis beatis, est summum bonum, non simpliciter, sed in genere bonorum participabilium a creatura."[111] But what does this prove except that some created proper good may be better than some created common good? He surely cannot be turning this conclusion against me when I have so plainly exposed this particular brand of sophistry.

> La plupart de ces objections jouent donc sur la transgression des genres, elles exploitent le *par accident.* De ce que quelque bien privé est meilleur que quelque bien commun, comme c'est le cas de la virginité meilleure que le mariage, on conclut que quelque bien privé pris comme bien privé est meilleur que quelque bien commun pris comme bien commun; que le bien privé comme tel peut avoir une éminence qui échappe au bien commun comme tel; qu'on peut dès lors préférer un bien privé à un bien commun, parce qu'il est privé. Nier par cette voie tous les premiers principes, quoi de plus facile? (*BC*, 71)

The good which we maintain is greater than the personal good of the Blessed is not a common good of an inferior order but the common good of objective beatitude.

> . . . Il est très vrai que la vie spéculative est solitaire, mais il reste vrai aussi que, même la béatitude souveraine qui consiste dans la vision de Dieu, est essentiellement bien commun. Cette apparente opposition entre la vie solitaire et le bien commun qui est l'objet de cette vie s'explique du fait que cette félicité peut être considérée, soit de la part de ceux qui en

> jouissent, soit de la part de l'objet même de cette félicité. Or, cet objet est, de soi, communicable à plusieurs. (*BC*, 62)

If Father Eschmann should claim that this is not his argument against me, he would have but one alternative: an even more crass sophism. I maintain that objective beatitude can never be other than a common good of the supernaturally elevated person. He tries to prove that it is not. What is the reason he offers? Formal beatitude is a strictly proper good of the person. Does he infer from this that objective beatitude is also a proper good of the created person? This would be a wretched sophism begging a real identity of our formal and objective beatitude; their distinction would be one of reason only. It would mean that the formal beatitude of the creature is wholly commensurate with its objective beatitude; that its formal beatitude is identical with the formal beatitude of God Himself, and that in the beatific vision God and the creature are identified. Then, indeed, God would not be a *bonum universale in causando* in the strict sense of cause. His essence would not inform the created intellect as a formal extrinsic cause; His goodness would not be "realiter causa finalis" of the created will; He would not even be "quoddam bonum commune" in Father Eschmann's sense.

The reader will recall my Opponent's indictment:

> In setting up a "principle of the New Order" Professor De K. has done a work which is—shall we say—surprisingly radical and daring: he has at the same time taken in his stride a new foundation of Christian ethics and moral theology. (*DM*, 196)

Should any doubt remain in the reader's mind, or should he feel, perhaps, that the absurd positions to which we have reduced our Opponent's principles are merely laborious inferences, let him weigh the following paragraphs:

> *Objectively*, i.e. viewed from the part of its uncreated object, the vision is not a common good; it is not even God as Common Good (to speak of common good in a proper and adequate language) but it is God Himself, the *Bonum universale in essendo*, as has been shown above.
>
> *Formally*, i.e. viewed as a created act and good, the vision is that supreme, personal good by which a created intellect, elevated by the light of glory, is most intimately united with, and most perfectly likened to, God.

> With these two elements the *essence* of the vision and of final beatitude is fully circumscribed. No further element needs to be added. No further element pertains to the *intrinsic nature* of final beatitude.
>
> *Extrinsically,* however, i.e. in virtue of the fact that there is a multitude of the Blessed sharing, as it were, in the same good, the vision might be called a certain common good which, then, is the constitutive of a certain "society," a society which St. Augustine has called *societas fruendi Deo et invicem in Deo.*[112] With regard to this "society" all that St. Thomas has to say is that it *quasi concomitanter se habet . . . ad perfectam beatitudinem*[113] because, speaking of the essence of things, every single "member" of it has his full sufficiency in God and in God alone. (*DM,* 202–3)

A simplistic idea could hardly be more simply stated.

VII. "The Chief 'Personalist' Text"

Let us examine closely what Father Eschmann calls "the chief 'personalist' text" (*DM,* 206), "the most concise and the most explicit statement of what we now call Personalism" (*DM,* 197): *quae est multo major assimilatio.* We must attempt to determine, first why this text is produced; why the parenthesis is inserted; what the composite of quotation and parenthesis proves; and, finally, how it may be taken to contradict my position. Father Eschmann desires to show that God, as the object of beatitude, cannot be a common good. Now, if such is to be his conclusion from the quotation and parenthesis, it can follow only from an argumentation which, simplified to its utmost, will go something like this:

I. The object of the practical intellect is an operable good.[114] But, the common good is the highest object of the practical intellect. Therefore, the common good is an operable good.

II. The operable good is not an object of the speculative intellect. But the common good is an operable good. Therefore, the common good is not an object of the speculative intellect.

III. The common good is an operable good. But God is not an operable good.[115] Therefore, God is not a common good.

IV. The assimilation of the speculative intellect to God is not a common good. But beatitude is "assimilatio intellectus speculativi ad Deum." Therefore, beatitude is not a common good.

Our answer will be brief. We distinguish the minor of the first two arguments and contradistinguish their conclusion: The common good which is the highest object of the practical intellect is the common *operable* good, not the common good which is an intelligible end.[116] The same distinction applies to the major of the third argument, and to its conclusion; the minor of this argument we concede. We concede the major of the last argument, and contradistinguish the minor and the conclusion: if beatitude is taken as it is in the major, i.e. formal beatitude, we agree; if taken to mean the objective beatitude of the creature, we deny.

Father Eschmann may object to the form in which the minor of the first argument is cast: for it states the common good to be the highest object of the practical intellect, whereas his parenthesis ran: "the highest object of the practical intellect is a common good." But the point is that unless he accepts this statement of his premise, he cannot possibly reach that conclusion. It is the interpretation he must put upon his own words.

The text in question might be used to show that formal beatitude is a proper personal good, since the assimilation to God is an operation of the intellect of the Blessed.[117] But if my Opponent merely intends to prove that "this act and good of the speculative intellect [is] a personal good" (*DM,* 197), in using this text, he is following the most roundabout way one could imagine, and to no purpose, for no one has denied that formal beatitude is a purely personal inherent good of the Blessed. But this is not the end of the matter, for "Actus . . . noster non ponitur esse beatitudo, nisi ratione suae perfectionis, ex qua habet quod *fini exteriori* nobilissime conjungatur; et ideo nostrae beatitudinis non sumus nos causa, sed Deus."[118] It is that *finis exterior,* the formal and final cause of beatitude, that we are concerned to explain and defend. And it is this end which Father Eschmann does not want to be a common good.

While Father Eschmann's quotation proves absolutely nothing either for his position or against our own, there does exist a certain correspondence between this text and that on the preceding page which he takes from Ia, q. l 03, a. 4, c. But the correspondence is a disturbing one for my Opponent's method as well as for his argument. Let us place side by side the two passages concerned: first that from which is drawn the excerpt he gives on page 196, and secondly the entire *responsio* of St. Thomas from which he has selected the quotation on page 197. But first we shall do well to read the objection relatively to this *responsio:*

> . . . Videtur quod beatitudo consistat in operatione intellectus practici. Finis enim ultimus cujuslibet creaturae consistit in assimilatione ad Deum. Sed homo magis assimilatur Deo per intellectum practicum, qui est causa rerum intellectarum, quam per intellectum speculativum, cujus scientia accipitur a rebus. Ergo beatitudo hominis magis consistit in operatione intellectus practici quam speculativi.

And now the two above-mentioned texts:

Ia, q. 103, a. 4, c.	IaIIae, q. 3, a. 5:
Effectus igitur gubernationis potest accipi tripliciter. [a] Uno modo ex parte ipsius finis: et sic est unus effectus gubernationis, scilicet assimilari summo bono. [b]—Alio modo potest considerari effectus gubernationis secundum ea quibus ad Dei assimilationem creatura perducitur. Et sic in generali sunt duo effectus gubernationis. Creatura enim assimilatur Deo quantum ad duo: [i] scilicet quantum ad id quod Deus bonus est, inquantum creatura est bona; et [ii] quantum ad hoc quod Deus est aliis causa bonitatis, in quantum una creatura movet aliam ad bonitatem. Unde duo sunt effectus gubernationis: scilicet conservatio rerum in bono, et motio earum ad bonum. [c]—Tertio modo potest considerari effectus gubernationis in particulari: et sic sunt nobis innumerabiles.	Ad primum ergo dicendum quod [a] similitudo praedicta intellectus practici ad Deum, est secundum proportion alitatem; quia scilicet se habet ad suum cognitum, sicut Deus ad suum. [b] Sed assimilatio intellectus speculativi ad Deum, est secundum unionem vel informationem; quae est multo major assimilatio.—Et tamen dici potest, quod respectu principalis cogniti, quod est sua essentia, non habet Deus practicam cognitionem, sed speculativam tantum.

Clearly, the "similitudo secundum proportionalitatem" in part [a] of the second text is related to [b] of the first, and the practical intellect exemplifies what is said in its subdivision [ii]. It is also clear that the "assimilatio intellectus speculativi ad Deum secundum unionem vel informationem" in [b] of the second text is related to [a] of the first (the clause which our Opponent ignores) and not to subdivision [i] as Father Eschmann supposes when he says: "This is the likeness to God in *linea essendi*" (*DM*, 197).

Now, when my Opponent considers the likeness to God in *linea essendi* as opposed to what he understands by *in ordine causandi,*[119] he cannot formally consider the assimilation of the speculative intellect to God "secundum unionem vel informationem"; he cannot be considering it under the formality of union and information, but rather under the formal aspect of inherent perfection of the creature. This consideration is quite legitimate.[120] But when we do consider this aspect of the likeness to God, there cannot be, in this precise respect (that is, *in linea essendi*), any question of assimilation to God "secundum unionem vel informationem," although it is because of the union or information that there is a likeness *in linea essendi.* In other words, the likeness to God is, in this respect, only a "similitudo secundum proportionalitatem," as in the case of the practical intellect. True, even in this respect, it is a much more perfect likeness to God than that of the practical intellect, yet as being merely proportional it is confined to the same genus. But we hasten to add that there remains still the most essential difference between the speculative and the practical intellect: whereas the practical intellect can be only a likeness to God "secundum proportionalitatem," the speculative intellect can be, profoundly and uniquely, a likeness "secundum unionem vel informationem." This is, indeed, the basis for St. Thomas' distinction in the second text. And we now see just what is meant by *multo major assimilatio.*

When we confine ourselves then to the assimilation to God *in linea essendi* (thus prescinding from the formal extrinsic cause which is absolutely essential in the strictest sense when we speak of assimilation "secundum unionem vel informationem") we remain within the genus of likeness "secundum proportionalitatem." Thus Father Eschmann has rather missed the point. Whereas his intention was to show the radical difference between the speculative intellect and the practical, he, in fact, does not use the distinction he quotes from St. Thomas. Although he does not seem to realize it, having confined himself to the *linea essendi,* he can never reach anything higher than the genus of likeness "secundum proportionalitatem."

Personalists have been deeply moved by my Opponent's "quae est multo major assimilatio," printed in capital letters. Yet it is difficult to understand how this may be called "the chief 'personalist' text" and "the most concise and the most explicit statement of what we now call Personalism." It is very true that the "likeness to God *in linea essendi*" is a wholly personal good (such a likeness is common good only *in praedicando*). But this is completely beside the point. The question is: Is God a common good? Is objective beatitude a common good?

Yet, *per aciddens,* Father Eschmann has made a distinct contribution. In misinterpreting the *littera Sancti Thomae,* and in clouding the distinction between the object of beatitude and the act in its relevance to our problem, he has done a good deal to clarify the issue between "Personalism" and the primacy of the common good.

VIII. Beatitude, "The" Common Good

Section IV ("Professor De Koninck on Beatitude") of my Opponent's article is the one which breathes the most confidence, and which is obviously meant to deliver the *coup de grâce.* Given his notions of part and whole, of the good and of the common good, of charity and of beatitude, together with his remarkable ease in dealing with the *littera Sancti Thomae,* we can appreciate that it is difficult for him to feel anything but invincible. I venture to add, though, and for the same reasons, that it is equally difficult for him to read my text, much less explain it.

I might further suggest a possible oversight. The Historical Point of View draws its life-blood from the safe absence of the authors it expounds and judges, this being the most imperative condition of its freedom. I, however, am still present to disclose and to protest against my Opponent's distortion of the position I uphold. It may be doubted whether such assurance as his can admit this distinction. But in any case, the misrepresentation of my position is of small consequence in comparison with the doctrine he continues to advance in the name of St. Thomas.

The reader will recall that according to Father Eschmann it is contrary to the very nature of person to be a part of society because a person cannot be part of what is substantially one. But, although contrary to its nature, a person can yet, somehow, be such a part of society. I quote again from *DM*, 205:

> *Ratio partis contrariatur personae.*[121] Hence . . . the person, qua person, is not a part of society: and if a person is such a part, this "being part" will not be based upon the metaphysical formality and precision of "being person."

As has been pointed out already, the *ratio partis* St. Thomas indicates in the text my Opponent quotes, is that of a *natura per se.*[122] Our contention is that no person can be such a part because this, being contrary to the very nature of person, will imply an absolute contradiction. But what we should like to call attention to at this juncture is Father Eschmann's notion of society. If he understands and means what he says, he is maintaining that a society is substantially one. And however preposterous this may sound, it is nevertheless quite in keeping with his strange notion of common good,[123] namely, that it is a good not immediately and personally possessed by him who shares in it.

Such being the case, it is strange he does not emphasize that my own notion of common good is contradictory since I insist that the perfect common good is immediately shared by each person of the community, that the one and the other attain it in its very universality; while I yet uphold such a good as a true common good.

> . . . L'universalité même du bien est principe de béatitude pour la personne singulière. C'est, en effet, en raison de son universalité qu'il peut béatifier la personne singulière. (*BC*, 26)

Or at least he might have tried to show that the following statement is contradictory:

> L'indépendance des personnes les unes des autres dans la vision même n'exclut pas de l'objet cette universalité qui veut dire, pour toute intelligence créée, essentielle communicabilité à plusieurs. Loin de l'exclure, ou d'en faire abstraction, l'indépendance présuppose cette communicabilité. (*BC*, 62)

While he holds that my notion is contradictory (which means, of course, that it is not a notion at all) he has nowhere even remotely tried to show this to be so. However, he persistently confuses my interpretation with his own, and this makes it quite easy to impose upon me the contradictions which follow from his understanding. Section IV of Father Eschmann's article is a striking instance of this procedure. He assumes that what I mean by part, whole, society, and common good are what he means by these terms. As a result, my text, quite logically, is converted into a maze of contradictions.

The misinterpretations in this section IV deserve exposition in some detail. Its opening paragraph is the following:

> Ever since the days when Plato stated the problem of the philosophers and kings, every occidental theory of society has ultimately proved its truth and its value by the regard it has paid to, and the place it has left open for, that which is not society nor action, viz. solitude and contemplation. The modern problem which we are now accustomed to state in terms of Person and Society is nothing but the continuation of the age-old discussion of Philosophers and Kings. (*DM*, 197)

In this connection my Opponent has failed to inform his readers that, not to speak of repeated assertions in the essay he attacks, the second part of my book is entirely devoted to showing the disastrous consequences of *La négation de la primauté du spéculatif* (*BC*, 85–123). And even while quoting my own text, he will argue as if I denied these irrefutable truths: that *the ultimate end of the person consists in the vision of God*, that *the speculative life is solitary*, and that *the persons are independent of one another in the vision* (*BC*, 61–62). At times one wonders what type of reader Father Eschmann has in mind. If he presumes, as he surely must, that his reader knows what I actually say (he quotes the text), then the only reasonable thing for him to do would be to point out, simply and clearly, that in maintaining these essential truths I utterly contradict my own position concerning the primacy of the common good. He should not speak as if I denied them. Obviously, the proper course would have been harder to follow than that which he has chosen—it would have compelled him to face the notion of common good.

Immediately following the first paragraph of section IV my Opponent proceeds:

Professor De Koninck will already have surprised the attentive reader by the statement quoted above, that our beatitude is a common good ("le bien commun qu'est la béatitude," p. 26). Let us have a closer look into this statement.[124]

On page 61 the author composes (one might be tempted to say concocts) the following "objection" against his thesis of the absolute primacy of "the" common good:

> L'ordre pratique est tout entier ordonné à l'ordre spéculatif. Or, le bonheur parfait consiste dans la vie spéculative. Mais, la vie spéculative est solitaire. Donc, le bonheur pratique de la societé est ordonné au bonheur spéculatif de la personne singulière.

Professor De K.'s answer to this "objection" is as follows:

> Nous répondons que le bonheur pratique de la communauté n'est pas, par soi, ordonné au bonheur spéculatif de la personne singulière, mais au bonheur spéculatif de la personne en tant que membre de la communauté. [Here is quoted Petrus de Alvernia, *In VII Pol.*, lect. 2.][125] Il serait, en effet, contradictoire qu'un bien commun fût, de soi, ordonné à la personne singulière comme telle. Il est très vrai que la vie spéculative est solitaire, mais il reste vrai aussi que, même la béatitude souveraine qui consiste dans la vision de Dieu, est essentiellement bien commun. Cette apparente opposition entre la vie solitaire et le bien commun qui est l'objet de cette vie s'explique du fait que cette félicité peut être considérée, soit de la part de ceux qui en jouissent, soit de la part de l'objet même de cette félicité. Or, cet objet est, de soi, communicable à plusieurs. Sous ce rapport, il est le bien spéculatif de la communauté. Le bien commun pratique doit être ordonné à ce bien spéculatif qui s'étend comme bien commun aux personnes. L'indépendance des personnes les unes des autres dans la vision même n'exclut pas de l'objet cette universalité qui veut dire, pour toute intelligence créée, essentielle communicabilité à plusieurs. Loin de l'exclure, ou d'en faire abstraction, l'indépendance présuppose cette communicabilité.

Is this somehow "magisterial" *Nous répondons* in conformity with Master Thomas' famous *Respondeo. Dicendum?* (*DM*, 198)

I trust the reader will agree that in this passage I am stating as explicitly as I can that what I mean by the speculative good of the community is none other than the object of beatitude, and that the apparent opposition between the solitude of the speculative life and the community of its object is due to a failure to distinguish beatitude on the part of those who enjoy it from beatitude which is the very object. Father Eschmann, though he will mention the distinction, completely ignores its relevance to our problem.

Our formal felicity is not *beatitudo per essentiam,* but by participation and hence cannot be equal to its cause—objective beatitude. In its incommensurable communicability to many, objective beatitude is numerically one.[126] That it is actually communicated to many does not affect it intrinsically. Even for the creature, the respect of *excedens* et *excessum* remains entirely the same. It is for this reason that, as we have already shown, the divine good can only be compared to the creature as the good of the whole to the part, whether other creatures actually exist or not.

Let me attempt to convey this vital truth in terms more unmistakable still. When St. Thomas says that we must love God more than ourselves because He is the "*bonum commune omnium,*" he does not mean that we must love God more than ourselves because He happens to be also the good of this person and of that person, but because He *is,* by His own goodness, "the" common good. And that is why St. Thomas can say in a text prescinding from the actual existence of any neighbour that we must love God more than ourselves because He is the common good of all. It is for the same reason no created person dare think of the divine good as ordered to himself (which he most certainly should do if God were his proper good) but must rather see himself as ordered to God. Let me quote again the second answer from the article: *Utrum homo debeat ex caritate plus Deum diligere quam seipsum.* The objection is based on the assumption: "unumquodque diligitur inquantum est proprium bonum."

> Ad secundum dicendum quod bonum totius diligit quidem pars secundum quod est sibi conveniens: non autem ita quod bonum totius in se referat, sed potius ita quod seipsam refert in bonum totius.[127]

Does this not make it plain that our own personal speculative felicity must be ordered to its object as to the common good?

The divine good is a common good, not in that it is communicated, but in that it is communicable, to many. It would be unspeakably foolish to think that, because there are many persons participating in the divine good, the object of beatitude and charity is in any way divided or altered, or the immediacy in attaining and loving it is in any way interrupted. Indeed, only because we already love God as the "bonum commune omnium," shall we, consequently, love this and that neighbour. If we did not love our neighbour; if the fact that he too shared in the same numerical good, perhaps to a much greater extent, were either indifferent or repugnant to us, it could only be because we did not love the divine good as a common good, that is, because we would be placing above all else our singularity, and hence, the proper good.

This is the consideration which sustains my answer to the objection which my Opponent terms "concocted." It is an objection which has been often made, however, and which is supposedly based on book X of the *Ethics,* where Aristotle holds that the man of wisdom, "even when by himself, can contemplate truth, and the better the wiser he is" (c. 7, 1177a30) and also on the Thomistic acceptation of this teaching, which loyally follows the Philosopher. But the felicity in question here is formal felicity; while our problem turns on the one that Father Eschmann has chosen to ignore—objective felicity. Now the question to which we have been unceasingly trying to direct attention is simply, which of these two is the ultimate end? Must the person order himself to objective felicity, or objective felicity to himself? If he is to order himself to objective felicity, that will only be because the latter is not his proper good. If it is not a proper good and yet a good, it can only be a common good.

But when the principle is maintained that the person must order himself to his ultimate end as to a common good, in no way does it follow that this must be a mass movement, so to speak, by the community as a whole. Nothing obliges us to draw such an absurd conclusion, and I must once more protest against my Opponent's attribution of it to me. Throughout my work I have made it clear that our neighbour does not share in this ordering and that it is rather the task of each individual person. I leave it to my reader then to understand, if he can, how Father Eschmann, after my emphatic statement "Il est très vrai que la vie spéculative est solitaire" and my insistence on "l'indépendance des personnes les unes des autres dans la vision même" can proceed without making any distinction whatever (even after quoting my text!). He should at least, I repeat, allow that when I speak of solitude and independence I am right, although self-contradictory when I hold that these can have anything to do with the common good. But we shall search in vain for any such remark; he simply continues as if I maintained that the *assecutio* of this common good is an *assecutio communis* as opposed to the *assecutio singularis* of the speculative intellect (*DM*, 200).

IX. "Fidelissimus discipulus ejus"

Let us return to Father Eschmann's text.

> Is this somehow "magisterial" *Nous répondons* in conformity with Master Thomas' famous *Respondeo. Dicendum?*
>
> The "Thomistic" basis for the author's answer is not St. Thomas but Peter of Auvergne. The quotation from this continuator of St. Thomas' *Commentary on the Politics* is here all the more surprising since for the point in question a rich and authentically Thomistic documentation was at hand. It is, indeed, a fact as un-understandable to any serious Thomistic scholar as it is characteristic for Professor De K.'s scientific methods that at a juncture where the most proper and important point of the whole discussion is under debate—*hic Rhodus, hic salta!*—the author completely forgets about St. Thomas. The reader is avid to get good Thomistic bread, but he must content himself with *Ersatz.*
>
> Peter of Auvergne, as is well known, is a secular priest, a member, in the last decades of the thirteenth century, of the Faculty of Arts at the University of Paris and, at one time, a disciple of St. Thomas, whose lectures he attended in Paris, somewhere between 1269 and 1273. Although, because of his general doctrinal outlook, there is no doubt that he must be counted among the representatives of the oldest Thomist school, nevertheless, in every question of detail the quality of his Thomism is a matter, not of assumption, but of examination. For it is not impossible that the Averroistic atmosphere of the Parisian Artists might somehow have colored his doctrine, as it happened, not infrequently in those times, for instance and especially, in the case of another Parisian Artist, John Quidort, O.P. As long as the notion of a doctrinal *source* retains any

> proper and intelligible meaning, it is surely impossible to use Peter of Auvergne unqualifiedly as a Thomistic source; and, let it be noted, the same applies, of course, to Cajetan, John of St. Thomas, etc., commentators whom Professor De K. puts, without any distinction, on equal footing with St. Thomas himself. (*DM*, 198–99)

But, as I have mentioned already, my reply to that "concocted" objection is in no sense dependent on the authority cited in the footnote. To give the proper argument for a doctrine and to refer to an author as confirming it are not quite the same thing. It is probably what Father Eschmann has already done with the *littera Sancti Thomae* which now makes it necessary for him to believe that the text of Peter of Auvergne is the only possible Thomistic basis for my reasoning; nevertheless, it is my duty to humbly and stubbornly maintain that this same reasoning is founded on nothing other than the plain words of the Angelic Doctor.

The first two paragraphs of the passage I have quoted need no remark. But a word on Peter of Auvergne, and a few observations on my use of other commentators may perhaps be called for. Who was this Peter of Auvergne? Let it be noted that he was not just another disciple who attended the lectures of St. Thomas. Ptolemy de Luca, the man who was both disciple and confessor of St. Thomas, calls him, with reference, as it happens, to this same commentary on the *Politics, fidelissimus discipulus ejus.*[128] As to my dependence on other authors, a scrutiny of the essay which Father Eschmann attacks will reveal that it contains a single quotation from Cajetan (a mere paraphrase), and five quotations from John of St. Thomas. Of the latter, only two actually appear in my own text: the first being a passage which notes the obvious distinction between common good and alien good; the second to show even the temporal common good must be publicly ordained to God. But even if my references to such authorities were as numerous and as important as my critic seems to imply, I could not think myself obliged to apologize for them. Not only do I admit without hesitation a need for the assistance of these great minds, but in relying on them I think I am obeying an authority which not even Father Eschmann would be inclined to reject. In a later chapter we shall see why my Opponent has good reason for urging the reader to be on guard against these famous theologians.

Granted that I do not use the authority of Peter of Auvergne as the basis for my argument, there remains the question why I refer to him at all. The reason is simple enough. They who infer from Aristotle's *Ethics* that, since the speculative life is solitary, there can be no question of a common good of the speculative life, should be reminded of a passage in the *Politics*, VII, c. 3:

> But if these things are well said, and if happiness is to be defined as well-doing, the active life is the best life both for the whole state collectively and for each man individually. *But the active life is not necessarily active in relation to other men, as some people think, nor are only those processes of thought active that are pursued for the sake of the objects that result from action, but far more those speculations and thoughts that have their end in themselves and are pursued for their own sake;* for the end is to do well, and therefore is a certain form of action. And even with actions done in relation to external objects we predicate action in the full sense chiefly of the master craftsmen who direct the action by their thoughts. (1325b14–23)[129]

Yet this single passage remains obscure. A consideration made in chapter 2 of the same book will help to determine its meaning:

> On the other hand it remains to say *whether the happiness of a state is to be pronounced the same as that of each individual man,* or whether it is different. Here too the answer is clear: everybody would agree that it is the same; for all those who base the good life upon wealth in the case of the individual, also assign felicity to the state as a whole if it is wealthy; and all who value the life of the tyrant highest, would also say that the state which rules the widest empire is the happiest; and if anybody accepts the individual as happy on account of virtue, he will also say that the state which is the better morally is the happier. . . . Now it is clear that the best constitution is the system under which anybody whatsoever would be best off and would live in felicity; but the question is raised even on the part of those who agree that the life accompanied by virtue is the most desirable, whether the life of citizenship and activity is desirable or rather a life released from all external affairs, for example some form of contemplative life, which is said by some to be the only life that is philosophic. For it is manifest that these are the two modes of life principally chosen by the men most ambitious of excelling in virtue, both in past times and at the present day—I mean the life of politics and the

> life of philosophy. And it makes no little difference which way the truth lies; for assuredly the wise are bound to arrange their affairs in the direction of the better goal—and this applies to the state collectively as well as to the individual human being. (1324a5–35)

So we see that, in this work, while expressly repeating the doctrine of his *Ethics*, Aristotle nevertheless refers to a contemplative happiness of the community. He does not explain here how there can be such a thing; he does not state the principle. However, the principle which justifies his statement does exist, and it is my claim that upon that principle my own argument is founded.

Now, there exists a Thomistic commentary on this very text, a commentary by Peter of Auvergne, *fidelissimus discipulus* of St. Thomas whose lectures he attended. To my mind, it is distinctly "not impossible" that much of this commentary should reflect what he heard from St. Thomas himself.[130] Yet why should we go into such a matter? Whether the disciple is an authority or not is surely no question to detain us; our real task is simply to inquire if what he teaches makes good sense. Let us examine what he has to offer us:

> . . . Felicitas est operatio hominis secundum intellectum. In intellectu autem est considerare speculativum, cujus finis est cognitio veritatis, et practicum cujus finis est operatio. Et secundum hoc duplex felicitas assignatur hominis. Una speculativa quae est operatio hominis secundum virtutem perfectam contemplativam quae est sapientia. Alia autem practica quae est perfectio hominis secundum perfectam virtutem hominis practicam quae est prudentia. Est autem quaedam operatio secundum prudentiam et speculatio secundum sapientiam hominis secundum seipsum solum. Et est quaedam operatio prudentiae et speculatio totius civitatis; et ideo est quaedam felicitas practica et speculativa quaedam hominis secundum seipsum, et est quaedam felicitas practica totius civitatis et quaedam contemplativa totius civitatis. Felicitas autem speculativa secundum unum hominem melior est practica quae est secundum unum hominem, sicut evidenter docet Aristoteles in decimo *Ethicorum;* quoniam illa perfectio intellectus eligibilior est quae est respectu objecti magis intelligibilis, quia ratio perfectionis sumitur ex objecto; talis autem est speculativa. Felicitas enim est perfectio intellectus

respectu primi et maxime intelligibilis. Felicitas autem practica est perfectio intellectus respectu agibilis ab homine quod multo deficit a ratione intelligibilis primi; ergo felicitas contemplativa unius eligibilior est quam felicitas practica; et iterum magis est continua et sufficiens et delectabilis haec quam ilia. Et eadem ratione contemplativa totius civitatis eligibilior est quam politica seu civilis, et contemplativa totius civitatis simpliciter eligibilior est contemplativa quae est secundum unum; similiter civilis practica quae est secundum unum. Et hoc est quod intendebat dicere Aristoteles primo *Ethicorum;* si idem est uni et civitati, majusque et perfectius quod civitati videtur et suscipere et salvare. Amabile enim et uni; melius vero et divinius genti et civitati. Et ratio hujus potest esse, quia contemplativa et civilis civitatis comparantur ad contemplativam secundum unum, sicut totum ad partem: totum autem rationem magis perfecti et majoris boni habet quam pars; et ideo ista quam ilia. (lectio 2)

The reader will have noticed the most essential words of this text "ratio perfectionis sumitur *ex objecto;* talis autem est speculativa. Felicitas enim est perfectio intellectus respectu primi et maximi intelligibilis" as well as the quotation from I *Ethics,* c. l, 1094b8. To overlook these phrases, which show us the crucial importance of the object in any analysis of beatitude, is to be led of necessity into a hopeless misunderstanding of the entire passage. Now, throughout his discussion Father Eschmann has missed the relevance of the object; furthermore, as we have already seen, in rejecting the universality of the principle from *Ethics,* I, he does not appear to have quite grasped its meaning. It was inevitable, then, that he should have nothing but hard words for our *fidelissimus discipulus:*

> The clumsiness both of the notions themselves and of the whole procedure of combining and comparing them, is at once striking. (*DM,* 199)

Yet one cannot help thinking, however unlikely the possibility, that the writer of this harsh criticism cannot have read the whole of this second lesson. For Peter has already said:

> . . . Ab illis qui dicunt optimam vitam hominis esse in optima operatione virtutis, dubitatur utrum vita civilis quae consistit in communicatione

civili et activa quae consistit in directione vel ordine operationum quae sunt ad alterum, sit eligibilior, vel illa quae est absolute a turbatione civili et actionibus exterioribus magis, quam dicimus contemplativam, quam solam intendimus Philosophos intendere.

In an earlier paragraph he has stated:

Cum sint duae vitae hominis magis principales, scilicet practica et speculativa, quae istarum sit eligibilior: utrum illa quae consistit in communicatione civili in simul vivendo civiliter, scilicet activa, vel illa quae peregrina est et absolute ab hujusmodi communicatione civili, scilicet contemplativa. Vocat autem vitam contemplativam absolutem et peregrinam, quia principaliter consistit in applicatione hominis secundum intellectum ad primum objectum ejus et optimum, quae non potest esse sine sedatione motuum et perturbationum sine quibus non est vita civilis: et ideo oportet ipsam esse absolutam a communicatione civili; et per consequens peregrinam. Peregrinum enim dicitur quod longe ab habitudine consuete est. Magis autem consueta vita communiter est vita civilis.

There is also the following passage:

Optima autem vita hujusmodi injustitiam, quae magis accidit in civili communicatione, non habet, quia *optima vita nihil praeter rationem habet:* igitur optima vita non est civilis, sed absoluta magis.

And finally, in the same lesson, we read this admirable commentary on Aristotle's dictum that the free man is "cause of himself":[131]

. . . Sicut Philosophus dicit in primo *Metaphysicae,* liber est qui est suiipsius causa. Quod non potest intelligi sic quod aliquis sit causa suiipsius primo: nihil enim est causa sui: sed est intelligendum quod liber sit ille, qui secundum aliquid proprium sibi est causa sibi operandi. Et tunc veritatem habet quod liber est suiipsius causa in duplici genere causae: et in ratione agentis, et ratione finis. In ratione agentis, inquantum per aliquod principium quod est principale in eo operatur. In ratione autem finis, inquantum operatur ad finem sibi debitum secundum principium illud. Et quia homo maxime in esse constituitur per intellectum, est enim intellectus, vel maxime secundum intellectum secundum Aristotelem in decimo *Ethicorum,* et ideo homo liber dicitur, qui per virtutem intellectualem existentem in eo operatur non accipiens ab alio rationem operandi, nec impedimentum habens ex parte materiae; et qui operatur ad finem qui debetur ei secundum naturam praedictam. Et quanto magis natus est operari secundum illud quod perfectius est in intellectu in eo, et ad finem excellentiorem secundum idipsum, tanto liberior est. Et ideo qui simpliciter operatur secundum virtutem intellectualem, et ad finem secundum intellectum, perfectissime liber est.

Yet Father Eschmann, without an effort at the least distinction, can impose upon Peter, *fidelissimus discipulus* of St. Thomas, the stupid opinion that the contemplative life of the community is a "genuine social or common act," an "*assecutio communis*"! (*DM,* 200). But let us answer the question he raises:

> The clumsiness both of the notions themselves and of the whole procedure of combining and comparing them, is at once striking. For, what is this *operatio prudentiae totius civitatis?* And if, in spite of the manifest clumsiness of the terminology, an intelligible meaning might finally be discovered in this notion—what in the world can *speculatio totius civitatis* be? (*DM,* 199)

Since Peter's critic reluctantly allows that it is not impossible "an intelligible meaning might finally be discovered in this notion" of *operatio prudentiae totius civitatis,* we may pass at once to the second question.[132] Since Father Eschmann attacks Peter's *speculatio totius civitatis* chiefly in connection with supernatural beatitude, we too shall confine ourselves to this application. But what we shall find is that only St. Thomas' own doctrine explains its meaning.

The very letter of Peter's lesson 2 precludes an *assecutio communis* or "a genuine social or common act." In contemplation itself persons cannot share one another's ordination. The ultimate reason why such a thing is quite impossible must be found in the fundamental distinction between the practical and the speculative good: *Intellectus practicus ordinatur ad bonum quod est extra ipsum: sed intellectus speculativus habet bonum in seipso, scilicet contemplationem veritatis.*[133] In contemplation, considered as the act of the intellect, each person is more than anywhere else, *suiipsius causa,* as Peter explains. But the object of this act, be it noted, is not just any intelligible good, it is the very highest, the "primum et maxime intelligibile." Now, while the act of contemplation is proper to the knower, the object could not be proper to him, unless he himself were that object. Manifestly, this is out of the question; no finite intellect, not even the soul of Christ, could be thought of as adequate to the object of beatitude.[134]

And now we are again faced with a familiar problem: why did God in His goodness[135] and wisdom produce a manifold of intellects? The only acceptable reason is that He wished to communicate Himself abundantly, and that the communication of Himself to a single created intellect could not meet the greatness of His design. He has not chosen to manifest Himself merely to this person, but to many persons. In this respect it is the manifestation of Himself to the manifold which is His primary intention. This does not mean, however, that He manifests Himself to the manifold in such a way that, in this immediate manifestation, the many becomes, as it were one body reaching Him by an *assecutio communis* as opposed to *assecutio singularis,* for He obviously remains the object of this speculative intellect and that. But the Saints in their multitudes are not chosen by God merely that there may be a plurality. Each person is made for Him; no person is made for the other persons. He is not the good of a collectivist community. Yet He is the good of this and of that person. That He can be the good, the infinite good, of this and of that person is not accidental; it is His very nature, whether or not He makes only this person or only that. Hence, when He does make the one and the other, in no sense could He be called a common good *per accidens.* Yet such would be the case if the community of the divine good depended upon the existence of this *and* that person.

In domo Patris mei mansiones multae sunt.[136] The many mansions represent the formal beatitude of the Blessed. But the mansions are the chambers of a single heavenly home. In the passage which follows, St. Thomas shows us how this house of God may be understood in two ways.

> [a] . . . Cum uniuscujusque domus sit in qua habitat, illa dicitur domus Dei in qua habitat Deus; Deus autem habitat in sanctis; *Jer.* 14, vers. 9: *Tu in nobis es, Domine* etc. Sed in quibusdam quidem per fidem; II *Cor.*

6, vers.16: *Inhabitabo in illis, et inambulabo inter eos.* In quibusdam vero per fruitionem perfectam; I *Cor.* 15, 28: *Ut sit Deus omnia in omnibus.* Duplex est ergo domus Dei. Una est militans Ecclesia, scilicet congregatio fidelium; I *Tim.* 3, 15: *Ut scias quomodo oporteat te in domo Dei conversari, quae est Ecclesia Dei vivi.* Et hanc inhabitat Deus per fidem; *Apoc.* 21, 3: *Ecce tabernaculum Dei cum hominibus, et habitabo in illis.* Alia est triumphans, scilicet sanctorum collectio in gloria Patris; *Ps.* 64, 6: *Replebimur in bonis domus tuae. Sanctum est templum tuum, mirabile in aequitate.*

[b] . . . Sed domus Patris dicitur non solum illa quam ipse inhabitat, sed etiam ipsemet, quia ipse in seipso est. Et in hac domo nos colligit. Quod autem ipse Deus sit domus, habetur II *Cor.* 5, 1: *Domum habemus a Deo, non manufactam, aeternam in caelis.* Et haec domus est gloriae, quae est ipse Deus; *Jer.* 17, 12: *Solium altitudinis gloriae tuae a principio, locus sanctificationis nostrae.* Manet autem homo in hoc loco, scilicet Deo, quantum ad voluntatem et affectum per fruitionem caritatis; I *Joan,* 4, 16: *Qui manet in caritate, in Deo manet, et Deus in eo*: et quantum ad intellectum per notitiam veritatis; infra 17, 17: *Sanctifica eos in veritate.* In hac ergo domo, idest in gloria, quae Deus est, *mansiones multae sunt,* idest diversae participationes beatitudinis ipsius; quia qui plus cognoscit, majorem locum habebit. Diversae ergo participationes divinae cognitionis et fruitionis, sunt diversae mansiones.[137]

The mansions are mansions of the same house both in the first meaning of house and in the second, and anyone's formal beatitude is but a single mansion. God Himself dwells in each mansion, yet His dwelling in the house is more abundant than in any single chamber. It is because of the narrowness of the individual mansion that it cannot receive Him in the fullness with which He is received in the house. When He said to His disciples: *Where there are two or three gathered together in my name, there I am in the midst of them,* He did not mean that He is not then present to Peter as Peter, or to John as John; yet He is more fully present then to Peter and John than to either Peter or John alone. And this is the reason why, even in the present life, that is, in the house in which God dwells according to faith, *if two of you shall consent upon earth concerning anything whatsoever they shall ask, it shall be done to them by my Father who is in heaven.*[138]

So it is in beatitude. Both Peter and John know that it is better that He be present to both together. They see the infinite greatness of God is such that, in truth, it can never be fully manifested either to one or the other, nor to both, nor even to all those whom He has chosen. Nor would they see God if they did not see that this goodness is incomprehensible, illimitable. They see that His indwelling in the house which is the Church, is, absolutely speaking, "eligibilior," because their viewpoint is truly divine. In seeing God, Peter sees what is greater than anything which could be his proper good for he knows that he is only Peter; he sees that God is infinitely more communicable than He is to Peter himself, and it is this infinity of goodness Peter loves, because he loves God in Himself and in that bounty which of its very nature is diffusive of itself. For this diffusion is not what proceeds from Him, "secundum operationem causa efficientis," it is His own goodness—"*prout est beatitudinis objectum.*" And if there be also John to share the vision, Peter cannot fail to rejoice, because the superabundance of the divine good is his joy. And if the share of John be greater than his own, Peter will again rejoice, for the prime measure of their happiness is neither Peter nor John, but the immeasurable liberality of the divine good. Yet the one soul does not need the other for their operation is their own; nor does the one aid the other to see, for God alone encompasses and draws them; and the very immediacy and freedom have their reason in the universality of divine truth and goodness. Even if Peter alone had been chosen, he would know that his is only a mansion in comparison with the House that is God Himself.

Now, the union between the mansions is twofold, according to the twofold meaning of the house of the Father. The first union is because of the identity of the House that is God. For, what the Blessed see and what they enjoy is the same. *Et in hac domo nos colligit.* This does not mean that the Blessed are present to one another in the primary object of the vision which can be nothing but God alone. Indeed their union would be infinitely less if the Deity Itself, *quantum ad id quod notum est sibi soli de seipso,* were not exclusively the primary object. This union is the most profound, for it is not merely an effect of the termination of the vision and love of the Blessed in the same object and good, but is founded also in the fact that the object and the good are attained and adhered to in their very universality.[139] Hence, the union because of the identity of the object is incomparably greater than any conceivable union dependent on the Blessed themselves.

The second union of the mansions is in that house which is the Church. Christ, Who according to His divinity is the separated good of the Church, is, according to His humanity the head, the principal member and part of the Church. The mansions are strictly part of this house; the members are strictly members of this body. And in this they communicate directly among themselves *extra Verbum*. Yet, the ultimate principle of this communion is still the separated good, for, as we read in the Encyclical *Mystici Corporis*:

> . . . In mystico, de quo agimus, Corpore conspirationi huic internum aliud adjungitur principium, quod *tam in universa compage, quam in singulis ejus partibus* reapse existens virtuteque pollens, talis est excellentiae, ut ratione sui omnia unitatis vincula, quibus vel physicum vel morale corpus copuletur, in immensum prorsus evincat. Hoc est, ut supra diximus, aliquid non naturalis, sed superni ordinis, immo *in semet ipso infinitum omnino atque increatum:* Divinus nempe Spiritus, qui, ut ait Angelicus, "*unus et idem numero, totam Ecclesiam replet et unit.*"[140]

We must note that, whether we compare the mansions to the House that is God or to the house that is the Church, in either case they are mansions, and the mansion of a house is a part. In the first comparison we have the *speculatio totius civitatis* as the greatest good which God has produced. Yet, we must not consider this *speculatio* or *felicitas* absolutely, that is, as the operation and inherent perfection of the created persons. We must consider it in relation to its object and cause. The unity of the divine City is to be sought, not in an absolute comparison of its parts or in their interrelations, but in the identity and universality of the divine good of the City. If we merely consider the parts in their formal beatitude, the good that is common to them is common only according to predication.

And this should suffice to show what is meant by *speculatio totius civitatis,* and to vindicate Peter of Auvergne who, in this lesson 2 of VII *Politics,* is a most faithful disciple of St. Thomas. To see that the common good of the entire heavenly city is "eligibilior," our eye must be fixed on God and on His purpose in creating and choosing, not Peter alone, but Peter and John, and all the mighty host of the Elect.

And now may I be forgiven if I set down once more a passage which Father Eschmann (*DM,* 193) has spared no effort to ridicule:

L'universalité même du bien est principe de béatitude pour la personne singulière. C'est, en effet, en raison de son universalité qu'il peut béatifier la personne singulière. Et cette communication au bien commun fonde la communication des personnes singulières entre elles *extra verbum:* le bien commun en tant que bien commun est la racine de cette communication qui ne serait pas possible si le bien divin n'était déjà aimé dans sa communicabilité aux autres: "praeexigitur amor boni communis toti societati, quod est bonum divinum, prout est beatitudinis objectum." (*BC,* 26–27)

X. "Unusquisque seipsum in Deum ordinatur sicut pars ordinatur ad bonum commune"

In large capital letters Father Eschmann repeats a text from St. Thomas: "ipse solus dirigitur in speculationis finem" (*DM,* 200). Let us first read the text as he reproduces it. Immediately following his question ". . . what in the world can *speculatio totius civitatis* be?" he has this:

> It is exactly this notion which, most unfortunately, Professor De K. has picked out to be the cornerstone of his answer.
>
> St. Thomas speaks quite a different language:
>
> > Sicut bonum unius consistit in actione et contemplatione, ita et bonum multitudinis, secundum quod contingit multitudinem contemplationi vacare. Hoc est verum, quod . . . assecutio finis quem intellectus practicus intendit, potest esse propria et communis, inquantum per intellectum practicum aliquis [!] se et alios dirigit in finem, ut patet in rectore multitudinis [!] Sed aliquis ex hoc, quod speculatur, ipse solus dirigitur in speculationis finem. Ipse autem finis intellectus speculativi tantum praeeminet bono intellectus practici, quantum singularis assecutio ejus excedit communem assecutionem boni intellectus practici. Et ideo perfectissima beatitudo in intellectu speculativo consistit.[141]

How conscientious, how realistic a thinker is young St. Thomas who wrote these passages already in or about 1255 to 1256! He, indeed, never indulges in combining his notions merely for the sake of obtaining some neat scheme, but he examines them with regard to their inner possibility and truth. In the first passage it seems to be evident that St. Thomas somehow inclines towards something like Peter of Auvergne's *speculatio totius civitatis.* Yet Aquinas at once checks himself by adding, with remarkable finesse: *secundum quod contingit multitudinem contemplationi vacare.* Is contemplation, as a genuine social or common act, possible at all? In the second text to the *assecutio communis finis intellectus practici* the right, personal subject is assigned, namely the *rector multitudinis* (cf. II-II, 47. 12). And St. Thomas now vigorously sets in relief the inner impossibility of an *assecutio communis* of the end of the speculative intellect. The words IPSE SOLUS DIRIGITUR IN SPECULATIONIS FINEM and the subsequent statement of the absolute preeminence of the SINGULARIS ASSECUTIO of the speculative good—deserve to be written as a motto at the head of a treatise of Thomistic social philosophy. And be it noted that this whole statement is the Thomistic answer to the following *argumentum in contrarium* which most exactly states the problem of the pretended absolute pre-eminence of the common good:

> Videtur quod beatitudo magis consistat in actu intellectus practici quam speculativi. Quanto enim aliquod bonum est communius, tanto est divinius, ut patet in I Eth. Sed bonum intellectus speculativi est singulariter ejus qui speculatur. Bonum autem intellectus practici potest esse commune multorum. Ergo magis consistit beatitudo in intellectu practico quam speculativo. (*DM,* 199–200)

Overlooking the paper shortage a second time, we shall reproduce Father Eschmann's quotations in the context and order they have in St. Thomas. The first sentence of his first citation is taken from *In III Sententiarum,* d. 35, q. l, a. 4, sol. l, ad 2. First let us read the objection, and then the answer:

> *Bonum gentis divinius est quam bonum unius.* Sed vita contemplativa consistit in bono unius hominis, vita activa in bono multorum. Ergo vita activa est nobilior quam contemplativa.
>
> Ad secundum dicendum quod sicut bonum unius consistit in actione et contemplatione; ita et bonum multitudinis, secundum quod contingit multitudinem contemplationi vacare. Sed ad bonum multitudinis pervenitur per regimen activae vitae; unde ex hoc non probatur quod activa sit dignior, sed utilior.

The important point about that reply is that St. Thomas does not distinguish the major. To do so would have been simple; he need only have conceded the major is true of the practical good, and denied that it applies to the speculative. But he did not do so. Why not? Father Eschmann will say that the major of the *argumentum in contrarium* is a *dictum authenticum* which, "to a medieval writer, is always true" (*DM*, 200, n. 27). We agree that in some circumstances St. Thomas will often concede a statement which he himself would not express in those terms and which, if understood in his own technical language, would mean something quite different. But it would be preposterous to believe that this caution applies here. For, if St. Thomas did not make the distinction defended by Father Eschmann, then, even from Father Eschmann's point of view, it could only be because of the "authority" of this major. Now: what accepted truth can this proposition have been thought to convey? What could it have been taken to mean? There seems no escape from the conclusion that it was understood in St. Thomas' day as meaning that the supremacy of the common good applies both to the practical and the speculative order. Not much remains, then, in the accusation that the "surprisingly radical and daring" work of providing "a new foundation of Christian ethics and moral theology" (*DM*, 196) by applying this principle to both the practical and the speculative was undertaken only toward the middle of the twentieth century.

And now let us revert to the second part of my Opponent's quotation. This time the text is complete. But he quotes first the second part of St. Thomas' reply to an objection quoted in second place, and finally, in a footnote, he produces the first part of the same reply. For the sake of convenience we will reproduce the text in its original order.

> Videtur quod beatitudo magis consistat in actu intellectus practici quam speculativi. Quanto enim aliquod bonum est communius tanto est divinius, ut patet in I *Ethic.*, cap. 1. Sed bonum intellectus speculativi est singulariter ejus qui speculatur; bonum autem intellectus practici potest esse commune multorum. Ergo magis consistit beatitudo in intellectu practico quam speculativo.
>
> Ad primum ergo dicendum, quod bonum cui intellectus speculativus conjungitur per cognitionem, est communius bono cui conjungitur intellectus practicus, inquantum intellectus speculativus magis separatur a particulari quam intellectus practicus cujus cognitio in operatione

> perficitur, quae in singularibus consistit. Sed hoc est verum quod assecutio finis, ad quem pervenit intellectus speculativus, inquantum hujusmodi, est propria assequenti; sed assecutio finis quem intellectus practicus intendit, potest esse propria et communis, inquantum per intellectum practicum aliquis se et alios dirigit in finem, ut patet in rectore multitudinis; sed aliquis ex hoc quod speculatur, ipse singulariter dirigitur in speculationis finem. Ipse autem finis intellectus speculativi tantum praeeminet boni intellectus practici quantum singularis assecutio ejus excedit communem assecutionem boni intellectus practici; et ideo perfectissima beatitudo in intellectu speculativo consistit.[142]

Again St. Thomas avoids distinguishing the major ("Quanto aliquod bonum est communius tanto est divinius"). On the contrary, he shows that the *dictum authenticum* applies more perfectly to the good of the speculative intellect than to that of the practical. And we must note carefully that St. Thomas calls "communius," not the good which consists in the *act* of the speculative intellect, but the "bonum *cui* intellectus speculativus *conjungitur* per cognitionem," and this is objective beatitude. The good of the speculative intellect as such is more common because it is formally more abstract, more separated from the singularity of the operable which involves potentiality, and hence more communicable.

His position having been plainly contradicted by the *littera Sancti Thomae,* here is how Father Eschmann behaves: the embarrassing sentences are confined to a footnote, and in the footnote their meaning is also taken care of:

> To understand this and similar texts (one of which is quoted by De K., p. 172, note 13)[143] it must be noted, first, that the notion of common good is an analogical notion which St. Thomas has not always used in the same nor in its proper sense; and, secondly, that the Thomistic discussion of the primacy of the common good is frequently not, in the first place, a discussion of a doctrine, but of an "authority." A *dictum authenticum,* to a medieval writer, is always true. The only thing, therefore, that can be done about it, is to sustain it and to interpret it. A student of the Thomistic primacy of the common good must first of all know the characteristic medieval techniques of how to deal with a *dictum authenticum.* (*DM,* 200–1)

Thrusting into the back of our minds the nightmarish vision of a great doctrine degenerating into interminable historical controversies on the historical use of formulae, we shall turn our attention to Father Eschmann's unusual notion of analogy, with which we already have some acquaintance. When an analogical notion is not being used in its proper sense, in what precise sense is it being used? The good St. Thomas speaks of is surely a good in the strict sense. There could be little doubt on this point. Father Eschmann's Latin, it is true, might allow him to object that beatitude is a good only "in a certain sense," since he could point out that St. Thomas' clear and precise *littera* expressly states: "beatitudo est *quoddam* bonum excedens naturam creatam."[144] The very definition of beatitude ("bonum perfectum intellectualis naturae") would be destroyed, of course, but further historical research of this kind could always manage to break down even this definition into a mere *dictum authenticum.*

Is St. Thomas' "bonum communius" really a common good? Not in the proper sense, Father Eschmann might reply. But this means raising a question of sophistry, for the *bonum commune* of the practical intellect is a common good in the strict sense, whereas the common good of the speculative intellect would be understood in an improper sense, and yet the latter is to be called "more common" than the former. Perhaps this should embarrass no one. When faced with a *dictum authenticum,* a *sophisma aequivocationis* is doubtless quite in order.

The reader will have noticed that the first proposition of the *argumentum in contrarium* was: "quod beatitudo magis consistat *in actu* intellectus practici quam speculativi." This means that, in this phrase, we have to do with formal felicity, which consists in the *adeptio finis.* The second part of St. Thomas' answer is also concerned with this operation and good: "Sed hoc est verum quod assecutio finis. . . ." The speculative *assecutio* is proper to the intellect of the individual person. St. Thomas now compares this *assecutio singularis* to the *assecutio* of the practical good which is outside the intellect. And this *assecutio* may be either *singularis* or *communis* as in the one who directs both himself and the multitude toward the good. It is called common, not formally because of the community of the good involved (as my Opponent supposes when he defines the common good by the *assecutio communis*) but "inquantum per intellectum practicum aliquis se et alios dirigit in finem," which is quite a different matter.

But these distinctions are of slight importance to Father Eschmann. Just as from the fact that formal beatitude is a proper good he inferred that objective beatitude is also a proper good of the person, so now, from the *assecutio singularis,* that is, from the same formal beatitude, he infers, without troubling to explain how, that the good attained by this *assecutio* cannot be a common good.

On page 203, Father Eschmann again indulges in another historical observation. He is about to quote a text from chapter 13 of St. Thomas' opuscule *De perfectione vitae spiritualis.* Between parentheses he makes the following remark:

> (Let us note that this work was written against the pragmatism of Gerald of Abbéville whose main mistake was to have turned the relative primacy of the common good into an absolute one!)[145]

My Opponent's understanding of St. Thomas' words "ipse solus dirigitur in speculationis finem" is quite obviously opposed to my central position, namely, that the intellectual creature is directed to God as a part to the good of the whole. Now, it is quite remarkable that in the very chapter Father Eschmann refers to, St. Thomas says just that. Speaking of the love of our neighbour, St. Thomas says:

> [Dilectio proximi sancta dicitur] ex hoc quod . . . ordinatur in Deum: sicut enim homines qui sunt unius civitatis consortes in hoc conveniunt, quod uni subduntur principi, cujus legibus gubernantur, ita et omnes homines in quantum naturaliter in beatitudinem tendunt, habent quamdam generalem convenientiam in ordine ad Deum, sicut ad summum omnium principem et beatitudinis fontem et totius justitiae legislatorem. Considerandum est autem, quod *bonum commune secundum rectam rationem est bono proprio praeferendum:* unde unaquaeque pars naturali quodam instinctu ordinatur ad bonum totius. Cujus signum est, quod aliquis percussioni manum exponit, ut cor vel caput conservet, ex quibus totius hominis vita dependet. *In praedicta autem communitate qua omnes homines in beatitudinis fine conveniunt, unusquisque homo, ut pars quaedam consideratur, bonum autem commune totius est ipse Deus, in quo omnium beatitudo consistit. Sic igitur secundum rectam rationem et*

> *naturae instinctum unusquisque seipsum in Deum ordinat sicut pars ordinatur ad bonum totius, quod quidem per charitatem perficitur, qua homo seipsum propter Deum amat.* Cum igitur aliquis etiam proximum propter Deum amat, diligit eum sicut seipsum, et per hoc dilectio sancta efficitur. Unde dicitur, *I Joan.* 4: *Hoc mandatum habemus a Deo, ut qui diligit Deum, diligat et fratrem suum.*

The position Father Eschmann attacks could not be more clearly stated.

Now for the text my Opponent quotes from this same chapter of the opuscule. Immediately following the parenthesis given above he quotes:

> "Proximus autem noster non est universale bonum supra nos existens, sed particulare bonum infra nos constitutum." Will Professor De K. be able to give us a fitting explanation of this "infra nos" of St. Thomas? I cannot help but think that, he will not. According to the suppositions of his system he will protest (in fact, he does so, on similar occasions) that this is the "base abomination of egoism." We have no reason to recede even one iota from the clear and precise *littera Sancti Thomae.* That it contains no egoism at all is clear to everyone who, with St. Thomas, knows how to distinguish between *amor sui ordinatus* and *amor sui inordinatus.* (*DM,* 203)

Does Father Eschmann mean that I hold our neighbour to be "bonum universale supra nos existens," and not "bonum particulare infra nos constitutum"? My Opponent's question reveals such "remarkable finesse" in dealing with this problem and such scrupulous care in reading my book that I feel quite speechless. Yet, lest the reader believe I concede Father Eschmann's interpretation of this text (his interpretation is unmistakably clear from his general doctrine and from his purpose in quoting it against me) I should like to add that it represents inescapably that position which I do not hesitate to qualify in those very terms he quotes from my essay.

Why should we love ourselves more than our neighbour? Obviously the reason cannot be that, absolutely speaking, we are better than our neighbour. He who would not be content to be the last to leave purgatory, to be the last in the kingdom of heaven and therefore the least of all the Blessed, would stand small chance of ever getting there. The reason why we must love ourselves more than our neighbour is not that we are better than our neighbour, but because, as St. Thomas says,

> licet proximus melior sit Deo propinquior, *quia tamen non est ita propinquus caritatem habenti sicut ipse sibi,* non sequitur quod magis debeat aliquis proximum quam seipsum diligere.[146]

It is essential for each one of us to realize in a most practical manner that certainly many, and possibly every one of our neighbours is better than our own person, and by "better" I mean better in the eyes of God and more lovable to Him. If we cannot love them according to their own, absolute amiability, it is because we cannot love them as God loves them.[147]

In the same article 4 (IIaIIae, q. 26) St. Thomas adds something which, according to Father Eschmann, would be in open contradiction with the passage we have just quoted:

> Ad tertium dicendum quod, sicut Augustinus dicit, in *Regula, quod dicitur, Caritas non quaerit quae sua sunt, sic intelligitur quia communia propriis anteponit.* Semper autem commune bonum est magis amabile unicuique quam proprium bonum: sicut etiam ipsi parti est magis amabile bonum totius quam bonum partiale sui ipsius, ut dictum est. (Scil. a. 3)[148]

Since Father Eschmann expressly maintains that my understanding of the primacy of the common good is something unheard of in "Christian ethics and theology," something "suprisingly radical and daring," I might quote an opinion of a seventeenth-century theologian, perhaps the last of the greater disciples of St. Thomas:

> Post Deum autem, unusquisque magis diligit se, quam proximum, debet enim diligere alios, sicut seipsum, unde ipsemet est quasi exemplar primum et diligendorum, quia se ut participem gloriae divinae, alios ut socios in participando. *Excipio tamen Christum Dominum, etiam ut hominem, et Beatissimam Virginem matrem,* eo quod participant quamdam rationem principii communicantis nobis gratiam, et beatitudinem, est enim Christus ut homo caput gloriae, et Beatissima Virgo mater capitis,

> et collum per quod derivatur gratia, et ideo magis debemus ipsos diligere, quam nos.[149]

XI. "Civitas homini, non homo Civitati existit"

At the very beginning of his article, Father Eschmann quoted (*DM,* 86) a passage from the Encyclical *Divini Redemptoris.* The line most relevant to our problem is the following: *Civitas homini, non homo Civitati existit.* My Opponent's reader is, presumably, to understand that this text implies a negation of the primacy of the common good. The application is apparently so inevitable that Father Eschmann does not feel obliged to inform the reader that I had answered a current objection drawn from this very text (*BC,* 67–71). It is surely clear, even from my Opponent's own paper, that I hold God to be the supreme common good. At the same time, quite inexplicably, Father Eschmann will speak as if I held the supreme common good, to which all else must be subordinated, was none other than the common good of mere political society. So again, in the section of his article we are now examining, we find him making this same implication. When, because of an emergency, the contemplative are called upon to share more fully in the active life, why do they obey? My Opponent asks,

> Is it because they have been enjoying the *dulcedo contemplationis* as parts of the community, and thus, already, in subordination to its interests and laws? (*DM,* 202)

I do not think that at this stage, it will be necessary to comment upon the above statement. We already know how freely Father Eschmann skips from one order to the other, as if the notion of common good were a univocal one. We have learned that his common good is a strange thing indeed: an efficient cause; an object of the practical intellect; and now, after the admission that there is a respect in which God is "a common good properly speaking" (*DM,* 196), we learn that the common good "is essentially a *bonum utile,* the highest *bonum utile,* but nothing more" (*DM,* 202). May we be pardoned if we cannot help murmuring like another: "Would it not be desirable that an author who uses traditional notions knew exactly what they mean?" (*DM,* 189, n. 12).

The main reason why many a personalist has been irked by my essay is that it took him off-guard. Instead of discussing the problem in terms of "person" and "society," I approach it in the fundamental terms of "proper good" and "common good." Ultimately, person and society are not to be judged by what they *are* absolutely, but by what is their perfection, i.e. by what is their good; that is the only way in which Aristotle and St. Thomas ever discussed this problem. To look upon the absolute comparison of person and society as the most basic consideration is distinctly modern. It is also distinctly modern to accord absolute priority to the subject and to believe, with Spinoza (who, in this respect, follows in the footsteps of David of Dinant) that "to be absolutely" is "to be good absolutely," i.e. that "ens simpliciter" is "bonum simpliciter."[150] From this identification it follows quite logically: "Per finem, cujus gratia aliquid facimus, appetitum intelligo."[151] *Finis cui* becomes *finis qui.* From such a point of view, the problem of person and society quite naturally becomes the question: is the person better than society? instead of: is the proper good of the person better than his common good? When the problem itself has been so distorted, what can be expected in the solution?

The totalitarian solution is that the individual person is ordered and subjected to society. We are inclined, in rejecting this doctrine, to swing to the opposite extreme; but if we prescind from the common good of the persons which is the final, and therefore first cause of society, we are left with a mere aggregate of individuals. Now, in this formal consideration, each and every one of that group could never be more than an *alter ego,*[152] and the group itself could never be more than an aggregate, a mere *unum coacervatione* of *alter egos.* Hence, in this perspective, the whole question of our relation to the common good and to our particular good becomes a problem entirely different from that over which the battle has raged until now, resolving itself into the simple question: must one love oneself more than one's neighbour? There is not the slightest doubt that we must love ourselves more. Even Aristotle expressly taught this fundamental truth. Indeed, St. Thomas used the Philosopher's doctrine for an objection against the primacy of the common good, which was reproduced by me:

> On pourrait, tout en s'appuyant sur le Philosophe (IX *Ethic.*, cc. 4 et 8), pousser plus loin l'objection: "Les témoignages d'amitié que l'on rend

> aux autres ne sont que des témoignages d'amitié rendus à soi-même."— A cette objection saint Thomas répond "que le Philosophe parle ici des témoignages d'amitié rendus à un autre chez qui le bien qui est objet de l'amitié se trouve selon un certain mode particulier: il ne parle pas des témoignages d'amitié rendus à un autre chez qui le bien en question se trouve sous la raison de bien du tout." (*BC*, 60–61)[153]

To love our neighbour more than ourselves would be contrary to nature, since we are more one with ourselves than with our neighbour.[154] Nor is there any doubt that for the very same reason we must love ourselves more than any society so considered. Only God, Christ even as man, and Mary, who as truly universal sources dispense to us the divine good, can be loved by us more than ourselves.

Throughout my essay I repeatedly called attention to the simplistic confusion of these two problems; in fact, it was written mainly to dissipate the false assumption that the common good is an alien good, that is, either a personal good of our neighbour or the sum total of proper goods. When we state the fundamental problem in terms of person and society, it is quite natural that the subordination of the personal good to the common good should be interpreted as the collectivist and totalitarian subjection of the individual to the mass. But the truth is, as I sought to explain in the third chapter of my essay, that personalism and totalitarianism proceed from the same assumption (*BC*, 73–79). When we say, in opposition to the personalists, that the individual person is subordinated to society, we do not mean, as they would have us mean, that the person and his proper good are subordinated to society considered absolutely, that is, to a mere aggregate of proper goods in which no aspect of a real common good is to be seen. We mean that, within a given order, the good of the individual person is subordinated to the common good of the community. If the political community has the right to execute a criminal citizen, it is not formally because it represents a number of persons, but rather "ut bonum commune conservetur."[155] The condemned man does not become the victim of mob violence; he is destroyed because he has proved a responsible menace to the common good.

The reader can now see that Father Eschmann does not seem to be aware of the real problem, and therefore can easily impose upon me the ignominious positions which follow from his own misunderstandings. The very opening paragraph of my book was:

> La société humaine est faite pour l'homme. Toute doctrine politique qui ignore la nature raisonnable de l'homme, qui nie, par conséquent, sa dignité et sa liberté, est viciée à la racine et soumet l'homme à des conditions inhumaines. C'est donc à bon droit qu'on s'insurge contre les doctrines totalitaires au nom de la dignité de l'homme. (*BC*, 1)

And on the following page I said:

> On peut à la fois affirmer la dignité de la personne et être en fort mauvaise compagnie. Suffrait-il d'exalter la primauté du bien commun? Non plus. Les regimes totalitaires saisissent le bien commun comme prétexte pour asservir les personnes de la facon la plus ignoble. Comparée à l'esclavage où ils menacent de nous soumettre, la servitude des bêtes est liberté. Commettrons-nous la lâcheté de concéder au totalitarisme ce pervertissement du bien commun et de sa primauté?

If no more than Father Eschmann's misunderstandings and accusations were at stake, we should have had little to reply; but when he invokes "the clear and precise *littera Sancti Thomae*" in support of a doctrine which, as far as we can see, is indistinguishable from the *amor sui inordinatus*, we must surely consider it a duty to rally to the defence of true Thomistic principles. Let it be added at once that we have no right to assume that Father Eschmann is conscious of all the implications of his position. And that this should be so is the more understandable because of the fact that his viewpoint, even as regards the primary notion of the good, is the modern one. It is not easy to escape erroneous tendencies when they are those of the age in which we actually live. *Intus existens prohibet extraneum.*

And now let us return to the Encyclical *Divini Redemptoris*. Pius XI precisely denounces the totalitarian conception of person and society. In the very next phrase he adds (and Father Eschmann himself quotes the passage):

> Id tamen non ita intelligendum est, quemadmodum ob suam *individualismi* doctrinam *Liberales*, quos vocant, asseverant; qui quidem communitatem immoderatis singulorum commodis inservire jubent. . . .[156]

Since my Opponent has so discreetly overlooked my answer to the objection drawn from this text, I will reproduce it in full.

La cité existe pour l'homme, l'homme n'existe pas pour la cité.

Pour convertir ce texte en objection contre notre position, il faudrait le traduire: "Le bien commun de la cité existe pour le bien privé de l'homme." Nous pourrions, alors, citer la suite immédiate de ce même texte: "Ce qui ne veut point dire, comme le comprend le libéralisme individualiste, que la société est subordonnée à l'utilié égoïste de l'individu."

La cité existe pour l'homme. Cela doit s'entendre de deux manières. Premièrement, la cité, quand nous l'envisageons comme organisation en vue du bien commun, doit être entièrement soumise à ce bien en tant qu'il est commun. Envisagée sous ce rapport, elle n'a d'autre raison d'être que le bien commun. Or, ce bien commun lui-même est pour les membres de la société; non pas pour leur bien privé comme tel; il est pour les membres en tant que bien commun. Et, comme il s'agit d'un bien commun de natures raisonnables, il doit être conforme à la raison, il doit regarder les natures raisonnables en tant qu'elles sont raisonnables. La cité n'est pas, ou ne peut pas être, un 'pour soi' figé et refermé sur soi, opposé comme un singulier à d'autres singuliers: son bien doit être identiquement le bien de ses membres. Si le bien commun était le bien de la cité en tant que celle-ci est, sous un rapport accidentel, une sorte d'individu, il serait du coup bien particulier et proprement étranger aux membres de la société. Il faudrait même accorder à l'organisation ainsi ravie à ses membres, intelligence et volonté. La cité serait alors comme un tyran anonyme qui s'assujettit l'homme. L'homme serait pour la cité. Ce bien ne serait ni commun ni bien de natures raisonnables. L'homme serait soumis à un bien étranger.—Deuxièmement, la cité, comme le bien commun de la cité, est pour l'homme en tant que celui-ci comprend des formalités qui l'ordonnent à des biens communs supérieurs, formalités qui sont, dans l'homme, supérieures à celle qui l'ordonne au bien commun de la cité. Or, l'identité du sujet de ces diverses formalités peut prêter à confusion. Le bien privé et le bien commun sont l'un et l'autre biens de l'homme. Et pourtant, tout bien de l'homme n'est pas bien de l'homme purement homme. Le bien de l'homme purement homme, d'après le sens que lui accorde saint Thomas dans les textes déjà cités, n'est autre chose que le bien qui lui convient en raison de l'individu. Le bien commun ne peut jamais être subordonné à cet homme purement homme.

La formalité 'homme purement homme' ne peut pas être identifiée à la formalitè 'citoyen', comme elle ne peut l'être au sujet 'homme'. Dès lors, quand nous disons un bien commun subordonné à l'homme, ce ne peut être qu'en raison d'une formalité qui regarde un bien commun supérieur. Seul le bien commun le plus parfait ne peut être subordonné à l'homme.

De plus, quand nous disons que le bien commun ne peut jamais être considéré comme une pure extension du bien de l'homme dans la ligne de son bien singulier, en sorte que le bien commun ne serait qu'un détour pour rejoindre le bien singulier, nous n'entendons pas par là que le bien singulier est méprisable, qu'il est néant, qu'il ne doit pas être respecté ou qu'il n'est pas en lui-même respectable. Cependant, un respect plus grand est dû à la personne quand nous envisageons celle-ci dans son ordination au bien commun. Même le bien singulier de la personne est meilleur quand nous le considérons comme ordonné au bien commun de la personne. Du reste, une cité qui ne respecte pas le bien privé ou le bien des familles, agit contrairement au bien commun. De même que l'intelligence dépend du sens bien disposé, ainsi le bien de la cité depend de l'intégrité de la famille et de ses membres. Et de même qu'une nature sensible bien soumise à la raison est plus parfaite dans la ligne même de la nature sensible, de même, dans une cité bien ordonnée, le bien singulier de l'individu et le bien commun de la famille doivent être plus parfaitement réalisés et assurés. Cependant, si le bien commun de la cité était subordonée à ces derniers, il ne serait pas leur bien commun et l'homme serait privé de son bien temporel le plus grand; la cité ne serait pas cité. Elle serait comme une intelligence subordonnée au sens et réduite à la condition d'instrument pour le bien privé. (*BC*, 67–71)

The following objection and answer are related to the same problem:

". . . L'homme n'est pas ordonné à la société politique selon tout lui-même et tout ce qui est sien."[157]

On a voulu conclure de ce texte isolé que la société politique est en dernière instance subordonnée à la personne singulière prise comme telle. Et quiconque ose contredire cette grossière inférence tournée en faveur du personnalisme, se fait traiter de totalitaire. Or, ainsi que nous l'avons vu, il est contraire à la nature même du bien commun d'être, comme tel, subordonné à un singulier, à moins que ce singulier n'ait lui-même raison de bien commun. Saint Thomas veut dire seulement que l'homme n'est pas ordonné à la seule société politique. Il n'est pas selon tout lui-même partie de la société politique, puisque le bien commun de celle-ci n'est qu'un bien commun subordonné. L'homme est ordonné à cette société en tant que citoyen seulement. Bien que l'homme, l'individu, le membre de famille, le citoyen civil, le citoyen céleste, etc., soient le même sujet, ils sont formellement différents. Le totalitarisme identifie la formalité homme à la formalité citoyen. Pour nous, au contraire, non seulement ces formalités sont distinctes, mais elles sont subordonnées les unes aux autres selon l'ordre même des biens. Or, c'est l'ordre des biens, causes finales et premières, et non pas l'homme purement homme, qui est principe de l'ordre de ces formalités d'un même sujet. Le personnalisme renverse cet ordre des biens: il accorde le plus grand bien à la formalité la plus inférieure de l'homme. Ce que les personnalistes entendent par personne, c'est, en vérité, ce que nous entendons par pur individu, tout matériel et substantiel enfermé en soi, et ils réduisent la nature raisonnable à la nature sensible qui a pour objet le bien privé.

L'homme ne peut pas s'ordonner au seul bien de la société politique; il doit s'ordonner au bien du tout parfaitement universel, auquel tout bien commun inférieur doit être expressément ordonné. Le bien commun de la société politique doit être expressément ordonné à Dieu, tant par le citoyen-chef que par le citoyen-partie, chacun à sa manière. Ce bien commun demande, lui-même, cette ordination. Sans cette ordination expresse et publique, la société dégénère en Etat figé et refermé sur soi. (*BC*, 66–67)

XII. The Private Law of the Holy Ghost

There remains one more objection deserving of our attention before we enter upon the final chapter of this article. This argument against our doctrine, while hardly formidable in itself, makes a vivid appeal to one's piety and so has its danger for the person who may not have the leisure to examine it thoroughly. For the principle which inspires it Father Eschmann turns to the *Canon Urbani*:

> To use (and extend) the language of an old and venerable papal document of the eleventh century, the so-called *Canon Urbani*—a document which has played an important role in the medieval canonist and theological discussions of our problem[158]—it is not the personalist contention that nobody dare resist the caprices of any given individual person, of Tom, Dick and Harry, but that nobody dare resist the Holy Ghost (Act. 7:51). (*DM*, 207)

We shall examine the actual words of the document in a moment or two; for the present let us merely note that from it our Opponent draws an argument based on the supremacy of the law of the Holy Ghost as written in the heart of the individual person, over any possible public law.

The whole objection, then, turns on the notion of law, and yet reveals a curious failure to grasp what is most fundamental in that notion. Indeed we have only to bring our Opponent to admit that this private law of the Holy Ghost is truly a law to quite destroy his reasoning. For if it be true law, it must have what is essential to any law (meaning simply that without which no law would be a law); and St. Thomas most uncompromisingly tells us that this "nihil est aliud quam quaedam rationis ordinatio ad *bonum commune*, ab eo qui curam communitatis habet, promulgata."[159] If the private law of the Holy Ghost is a law, then, like any other law, it is a *rationis ordinatio ad bonum commune.*

Since any law "proprie, primo et principaliter respicit ordinem ad bonum commune"[160] my Opponent could hardly have chosen a better example to defeat his own position. How untenable his position is might be best shown by reference to the second article of IaIIae, q. 90, which answers the question: *Utrum lex ordinetur semper ad bonum commune.* The whole article places us right in the middle of our problem. If anyone entertained the slightest doubt as to the strict meaning of the *bonum commune,* which is the end of the law, he may read in the reply to the second objection that St. Thomas means "bonum commune, non quidem communitate generis vel speciei, sed *communitate causae finalis,* secundum quod bonum commune dicitur finis communis." Here is the body of the article:

> Respondeo dicendum quod ... lex pertinet ad id quod est principium humanorum actuum, ex eo quod est regula et mensura. Sicut autem ratio est principium humanorum actuum, ita etiam in ipsa ratione est

> aliquid quod est principium respectu omnium aliorum. Unde ad hoc oportet quod principaliter et maxime pertineat lex.—Primum autem principium in operativis, quorum est ratio practica, est finis ultimus. Est autem *ultimus finis humanae vitae felicitas vel beatitudo, ut supra habitum est. Unde oportet quod lex maxime respiciat ordinem qui est in beatitudinem.*—Rursus, *cum omnis pars ordinetur ad totum sicut imperfectum ad perfectum; unus autem homo est pars communitatis perfectae: necesse est quod lex proprie respiciat ordinem ad felicitatem communem.* Unde et Philosophus, in praemissa definitione legalium, mentionem facit et de felicitate et communione politica. Dicit enim, in V *Ethic.*, quod *legalia justa dicimus factiva et conservativa felicitatis et particularum ipsius, politica communicatione:* perfecta enim communitas civitas est, ut dicitur in 1 *Polit.*
>
> In quolibet autem genere id quod maxime dicitur, est principium aliorum, et alia dicuntur secundum ordinem ad ipsum: sicut ignis, qui est maxime calidus, est causa caliditatis in corporibus mixtis, quae intantum dicuntur calida, inquantum participant de igne. Unde oportet quod, *cum lex maxime dicatur secundum ordinem ad bonum commune, quodcumque aliud praeceptum de particulari opere non habeat rationem legis nisi secundum ordinem ad bonum commune. Et ideo omnis lex ad bonum commune ordinatur.*

To the central doctrine conveyed in these words we may add the main divisions of law laid down by St. Thomas in this same treatise, before proceeding to the actual document which our Adversary has invoked. By the eternal law St. Thomas means "ipsa ratio gubernationis rerum in Deo sicut in principe universitatis existens,"[161] and the end of this divine government is "ipse Deus, nec ejus lex est aliud ab ipso."[162] The natural law is a "participatio legis aeternae in rationali creatura."[163] Now, since the precepts of natural law are very general—e.g. *commit no evil*—whereas action is in the singular, human reason must derive more particular directives from these naturally known principles, either by way of conclusion, e.g. *one must not kill,* or by way of further determination, e.g. life imprisonment for murder. Such conclusions or determinations constitute human law.[164] But because man is ordained to a supernatural end, "ideo superadditur lex divinitus data, per quam lex aeterna participatur altiori modo."[165] Finally, when a human law is contrary to human good, it does not bind in conscience, "nisi forte propter vitandum scandalum vel turbationem, propter quod etiam homo juri suo debet cedere";[166] when opposed to the divine good, however, "nullo modo licet observare: quia sicut dicitur *Act.* 5, *obedire oportet Deo magis quam hominibus.*"[167]

Now, what does the *Canon Urbani* mean by the private law of the Holy Ghost? Let us look into the text as Father Eschmann himself quotes it:[168]

> Duae sunt, inquit (i.e. Urbanus Papa), leges: una publica, altera priuata. Publica lex est, que a sanctis Patribus scriptis est confirmata, ut lex est canonum, que quidem propter transgressiones est tradita. Verbi gratia: Decretum est in canonibus, clericum non debere de suo episcopatu ad alium transire sine commendatitiis litteris sui episcopi, quod propter criminosos constitutum est, ne uidelicet infames ab aliquo episcopo suscipiantur personae. Solebant enim offitia sua, cum non in suo episcopatu poterant, in alio celebrare, quod iure preceptis et scriptis detestatum est. 1. Lex uero priuata est, que instinctu S. Spiritus in corde scribitur, sicut de quibusdam dicit Apostolus: "Qui habent legem Dei scriptam in cordibus suis," et alibi: "Cum gentes legem non habeant, si naturaliter ea, que legis sunt, faciunt, ipsi sibi sunt lex." Si quis horum in ecclesia sua sub episcopo populum retinet, et seculariter uiuit, si afflatus Spiritu sancto in aliquo monasterio uel regulari canonica salvare se (Variante: salvari se) uoluerit, quia (Variante: qui enim) lege priuata ducitur, nulla ratio exigit, ut a publica lege constringatur. Dignior est enim lex priuata quam publica. Spiritus quidem Dei lex est, et qui Spiritu Dei aguntur lege Dei ducuntur; et quis est, qui possit sancto Spiritui digne resistere? Quisquis igitur hoc Spiritu ducitur, etiam episcopo suo contradicente, eat liber nostra auctoritate. Iusto enim lex non est posita, sed ubi Spiritus Dei, ibi libertas, et si Spiritu Dei ducimini, non estis sub lege.

If this text is to furnish an argument against the primacy of the common good, it can only be on condition public law is taken to mean a law which is ordained to the common good, whereas the private law of the Holy Ghost would be that ordained to the private good. Such an interpretation seems inconceivable and yet if it be not that which my Opponent intends, what possible alternative can he find? Father Eschmann seems to have discovered an entirely new kind of law—a law which is neither eternal, nor natural, nor human, nor divine, nor even law in any strict sense of the word; it is an entity resembling, perhaps, his "quoddam bonum commune."

St. Thomas, like Pope Urban himself, has a quite different understanding of the distinction between public law and the private law of the Holy Ghost. In article 5 (*Utrum omnes subjiciantur legi*), q. 96, he formulates the following objection:

> 2. Praeterea, Urbanus Papa dicit, et habetur in *Decretis,* XIX, qu. ii *Qui lege privata ducitur, nulla ratio exigit ut publica constringatur.* Lege autem privata Spiritus Sancti ducuntur omnes viri spirituales, qui sunt filii Dei; secundum illud *Rom.* 8 *Qui Spiritu Dei aguntur, hi filii Dei sunt.* Ergo non omnes homines legi humanae subjiciuntur.

His reply is:

> Ad secundum dicendum quod lex Spiritus Sancti est superior omni lege *humanitus posita.* Et ideo viri spirituales, secundum hoc quod lege Spiritus Sancti ducuntur, non subduntur legi, quantum ad ea quae repugnant ductioni Spiritus Sancti. Sed tamen hoc ipsum est de ductu Spiritus Sancti, quod homines spirituales legibus humanis subdantur; secundum illud I *Pet.* 2: *Subjecti estote omni humanae creaturae, propter Deum.*

In formulating this objection, Father Eschmann seems to have been under the impression that my obsession with the common good was such as to lead me to teach that the person, in his subjection to an inferior good, may obey a law which is opposed to the divine good. It is this which compels me to recall a rather lengthy paragraph from my essay:

> On pourrait encore objecter que si la dignité de la créature raisonnable est liée à sa subordination à Dieu d'où la personne tient tout ce qu'elle est, sa dignité n'est pas liée à sa subordination à d'autres fins si supérieures soient-elles. Dès lors, cette dignité est antérieure à toute subordination autre qu'à Dieu, et indépendante de l'ordre dans les choses créées. En effet, "quand le bien propre d'un être est subordonné a plusieurs biens supérieurs, l'agent doué de volonté est libre de sortir de

> l'ordre qui se rattache à l'un de ces êtres supérieurs et de rester dans l'ordre qui se termine à un autre, que ce dernier soit plus ou moins élevé."[169] A cela nous répondons que quand un agent doué de volonté doit subordonner son bien propre à un bien créé supérieur, ce ne peut être qu'en tant que celui-ci est lui-même conforme à l'ordre divin. Dès lors, quand l'inférieur doit se soustraire à ce qui lui est supérieur, c'est que ce supérieur s'est écarté de l'ordre où il devait lui-même se tenir. Mais, tant que ce supérieur se tient dans l'ordre, il est un bien supérieur auquel l'inférieur doit se soumettre. "Par exemple, le soldat qui est soumis au roi et au général de l'armée peut subordonner sa volonté au bien du general et non à celui du roi, et inversement; mais dans le cas où le général transgresserait l'ordre donné par le roi, la volonté du soldat serait bonne, s'il la détachait de la volonté du général pour la soumettre au roi; elle serait mauvaise s'il exécutait la volonté du général contrairement à la volonté du roi; car l'ordre d'un principe inférieur dépend de l'ordre du principe supérieur."[170] Toutefois, "il y aurait péché dans les substances séparées si quelqu'une d'un rang inférieur sortait de l'ordre d'une substance supérieure qui reste soumise à l'ordre divin." Dès lors, la révolte de l'inférieur contre un supérieur insoumis est une révolt contre le désordre. (*BC*, 47–48)

XIII. "The Term 'Personalism' (in Itself, No Doubt, a Bad One)"

Some people call themselves personalists but, when one brings to their attention what that term usually emphasizes, they will hasten to add that they do not mean it in such a sense. In their special acceptance of it, the term may represent nothing objectionable, but it is doubtful if that be enough to justify its common use. In a certain class of Catholic writers there has appeared a tendency to effect something like a theft of the adversary's thunder by using his own vocabulary in applications which, in the end, turn out to be quite different from the impositions given them in the original.[171] The result is, of course, an ambiguity sufficient to mislead the most well-intentioned of readers.

That the tendency I mention is a pernicious one may be convincingly demonstrated by the case of personalism. The writers who represent this theory reach an audience that is both large and important in the world of

Catholic education. Whether or not their books are being correctly interpreted by this circle of readers is not the question which concerns us at the moment; the point is that a considerable number of people holding responsible posts in our institutions of higher learning are clearly taking the personalism expounded in such works to imply the negation of the primacy of the common good. In giving approval to an article like that of Father Eschmann, not only do they, quite unconsciously, welcome the central thesis of personalism in its most abject form, but they also prove the dangerous fruitfulness of ambiguity.

Had we read and obeyed the *littera Sancti Thomae* we would have been spared this disastrous and widespread condition:

> . . . Cum infidelibus nec nomina debemus habere communia, ne ex consortio nominum possit sumi erroris occasio; nomine *fati* non est a fidelibus utendum, ne videamur illis assentire qui male de fato senserunt, omnia necessitati siderum subjicientes. Unde Augustinus dicit, in V *de Civitate Dei: Si quis voluntatem vel potestatem Dei fati nomine appellat, sententiam teneat, linguam corrigat.* Et Gregorius, secundum eundem intellectum, dicit: *Absit a fidelium mentibus ut fatum aliquid esse dicant. . . .*[172]

> . . . Sicut Hieronymus dicit, *ex verbis inordinate prolatis incurritur haeresis.* Unde cum haereticis nec nomina debemus habere communia: ne eorum errori favere videamur.[173]

He who rightly believes that every human being is a person capable of and immediately ordained to the supreme immutable common good and that in this consists his dignity, let him not assume that he must therefore call himself a personalist—*sententiam teneat, linguam corrigat.* Equivocation implies a grave risk and no matter how unwittingly one may have employed it an inescapable duty may ensue. The following passage from a sermon of St. Thomas may be read in this connection:

> Inveniuntur aliqui qui student in philosophia, et dicunt aliqua qua non sunt vera secundum fidem; et cum dicitur eis quod hoc repugnat fidei, dicunt quod philosophus dicit hoc, sed ipsi non asserunt, imo

> solum recitant verba philosophi. Talis est falsus propheta, sive falsus doctor, quia idem est dubitationem movere et eam non solvere quod eam concedere; quod signatur in *Exod.* (21, 33), ubi dicitur quod si aliquis foderit puteum, et aperuerit cisternam et non cooperuerit eam, veniat bos vicini sui, et cadat in cisternam, ille qui aperuerit cisternam teneatur ad ejus restitutionem. Ille cisternam aperuit, qui dubitationem movet de his quae faciunt ad fidem. Cisternam non cooperit, qui dubitationem non solvit, etsi ipse habeat intellectum sanum et limpidum, et non decipiatur. Alter, tamen qui intellectum non habet ita limpidum bene decipitur, et ille qui dubitationem movit tenetur ad restitutionem, quia per eum ille cecidit in foveam.[174]

May I also remind the reader that the personalist conception of marriage[175] has been condemned by the *Suprema Sacra Congregatio S. Officii,* in a decree published at the order of Pius XII, on April 1, 1944.[176]

I should like it understood that I do not at all accuse Father Eschmann of using the term "personalism" ambiguously. On the contrary, he employs it in its strict meaning, as may be clear from his fundamental position:

> *Objectively,* i.e. viewed from the part of its uncreated object, the vision is not a common good; it is not even God as Common Good (to speak of common good in a proper and adequate language) but it is God Himself, the *Bonum universale in essendo,* as has been shown above. (*DM,* 202)

Nor could he have ever attacked my essay had I not been clear about what I mean by personalism and what I have against it.

XIV. The Devil and the Common Good

The rather flamboyant title of this final chapter might suggest that it is to contain doctrine of a novel and startling kind; its aim, however, is exactly the contrary. The fitting and proper close for any discussion in Christian Theology will always be an appeal to traditional and ancient teaching, and it is this indispensable support which we propose to seek in our last pages. If, prescinding now from the explicit *littera Sancti Thomae,* there were any truth in the accusation, that the primacy of the divine common good is a modern innovation, then for that reason alone he who held it should feel uneasy. But we shall leave it to the reader to judge, after reading the testimony now to be described, whether or not our position draws its strength from the roots of tradition.

In a paragraph of my book which arouses Father Eschmann's amusement as recalling "by its style and bearing the 'heroic' ages of baroque-Scholastic controversy" (*DM*, 184), I refer to John of St. Thomas in support of my position. The note to this passage presented in full the actual words of this recognized theologian. This citation will receive fuller notice in a moment, but for the present let it be noted that the text of John of St. Thomas (1589–1644) is based in turn directly on the authority of St. Augustine (354–430) "a superiore communi omnium beatifico bono [mali angeli] ad propria defluxerunt"; and also on the authority of Pope St. Gregory (c. 540–604) "Dum [Leviathan] privatam celsitudinem superbe appetiit, jure perdidit participatam";[177] and again on the authority of St. Bernard (1090–1153) "[Homines] infirmiores sunt, inquit [diabolus], inferioresque natura, non decet esse concives, nec aequales in gloria";[178] on the authority of St. Thomas (1224/5–1274) "affectavit [diabolus] excellentiam singularem."[179] The innovator Father Eschmann denounces was born in 1906.

The extraordinary thing is that all these mighty witnesses are as one in recognizing the denial of the Common Good as being the peculiar crime of Satan. Yet it is understandable that this be so, for the temptation could be for none more alluring than for persons as glorious as Lucifer and his followers before their fall. John of St. Thomas explains this in the passage I quoted.

> . . . Quia videntes dignitatem suam, appetierunt singularitatem, quae maxime est propria superborum . . . (recusat diabolus beatitudinem supernaturalem) habere sine singularitate propria, sed communem cum hominibus; ex quo consecutum est quod voluerit specialem super eos habere praelationem potius quam communicationem, ut etiam Divus Thomas fatetur in hac quaestione LXIII, a. 3, in calce. Accedit ad hoc auctoritas S. Gregorii papae: 'Angelos perdidisse participatam celsitudinem, quia privatam desideraverunt', id est, recusarunt coelestem beatitudinem, quia participata, et communis erat multis, et solum voluerunt privatam, scilicet quatenus privatam, et propriam, quia prout sic habebat

> duas conditiones maxime opportunas superbiae, scilicet singularitatem, seu nihil commune habere cum inferioribus, quod ipsis vulgare videbatur, etiamsi esset gloria supernaturalis, et non habere illam ex speciali beneficio, et gratia, et quasi precario: hoc enim maxime recusant superbi, et maxime recusavit angelus. Et ad hoc pertinet parabola illa *Lucae* xiv, de homine qui fecit coenam magnam, et vocavit multos, et cum vocasset invitatos coeperunt se excusare; ideo enim fortassis recusaverunt ad illam coenam venire, quia magna erat, et pro multis, dedignantes consortium habere cum tanto numero, potiusque eligerunt suas privatas commoditates, licet longe inferiores, utpote naturalis ordinis, iste quia villam emit, ille quia juga boum, alius quia uxorem duxerat, unusquisque propriam excusationem praetendens, et privatum bonum, quia proprium, recusans vero coenam, quia magnam, et multis communem. Iste est propriissime spiritus superbiae.[180]

Could one state more clearly that the fallen Angels refused supernatural beatitude because it can be achieved only as a common good and because they had to seek it *qua* common good? Yet, by their faith and their most perfect natural knowledge the Angels, who cannot err in matters of speculative science, knew, incomparably better than we, that the *adeptio finis* is an *assecutio singularis.* They knew that God Himself and God alone is the primary object of this happiness and that the vision is in no way interrupted by the existence of any neighbour nor by any number of them. Yet they prefer that lower good which is possessed as a privilege of their angelic nature or as wholly personal, to a good common to many and dispensed according to the free choice of God Himself Who can make the last first and the first last.[181] They may be compared to those who refused to attend the great supper, simply because it was a great one to which many were invited, and they scorned to take part with such a crowd. They preferred, accordingly, to turn to private affairs, even though these were far inferior and of a quite earthly nature. Nothing could be more characteristic of the proud. The Angels well knew the object of heavenly beatitude is the proper good of God alone which to angel or man can be only a common good. And in desiring to confine themselves to their personal good, we may think of them as pleading with great show of argument that, in so doing, they were only striving generously to be like to God in a more unique and personal fashion, since in this they would be less dependent upon His grace and favor, possessing their good by way of a strictly personal appropriation.[182] In other words, they sought to be assimilated to God only with regard to this that God is good, thus aiming to be most like to Him by being good in themselves, instead of seeking the assimilation *secundum unionem vel informationem* to an object which is common and impossible to attain as a proper good. And so, as St. Augustine (354–430) says, from that higher and beatific good which was common to all, they lapsed to this private good of their own:

> Angelorum bonorum et malorum inter se contrarios appetitus non naturis principiisque diversis, cum Deus omnium substantiarum bonus auctor et conditor utrosque creaverit; sed voluntatibus et cupiditatibus exstitisse, dubitare fas non est; dum alii constanter *in communi omnibus bono, quod ipse illis Deus est,* atque in ejus aeternitate, veritate, charitate persistunt; alii sua potestate potius delectati, velut bonum suum sibi ipsi essent, *a superiore communi omnium beatifico bono ad propria defluxerunt,* et habentes elationis fastum pro excelsissima aeternitate, vanitatis astutiam pro certissima veritate, studia partium pro individua charitate, superbi, fallaces, invidi effecti sunt.[183]

That, I believe, was authentic personalism in high places. Yet, it is quite different from contemporary doctrine. Before explaining what we mean let us quote from the second page of Father Eschmann's article:

> This is the personalism which is at issue in a passage on page three of Professor De Koninck's book, a passage which recalls by its style and bearing the "heroic" ages of baroque-Scholastic controversy:
>
>> Le péché des anges fut une erreur pratiquement personnaliste: ils ont préféré la dignité de leur propre personne à la dignité qui leur serait venue dans la subordination à un bien supérieur mais commun dans sa supériorité même.
>>
>> L'hérésie pélagienne, dit Jean de Saint Thomas, peut être considérée comme une étincelle de ce péché des anges. Elle n'en est qu'une étincelle, car, alors que l'erreur des anges fut purement pratique, l'erreur des pélagiens était en même temps spéculative. Nous croyons que le personnalisme moderne n'est qu'une réflexion de cette étin-

> celle, spéculativement encore plus faible. Il érige en doctrine spéculative une erreur qui fut a l'origine seulement pratique. . . .
>
> . . . Nous n'entendons pas soutenir ici que l'erreur de tous ceux qui se disent aujourd'hui personnalistes est plus que spéculative. Qu'il n'y ait là-dessus aucune ambiguïté. Sans doute notre insistance pourra-t-elle blesser ceux des personnalistes qui ont identifié cette doctrine à leur personne. C'est là leur responsabilité très personnelle. Mais il y a aussi la nôtre—nous jugeons cette doctrine pernicieuse à l'extrême. (*DM,* 184)

The squib "How many Angels can dance on a pin-point?" has been, perhaps, not without its effect even on learned Catholic circles. It is considered in bad taste to even mention the pure spirits—except, of course, in "objective" Historical Point of View research. In fact, we are led to wonder why God bothered to tell us of them, and why He has repeatedly warned us against those that move in the darkness. Yet, He seems to be of the opinion that Angel and Devil play a rather prominent role in His universe and concern us more than the Evil One would like us to believe. And is it not He who tells us: *invidia autem diaboli mors introivit in orbem terrarum: imitantur autem illum qui sunt ex parte illius?*[184] Christ Himself has said: *Vos ex parte diabolo estis: et desideria patris vestri vultis facere. Ille homicida erat ab initio, et in veritate non stetit.*[185] We are warned that Satan will *seduce the nations,*[186] and in daily evening prayers the Church repeats the words of St. Peter: *Sobrii estote et vigilate: quia adversarius vester diabolus tamquam leo rugiens circuit, quaerens quem devoret.*[187] The invidious personalism of the Devil is our concern, and on highest authority we must fear him and pay no heed to those who smile at our solicitude.

Now, in the first sentence of the citation which recalled to my Opponent, "by its style and bearing the 'heroic' ages of baroque-Scholastic controversy,"[188] it is stated that the Angels could commit no speculative error—even Adam shared in this privilege.[189] The Angels knew, therefore, that divine beatitude could not possibly become their proper good. Their error could only have consisted in *ignorantia electionis,* a purely practical error. "The sin of the Angels was a practical personalist error: they preferred the dignity of their own person to that dignity which would come to them through their subordination to a good, higher but common in its very sovereignty" (*BC,* 3).

The Pelagians on the other hand, were guilty of simple speculative error when they taught that, absolutely speaking, the natural powers were adequate to achieve the supernatural end of man. To maintain that the Angels could entertain such crass speculative ignorance would be to ignore the power of their intelligence.

> Nec oportet [diabolo] attribuere errorem Pelagii de habendo merita condigna ex propria natura, quia ipsi non volebant consequi formaliter, et de facto gloriam per sua naturalia, sed recusabant habere illam, si per gratiam consequenda erat, ut vere erat. Unde non habuerunt errorem speculativam Pelagii, sed habuerunt maximam superbiam, unde erupit scintilla erroris Pelagii.[190]

Now the personalism I attack shows an even greater speculative debility than that of Pelagianism, since it mistakes not just the means of attaining supernatural beatitude, but bears directly on the nature of God Himself. It is deserving of more indulgence only because it is more stupid.[191]

That is what was meant by the first paragraph Father Eschmann quotes. The continuation of it is possibly even more "baroque-Scholastic" in style:

> L'asservissement de la personne au nom du bien commun est comme une vengeance diabolique à la fois remarquable et cruelle, une attaque sournoise contre la communauté du bien à laquelle le démon avait refusé de se soumettre. La négation de la dignité supérieure que l'homme reçoit dans la subordination de son bien tout personnel au bien commun assurerait la négation de toute dignité humaine. (*BC*, 3–4)

And by the negation of all human dignity we mean the fruits of personalism. Quite logically the inordinate exaltation of the human person has a principle and term in contempt of the other person.

> Ce refus de la primauté du bien commun procède, au fond, de la méfiance et du mépris des personnes. (*BC*, 19)

To grasp this we have only to recall what has been said already regarding the dual aspect, relative and absolute, in which we can consider the amiability of our neighbour. The created persons are amiable to God in the measure of the goodness He gratuitously bestows upon them. We, however, cannot love them according to their nearness to Him, but only according to their nearness to us. On the other hand, any created good, including ourselves, pales to nonentity before the divine good which in beatitude becomes our good—our common good. But to rejoice in the fact that our neighbour is only a particular good "infra *nos* constitutum" may be a rather doubtful attitude. Emphasis on the *nos* is definitely jeopardizing. Our Lord was rather insistent that some would be first and some last and what He said concerned beatitude. That is one point of view not to be ignored. In the end, His point of view must prevail. However, when *our* point of view (that of our neighbour's nearness to us) prevails over the former, then, of course, any common good, as well as any particular good except ourselves, becomes a mere *bonum utile*—i.e. a good only as a means, for the sake of that good which is our insatiable *ego*.

That the Devil exists, that he is envious of man, that he is a homicide, that in envy and revenge he craves our imitation of his initial deed, is uncreated truth. This may be called "baroque," yet it is truth divine. It is, absolutely speaking, more true than our own existence.

Such being the gravity of the error we attacked, the reader will understand why we were careful to add that in no way did we consider "the error of all those who call themselves personalists to be more than speculative. Let there be no ambiguity about that." For a man is good, not because of his science, but because of the rectitude of his appetite. Yet, we could hardly fail to disturb those personalists who have identified this speculative doctrine with their own person. And if personalism implied what we were certain it did imply (even before Father Eschmann made it rather explicit), the obligation was upon us to say just what that implication was. It would have been quite merciless not to say it.

And now let us turn to one more of my Opponent's statements. Immediately following the above quotation he proceeds:

> There is a proper and profound Thomistic doctrine of the relative superiority, within definite orders, of *their* respective common goods over the particular goods contained in those orders. It is this doctrine which Professor De Koninck has distorted into the contradictory and

> unintelligible position of the absolute superiority of "the" common good over all and everything. This will be shown later in detail. (*DM*, 184–85)

We believe our Opponent had a fair chance to show that our position is "contradictory and unintelligible." Yet, having carefully read his article to the end, if we accepted his conclusion it could only be on his word. We must, however, appreciate his predicament, for, when a *dictum authenticum* is also a *per se notum quoad sapientes*,[192] its rejection gives rise to endless difficulties.

Father Eschmnann's denunciation calls to our mind an aptly phrased indictment by that "ravenously affectionate uncle *Screwtape*":

> The whole philosophy of Hell rests on recognition of the axiom that one thing is not another thing, and, especially, that one self is not another self. My good is my good and your good is yours. . . .
>
> Now the Enemy's[193] philosophy is nothing more nor less than one continued attempt to evade this very obvious truth. He aims at a contradiction. Things are to be many, yet somehow also one. The good of one self is to be the good of another. This impossibility he calls *love*, and this same monotonous panacea can be detected under all He does and even all He is—or claims to be.[194]

The article I have written, long and difficult as it is, will doubtless tax the patience and energies of many of its readers. It has been composed with a threefold purpose: to vindicate the truth, to vindicate St. Thomas, and to utter a word in defence of the personage who so kindly wrote the preface of the little book which has been the occasion of so much controversy. While I hope my work will reveal a spirit of sincerity and devotion to truth, it is not difficult for me to believe that the task could have been done much better, that stronger arguments might have been found and, above all, that they might have been presented much more effectively. Still, I am convinced that the reasons here given are sufficient to establish the truth, and should they not succeed in convincing the adversaries, some other writer will surely appear with power to enlighten their ignorance.

NOTES

1. *In XII Metaphysicorum* (ed. Cathala), lectio 9, n. 2566.

2. "On the Common Good," in *The Review of Politics* 6, no. 4 (1944): 530–33.

3. *De la primauté du bien commun contre les personnalistes*, préface de S. E. le Cardinal Villeneuve (Québec: Editions de l'Université Laval; Montréal: Editions Fides, 1943). I shall use the initials *BC* in my references to this work. The number following indicates the page.

4. Yves Simon, "On the Common Good," 530.

5. Ibid., 531.

6. Ibid., 533.

7. Ibid., 532.

8. Ibid., 533.

9. Ibid., 531.

10. Ibid., 532–33.

11. Ibid., 533.

12. "In Defense of Jacques Maritain," *The Modern Schoolman* 22, no. 4 (May 1945): 183–208. I shall henceforth refer to this article by the initials *DM*.

13. Unless the reader is acquainted with Father Eschmann's own complete text he will hardly appreciate the directness of this reply.

14. "Ex hoc etiam patet diligenter consideranti dictum eorum qui posuerunt mundum semper fuisse: quia nihilominus ponunt eum a Deo factum, nihil de hac repugnantia intellectuum sentientes. Ergo illi qui tam subtiliter eam percipiunt, soli sunt homines, et cum eis oritur sapientia" (*De Aeternitate mundi contra murmurantes, Opuscula Omnia*, ed. Mandonnet, t. 1, p. 26).

15. "Si quis autem gloriabundus de falsi nominis scientia velit contra haec quae scripsimus aliquid dicere, non loquatur in angulis, nec coram pueris, qui nesciunt de causis arduis judicare, sed contra hoc scriptum scribat, si audet: et inveniet non solum me, qui aliorum sum minimus, sed multos alios, qui veritatis sunt cultores, per quos ejus errori resistetur, vel ignorantiae consuletur" (*De Unitate intellectus contra averroistas parisienses*, ibid., 69).

16. Section I of Father Eschmann's article bears the title, "On Censures, Insinuations, and Citations." See below, n. 191.

17. The reader is warned that he may find this paper difficult to follow because of its apparent lack of plan. In order to write a true rebuttal of my Opponent's attack I have felt obliged to forsake an order more in accordance with the nature of the subject and intend merely to follow him step by step through the pages of his own work. Now and then, to be sure, I may give a quotation from an earlier or later page when it seems to state more fully and clearly some point under discussion; and there are also certain passages towards the beginning of my Opponent's work (his handling of the words of the Encyclicals and his remark on "baroque-Scholastic controversy") which could be dealt with only at the close of my article for reasons the reader will discern by the time he reaches the last chapters. But my general procedure results inevitably in

overlapping and repetition, and in abrupt transitions from one subject to the next, so that a considerable demand will be made on the attention. It is scarcely necessary to add that anyone seriously interested in this question should first read Father Eschmann's work and indeed keep a copy of it at hand as he studies this reply. Since most of my Opponent's citations from St. Thomas are given in Latin only, I have not felt obliged to furnish translations of my own.

18. IIaIIae, q. 64, a. 2, c.

19. Ibid., q. 65, a. 1, c.

20. Indeed on p. 30 of my essay I said, "Bien sûr qu'on se révoltera contre cette conception si. . . ." If Father Eschmann believes that the object of "on se révoltera" may, in this instance, be rendered by "revolting," he is ill advised.

21. *In III Physicorum,* lectio 12, n. 2.

22. Ia, q. 65, a. 2, c.

23. ". . . Quod si totum aliquod non sit ultimus finis, sed ordinetur ad finem ulteriorem, ultimus finis partis non est ipsum totum, sed aliquid aliud. Universitas autem creaturarum, ad quam comparatur homo ut pars ad totum, non est ultimus finis, sed ordinatur in Deum sicut in ultimum finem. Unde bonum universi non est ultimus finis hominis, sed ipse Deus" (IaIIae, q. 2, a. 8, ad 2).

24. *Summa contra gentes,* II, c. 39.

25. Ibid., c. 45.

26. "Aliter dicendum est de productione unius particularis creaturae, et aliter de exitu totius universi a Deo. Cum enim loquimur de productione alicujus singularis creaturae, potest assignari ratio quare talis sit, ex aliqua alia creatura, vel saltem ex ordine universi, ad quem quaelibet creatura ordinatur, sicut pars ad formam totius. Cum autem de toto universo loquimur educendo in esse, non possumus ulterius aliquod creatum invenire ex quo possit sumi ratio quare sit tale vel tale; unde, cum nec etiam ex parte divinae potentiae quae est infinita, nec divinae bonitatis, quae rebus non indiget, ratio determinatae dispositionis universi sumi potest, oportet quod ejus ratio sumatur ex simplici voluntate producentis ut si quaeratur, quare quantitas caeli est tanta et non major, non potest hujus ratio reddi nisi ex voluntate producentis" (*Q.D. de potentia,* q. 3, a. 17, c.).

27. *In I ad Corinthios,* c. 12, lectio 3.

28. *Q.D. de veritate,* q. 21, a. 1, c.

29. Ibid., ad 4.

30. IIaIIae, q. 58, a. 9, ad 3.

31. Father Eschmann quotes this passage (*DM,* 198), but ignores its implication.

32. "Sciendum est autem, quod hoc totum, quod est civilis multitudo, vel domestica familia, habet *solam unitatem ordinis,* secundum quam non est aliquid simpliciter unum. Et ideo pars ejus totius, potest habere operationem, quae non est operatio totius, puta miles in exercitu habet operationem quae non est totius exercitus. Habet nihilominus et ipsum totum aliquam operationem, quae non est propria alicujus partium, sed totius, puta conflictus totius exercitus. Et tractus navis est operatio multitudinis trahentium navem. Est autem aliquid totum, quod habet unitatem non

solum ordine, sed *compositione;* aut *colligatione,* vel etiam *continuitate,* secundum quam unitatem est aliquid unum simplicter; et ideo nulla est operatio partis, quae non sit totius. In continuis enim idem est motus totius et partis; et similiter in compositis, vel colligatis, operatio partis principaliter est totius; et ideo oportet, quod ad eamdem scientiam pertineat talis consideratio et totius et partis ejus. Non autem ad eamdem scientiam pertinet considerare totum quod habet solum ordinis unitatem, et partes ipsius" (*In I Ethicorum* [ed. Pirotta], lectio 1, n. 5).

33. "Unusquisque seipsum in Deum ordinat sicut pars ordinatur ad bonum totius . . ." (*De perfectione vitae spiritualis,* c. 13).

34. *In I Sententiarum,* d. 19, q. 4, a. 1; d. 24, q. 1, a. 1, ad 4.

35. Ia, q. 3, a. 8, c. Cf. G. Théry, O.P., *Essai sur David de Dinant d'après Albert le Grand et saint Thomas, Mélanges Thomistes* (Kain, Belgium: Le Saulchoir, 1923), 361–408.

36. *Summa contra gentes,* II, c. 42.

37. Ibid., c. 44.

38. Ibid.

39. Ibid., I, c. 70.

40. Ibid., c. 71.

41. Ia, q. 22, a. 4, c.

42. Ibid., q. 49, a. 2., c.

43. *Q.D. de veritate,* q. 5, a. 3, c.

44. Ibid., ad 3.

45. Father Eschmann's note: *Eth.,* i, 1; 1094b 9; *Metaph.,* xii, 1075a 11.

46. *Q.D. de spiritualibus creaturis,* a. 8, ad 5.

47. *Summa contra gentes,* II, c. 45.

48. Ibid.

49. Ibid.

50. Ibid., III, c. 97.

51. Ibid.

52. *Q.D. de potentia,* q. 3, a. 16, c. See also *Compendium theologiae,* cc. 72, 73, 102.

53. Ia, q. 3, a. 7, ad 2.

54. The latitude of the terms *intensive* and *extensive* may be shown from the following text of St. Thomas, *In I Sententiarum,* d. 44, q. 1, a. 2, in answer to the question: *Utrum Deus potuerit facere universum melius.*

"Respondeo dicendum, quod secundum Philosophum, in XI *Metaphys.,* text 52, bonum universi consistit in duplici ordine: scilicet in ordine partium universi ad invicem, et in ordine totius universi ad finem, qui est ipse Deus; sicut etiam est in exercitu ordo partium exercitus ad invicem, secundum diversa officia, et est ordo ad bonum ducis, quod est victoria; et hic ordo est praecipuus, propter quem est primus ordo. Accipiendo ergo bonum ordinis qui est in partibus universi ad invicem, potest considerari, vel quantum ad partes ipsas ordinatas, vel quantum ad ordinem partium. Si quantum ad partes ipsas, tunc potest intelligi universum fieri melius, vel per additionem plurium partium, ut scilicet crearentur multae aliae species, et implerentur multi gradus bonitatis qui possunt esse, cum etiam inter summam creaturam et Deum infinita distantia sit; et sic Deus melius universum facere potuisset et posset: sed illud universum se haberet ad hoc sicut totum ad partem; et sic nec penitus esset idem, nec penitus diversum; et haec additio bonitatis esset per modum quantitatis discretae. Vel potest intelligi fieri melius quasi intensive, et hoc mutatis omnibus partibus ejus in melius, quia si aliquae partes meliorarentur aliis non melioratis, non esset tanta bonitas ordinis; sicut patet in cithara, cujus si omnes chordae meliorantur, fit dulcior harmonia, sed quibusdam tantum melioratis, fit dissonantia. Haec autem melioratio omnium partium, vel potest intelligi secundum bonitation accidentalem, et sic posset esse talis melioratio a Deo manentibus eisdem partibus et eodem universo; vel secundum bonitatem essentialem, et sic esset Deo possibilis, qui infinitas alias species condere potest. Sed sic non essent eadem partes, et consequens nec idem universum, ut ex praedictis patet. Si autem accipiatur ipse ordo partium, sic non potest esse melior per modum quantitatis discretae, nisi fieret additio in partibus universi; quia in universo nihil est inordinatum, sed intensive posset esse melior manentibus eisdem partibus quantum ad ordinem qui sequitur bonitatem accidentalem: quanto enim aliquid in majus bonum redundat, tanto ordo melior est. Sed ordo qui sequitur bonitatem essentialem, non posset esse melior, nisi fierent aliae partes et aliud universum. Similiter ordo qui est ad finem, potest considerari, vel ex parte ipsius finis; et sic non posset esse melior, ut scilicet in meliorem finem universum ordinaretur, sicut Deo nihil melius esse potest: vel quantum ad ipsum ordinem; et sic secundum quod cresceret bonitas partium universi et ordo earum ad invicem, posset meliorari ordo in finem, ex eo quod propinquius ad finem se haberent, quanto similitudinem divinae bonitatis magis consequerentur, quae est omnium finis.

"Ad sextum dicendum quod quamvis angelus absolute sit melior quam lapis, tamen utraque natura est melior quam altera tantum: et ideo melius est universum in quo sunt angeli et aliae res, quam ubi essent angeli tantum, quia *perfectio universi attenditur essentialiter secundum diversitatem naturarum,* quibus implentur diversi gradus bonitatis, et non secundum multiplicationem individuorum in una natura."

55. Ia, q. 47, a. 3, ad 2.

56. Ibid., a. 2.

57. *Q.D. de spiritualibus creaturis,* a. 8, ad 10; *Summa contra gentes,* II, c. 92.—"Intensive et *collective,*" a single created person is, *absolutely,* more perfect than the irrational parts of the universe; but this does not apply to a single person compared to the ensemble of other persons.

58. Ia, q. 50, a. 3.

59. *De divinis nominibus,* c. 13, lectio 2.

60. *Summa contra gentes,* II, c. 45.

61. Ia, q. 47, a. 1, ad 2.

62. *Q.D. de veritate,* q. 2, a. 4, ad 2.

63. Ia, q. 47, a. 2, ad 1.

64. Another explanation of this distinction (intensive, extensive) is to be found in the answer to the question: *Utrum Deus principalius incarnatus fuerit in remedium actualium peccatorum quam in remedium originalis peccati* (IIIa, q. 1, a. 4). We shall quote the relevant part of the body of the article as well as the third objection and answer.

"Tanto autem principalius ad alicuius peccati deletionem Christus venit, quanto illud peccatum maius est. Dicitur autem maius aliquid dupliciter. Uno modo, intensive: sicut est maior albedo quae est intensior. Et per hunc modum maius est peccatum actuale quam originale: quia plus habet de ratione voluntarii, ut in Secundo dictum est.—Alio modo dicitur aliquid maius extensive: sicut dicitur maior albedo quae est in maiori superficie. Et hoc modo peccatum originale, per quod totum genus humanum inficitur, est maius quolibet peccato actuali, quod est proprium singularis personae. Et quantum ad hoc, Christus principalius venit ad tollendum originale peccatum: inquantum *bonum gentis divinius quam bonum unius,* ut dicitur in I *Ethic.*"

The third argument *in contrarium* was: "Praeterea, sicut Chrysostomus dicit, in II *de Compunctione Cordis, hic est affectus servi fidelis, ut beneficia domini sui quae communiter omnibus data sunt, quasi sibi soli praestita reputet: quasi enim de solo loquens Paulus ita scribit, ad Galat. 2: Dilexit me, et tradidit semetipsum pro me.* Sed propria peccata nostra sunt actualia: originale enim est *commune peccatum.* Ergo hunc affectum debemus habere, ut aestimemus eum principaliter propter actualia peccata venisse.

"Ad tertium respondetur dicendum quod, sicut Chrysostomus ibidem inducit, verba illa dicebat Apostolus, *non quasi diminuere volens amplissima et per orbem terrarum diffusa Christi munera: sed ut pro omnibus se solum indicaret obnoxium. Quid enim interest si et aliis praestitit, cum quae tibi sunt praestita ita integra sunt et ita perfecta quasi nulli alii ex his aliquid fuerit praestitum?* Ex hoc ergo quod aliquis debet sibi reputare beneficia Christi praestita esse, non debet existimare quod non sint praestita aliis. Et ideo non excluditur quin principalius venerit abolere peccatum totius naturae quam peccatum unius personae. Sed illud peccatum commune ita perfecte curatum est in unoquoque ac si in eo solo esset curatum.—Et praeterea, propter unionem caritatis, totum quod omnibus est impensum, unusquisque debet sibi adscribere."

In his commentary on this article, Cajetan says: ". . . Auctor, explicando secundam conclusionem et non primam, et addendo secundae conclusioni rationem, scilicet, *quia bonum gentis divinius et eminentius est quam bonum unius,* insinuavit conclusionem responsivam quaesito simpliciter et absolute illa esse quam expressit in responsione ad tertium: *Principalius venit abolere peccatum totius naturae quam peccatum unius personae.*" See also F. C. R. Billuart, *Summa Sancti Thomae,* tract. de Incarnatione, dissert. 3, ad 4.

65. *In II Sententiarum,* d. 16, q. 1, a. 2, ad 3.

66. Father Eschmann's reference: *In Meta.,* xii, 12; *De Verit.,* 5. 5 ; *De Spirit. Creaturis,* art. 8.

67. Father Eschmann's reference: *Expos. in Ep. ad Rom.,* c. 8, lect. 6.

68. *In Ep. ad Romanos,* c. 8, lectio 6.

69. *Summa contra gentes,* III, c. 17.

70. The exclamation mark is Father Eschmann's.

71. See, for example, *'Bradley's Arnold' Latin Prose Composition,* ed. and rev. J. F. Mountford (London, New York, Toronto: Longmans, Green, 1938), 205, paragraph 361—or any elementary Latin grammar.

72. I quoted a text (*BC,* 24; 174, n. 5) which should have been a warning: "Est autem *quoddam bonum commune* quod pertinet ad hunc vel ad illum inquantum est

pars alicujus totius, sicut ad militem, inquantum est pars exercitus, et ad civem, inquantum est *pars civitatis* . . ." (*Q.D. de caritate,* a. 4, ad 2). Does St. Thomas mean that the common good of the citizen is a common good only in a certain sense? When St. Thomas says: "verum est quoddam bonum," does he mean that it is good only in a certain sense, that is, not properly? My Opponent's faulty Latin would destroy the entire Aristotelian and Thomistic doctrine of the speculative and practical intellect, and more particularly the absolute primacy of the speculative. Cf. Ia, q. 82, a. 3, ad 3 (ibid., Cajetan, nn. 21–24); ibid., a. 4, ad 1; *Q.D. de veritate,* q. 14, a. 4, c.; *In III Sententiarum,* d. 23, q. 2, a. 3, sol. 2, ad 3.

73. Here is St. Thomas' proof in the body of the article: "Sed falsitas huius opinionis (scil., angelus naturali dilectione plus diligit se quam Deum), manifeste apparet, si quis in rebus naturalibus consideret ad quid res naturaliter moveatur: inclinatio enim naturalis in his quae sunt sine ratione, demonstrat inclinationem naturalem in voluntate intellectualis naturae. Unumquodque autem in rebus naturalibus, quod secundum naturam hoc ipsum quod est, alterius est, principalius et magis inclinatur in id cuius est, quam in seipsum. Et haec inclinatio naturalis demonstratur ex his quae naturaliter aguntur; quia *unumquodque, sicut agitur naturaliter, sic aptum natum est agi,* ut dicitur in II *Physic.* Videmus enim quod naturaliter pars se exponit, ad conservationem totius: sicut manus exponitur ictui, absque deliberatione, ad conservationem totius corporis. Et quia ratio imitatur naturam, huiusmodi inclinationem invenimus in virtutibus politicis: est enim virtuosi civis, ut se exponat mortis periculo pro totius reipublicae conservatione; et si homo esset naturalis pars huius civitatis, haec inclinatio esset ei naturalis.

"Quia igitur bonum universale est ipse Deus, et sub hoc bono continetur etiam angelus et homo et omnis creatura, quia omnis creatura naturaliter, secundum id quod est, Dei est; sequitur quod naturali dilectione etiam angelus et homo plus et principalius diligat Deum quam seipsum.—Alioquin, si naturaliter plus seipsum diligeret quam Deum, sequeretur quod naturalis dilectio esset perversa; et quod non perficeretur per caritatem, sed destrueretur." The other answers to the arguments *in contrarium* are also relevant to our problem.

"Ad primum ergo dicendum quod ratio illa procedit in his quae ex aequo dividuntur, quorum unum non est alteri ratio existendi et bonitatis: in talibus enim unumquodque diligit naturaliter magis seipsum quam alterum, inquantum est magis sibi ipsi unum quam alteri. Sed in illis quorum unum est tota ratio existendi et bonitatis alii, magis diligitur naturaliter tale alterum quam ipsum; sicut dictum est quod unaquaeque pars diligit naturaliter totum plus quam se. Et quodlibet singulare naturaliter diligit plus bonum suae speciei, quam bonum suum singulare. Deus autem non solum est bonum unius speciei, sed est ipsum universale bonum simpliciter. Unde unumquodque suo modo naturaliter diligit Deum plus quam seipsum."

"Ad tertium dicendum quod natura reflectitur in seipsam non solum quantum ad id quod est ei singulare, sed multo magis quantum ad commune: inclinatur enim unumquodque ad conservandum non solum suum individuum, sed etiam suam speciem. Et multo magis habet naturalem inclinationem unumquodque in id quod est bonum universale simpliciter."

"Ad quartum dicendum quod Deus, secundum quod est universale bonum, a quo dependet omne bonum naturale, diligitur naturali dilectione ab unoquoque. Inquantum vero est bonum beatificans naturaliter omnes supernaturali beatitudine, sic diligitur dilectione caritatis."

74. *Quodlibetum* I, q. 4, a. 8, c.
75. IaIIae, q. 109, a. 3, c.
76. *Ethics,* IX, c. 4.
77. IIaIIae, q. 26, a. 3. See Cajetan's commentary.
78. 1 John 4:20.
79. See Cajetan's commentary, n. 6. Also IaIIae, q. 1, a. 4, ad 1, with commentary by the same.
80. IIaIIae, q. 58, a. 9, ad 3.
81. Father Eschmann's quotation begins here.
82. His quotation ends here.
83. IaIIae, q. 63, a. 4, c.
84. *Q.D. de caritate,* a. 5, ad 2.
85. Father Eschmann's quotation "principle of the New Order," is an allusion to the title of the second part of my book, which is mainly concerned with Marxist radicalism and nihilism as a logical outcome of the exaltation of the self, and it is this I call *le principe de l'ordre nouveau,* as opposed to the principle of the order of Redemption—humility and divine Mercy.
86. For these distinctions as well as for their application to God, see more particularly, Ia, q. 5, a. 1, ad 1 and 3; *De divinis nominibus,* c. 4, lectio 16; *In Boethii de Hebdomadibus,* o.; *Q.D. de veritate,* q. 21, *passim.*
87. A similar distinction applies even to our present knowledge of God. Metaphysics can reach God only insofar as He is knowable through the creatures, whereas faith and theology concern God as He is in Himself. "Sacra autem doctrina propriissime determinat de Deo secundum quod est altissima causa: quia non solum quantum ad illud quod per creaturas cognoscibile (quod philosophi cognoverunt, ut dicitur *Rom.* 1, *quod notum est Dei, manifestum est illis*): sed etiam quantum ad id quod notum est sibi soli de seipso, et aliis per revelationem communicatum" (Ia, q. 1, a. 6, c.).
88. *Q.D. de veritate,* q. 21, a. 3, ad 2.
89. Ibid., q. 21, a. 4, c.
90. Ibid., q. 22, a. 1, ad 7.
91. Ibid., q. 21, a. 5; Ia, q. 6, a. 3.
92. Ia, q. 103, a. 4, c.
93. *Q.D. de veritate,* q. 21, a. 4, c.
94. Ibid., a. 1, ad 4.
95. Father Eschmann cites *2 Sent.,* d. 26, 1, 1 ad 2.
96. He cites *ST,* I-II, 4. 8 ad 3.
97. He cites loc. cit.
98. He cites *In II Thess.,* c. 3, lectio 2.
99. IaIIae, q. 5, a. 3, ad 2.
100. Ia, q. 60, a. 5, ad 5.

101. IIaIIae, q. 26, a. 3.

102. *De perfectione vitae spiritualis,* c. 13.

103. Ludwig Feuerbach, *Das Wesen des Christenthums,* Dritte umgearbeitete und vermehrte Auflage (Leipzig: Wigand, 1849). In *Sämmtliche Werke,* t. 7, p. 212.

104. *Commentariorum ac Disputationum in Tertiam Partem Sancti Thomae,* Tomus Primus (Antverpiae: Petrum et Joannem Belleros, 1621), q. 25, d. 98, c. 2, p. 773.

105. *Cursus theologicus,* ed. Vivès (Paris, 1884), t. VII, d. 8, n. 12, p. 423

106. Ibid., n. 13, p. 424.

107. He cites *In Eth.* I, 2, ed. Pirotta, n. 30. And adds that the next quotation in the article is from the same place.

108. *In I Ethicorum,* lectio 2, n. 30.

109. "Sed adhuc alia differentia invenitur inter divinam bonitatem et creaturae; bonitas enim habet rationem causae finalis. Deus autem habet, rationem causae finalis cum sit omnium ultimus finis, sicut et primum principium; ex quo oportet ut omnis alius finis non habeat habitudinem vel rationem finis nisi secundum ordinem ad causam primam; quia causa secunda non influit in suum causatum nisi praesupposito influxu causae primae, ut patet in lib. de *Causis* (prop. 1); unde et bonum quod habet rationem finis non potest dici de creatura, nisi praesupposito ordine creatoris ad creaturam" (*Q.D. de veritate,* q. 21, a. 5, c.).

". . . Cum ens dicatur absolute, bonum autem superaddat habitudinem causae finalis; ipsa essentia rei absolute considerata sufficit ad hoc quod per eam dicatur aliquid ens, non autem ad hoc quod per eam dicatur aliquid bonum, sicut in aliis generibus causarum, habitudo secundae causae dependet ex habitudine causae primae; primae vero causae habitudo non dependet ex aliquo alio; ita est in causis finalibus, quia secundi fines participant habitudinem causae finalis ex ordine ad ultimum finem, ipse autem ultimus finis habet hanc habitudinem ex seipso; et inde est quod essentia Dei, quae est ultimus finis rerum, sufficit ad hoc quod per eam dicatur Deus bonus; sed essentia creaturae posita non dicitur res bona nisi ex habitudine ad Deum, ex qua habet rationem causae finalis. Et pro tanto dicitur quod creatura non est bona per essentiam, sed per participationem; uno modo scilicet in quantum ipsa essentia secundum rationem intelligendi consideratur ut aliud quid quam habitudo ad Deum, a qua habet rationem causae finalis, et ad quem ordinatur ut ad finem; sed secundum alium modum creatura potest dici per essentiam bona, in quantum scilicet essentia creaturae non invenitur sine habitudine ad Dei bonitatem; et hoc intendit Boetius in lib. de *Hebdom*" (ibid., a. l, ad 1).

110. *Metaph.,* XII, c. 7.

111. Ia, q. 26, a. 3, ad 1.

112. He cites *De Civ. Dei,* XIX, 13.

113. He cites *ST,* I-II, 4, 8 ad 3.

114. This proposition must be taken formally.—That which is formally the object of the practical intellect is the operable *qua* operable, for the operable may be also an object of speculative knowledge: "ut puta si aedificator consideret domum definiendo et dividendo et considerando universalia praedicata ipsius. Hoc siquidem

est operabilia modo speculativo considerare, et non secundum quod operabilia sunt" (Ia, q. 14, a. 16, c.). The term *operabile* must be taken for the genus divided into *factibile* and *agibile.*

115. When we say that God may be known practically, we do not mean, of course, that He can be a proper object of practical knowledge. As John of St. Thomas explains: ". . . Licet primarium objectum [theologiae], quod est Deus, non sit operabile operatione factiva, est tamen attingibile operatione morali per amorem, tamquam finis ultimus et regula actionum nostrarum, et sic practice cognoscibilis" (*Cursus theologicus* [ed. Solesm.], t. I, d. 2, a. 10, n. 18, p. 400).

116. For the expression *finis intelligibilis,* see IaIIae, q. 3, a. 4.

117. ". . . Regnum Dei, quasi antonomastice, dupliciter dicitur: quandoque congregatio eorum qui per fidem ambulant; et sic Ecclesia militans regnum Dei dicitur: quandoque autem illorum collegium qui jam in fine stabiliti sunt; et sic ipsa Ecclesia triumphans regnum Dei dicitur; et hoc modo *esse in regno Dei idem est quod esse in beatitudine. Nec differt secundum hoc, regnum Dei a beatitudine, nisi sicut differt bonum commune totius multitudinis a bono singulari uniuscujusque*" (*In IV Sententiarum,* d. 49, q.1, a. 2, sol. 5).

118. *In IV Sententiarum,* d. 49, q. 1, a. 2, sol. 2, ad 2.

119. I presume he opposes *in linea essendi* to *causandi* alone as understood in subdivision [ii] of the first text, and not to *operandi,* for, "cum Dei substantia sit ejus actio, summa assimilatio hominis ad Deum est secundum aliquam operationem. Unde, sicut supra dictum est [q. 3, a. 2], felicitas sive beatitudo, per quam homo maxime Deo conformatur, quae est finis humanae vitae, in operatione consistit" (IaIIae, q. 55, a. 2, ad 3).

120. ". . . bonum quod omnia concupiscunt, est esse, ut patet per Boetium in lib. iii *De Consol.* [pros. x, PL 63, col. 763 et seqq.]; unde ultimum desideratum ab omnibus est esse perfectum, secundum quod est possibile in natura illa. Omne autem quod habet esse ab alio, perfectionem sui esse ab alio habet: quia tanto perfectius esse recipit unumquodque, quanto verius conjungitur essendi principio; unde inferiora corpora propter longe distare a primo principio, esse corruptibile habet, ut patet *II De generatione.,* text. 59. Et ideo ultimus finis cujuslibet rei habentis esse ab alio est duplex: unus exterius, secundum scilicet id quod est desideratae perfectionis principium; alius interius, scilicet ipsa perfectio, quam facit conjunctio ad principium. Unde cum beatitudo sit ultimus hominis finis, duplex erit beatitudo: una *quae est in ipso, scilicet quae est ultima ejus perfectio,* ad quam possibile est ipsum pervenire; et haec est beatitudo creata; alia vero *est extra ipsum, per cujus conjunctionem praemissa beatitudo in ea causatur;* et haec est beatitudo increata, quae est ipse Deus" (*In IV Sententiarum,* d. 49, q. 1, a. 2, sol. 1).

121. Father Eschmann's reference 3 *Sent.,* d. 5, III, 2.

122. In the sentence immediately preceding the phrase *ratio partis contrariatur personae* (*In III Sententiarium,* d. 5, q. 3, a. 2), St. Thomas says: "anima est pars humanae naturae, et non natura *quaedam* per se." How are we to understand "natura quaedam per se" here? Does it mean *natura per se* "in a certain sense"?

123. When we speak of my Opponent's "notion of common good," we must, of course, prescind from the contradictory statements he makes about the very nature of common good, lest we allow him the impossibility of an "equivocal notion." See below, n. 124.

124. Father Eschmann here adds a footnote (*DM*, 198, n. 24): "Speaking of the Aristotelian *eudaimonia*, St. Thomas sometimes calls the felicity a common good: 'Felicitas autem est finis humanae speciei, cum omnes homines ipsam naturaliter desiderent. Felicitas igitur est quoddam commune bonum possibile provenire omnibus hominibus, nisi accidat aliquibus impedimentum quo sint "orbati"' (Arist. *Eth.* i. 1099a).—Of course this is not what Professor De K. means by *le bien commun qu'est la béatitude.* The Thomistic notion of common good is an analogical and very elusive notion."—Indeed it is not what I mean by "le bien commun qu'est la béatitude." Aristotle's *eudaimonia* is formal felicity and hence a purely personal good. When this is called common, the community is one of predication. What, exactly, was Father Eschmann's design in quoting this text? To show that the notion of common good is analogical? If so, his example is the worst he could have chosen, for, if the analogical notion is to embrace what is signified in this text by the expression "quoddam commune bonum," then he must understand this good to be a common good in the proper sense, and therefore a good in the proper sense. In other words, unless he is using the term "analogy" in an improper sense, he implies that *bonum commune in praedicando* is *perfectivum alterius per modum finis.*—That the community in the text he quotes is one of predication can be easily established. Presumably the passage is taken from *Summa contra gentes,* III, c. 39, where St. Thomas shows that human felicity cannot consist in the knowledge of God acquired by demonstration. His first argument is as follows: "Ea enim quae sunt alicujus speciei, perveniunt ad finem illius speciei ut in pluribus: ea enim quae sunt a natura, sunt semper vel in pluribus, deficiunt autem in paucioribus propter aliquam corruptionem. Felicitas autem est finis humanae speciei: cum omnes homines ipsam naturaliter desiderent. Felicitas igitur est quoddam commune bonum possibile provenire omnibus hominibus, nisi accidat aliquibus impedimentum quo sint *orbati.* Ad praedictam autem cognitionem de Deo habendam per viam demonstrationis pauci perveniunt, propter impedimenta hujus cognitionis, quae in principio Libri tetigimus. Non est igitur talis Dei cognitio *essentialiter ipsa humana felicitas.*" Hence, St. Thomas is speaking of *formal* felicity. It follows that the "commune bonum" is a *bonum commune in praedicando.* The sentence, "Felicitas igitur . . . sint *orbati,*" implies a reference to I *Ethic.*, 1099b, where Aristotle says (*versio antiqua* used by St. Thomas): "Erit autem utique et multum commune. Possibile enim existere omnibus non orbatis ad virtutem per quamdam disciplinam et studium." Aristotle is speaking of the felicity whose definition he had established in a preceding chapter, and which St. Thomas expresses in the following terms "felicitas est *operatio propria hominis* secundum virtutem in vita perfecta" (lectio 10, n. 130). In the passage "Erit autem utique . . ." the Philosopher proves that man himself is one of the causes of his own felicity. St. Thomas' commentary is as follows: "Ostendit idem [scil. tolerabiliter dici quod felicitas sit a causa humana] per hoc quod haec positio conveniat felicitati id quod pertinet ad finem alicujus naturae,

ut scilicet sit commune aliquod his quae habent naturam illam. Non enim natura deficit ab eo quod intendit, nisi in paucioribus. Et ita si felicitas est finis humanae naturae, oportet quod possit esse communis omnibus vel pluribus habentibus humanam naturam. Et illud salvatur si sit ex causa humana. Quia si sit per quamdam disciplinam et studium, poterit provenire omnibus non habentibus aliquod impedimentum ad operandum opera virtutis, vel per defectum naturae sicut qui sunt naturaliter stulti, aut per malam consuetudinem quae imitatur naturam" (lectio 14, n. 170). All these texts concern the felicity which is an inherent, proper good. When this good is called something common, or "quoddam commune bonum," the community is one of predication, not of causality. But we have already learned that to Father Eschmann this distinction does not seem very important.—Regarding his last remark in the footnote we have quoted, we might suggest that the analogical notion of common good is even more elusive than he seems to realize. Unless he is using the term analogy in an improper sense, the analogical notion of common good could not possibly comprise both *bonum commune in causando* and *bonum commune in praedicando,* since the latter is not formally a good. When used for the one and for the other, the expression 'common good' is equivocal, not analogical.

125. To the reader unacquainted with my book, I should like to point out that Petrus de Alvernia is not quoted in the body of my writing, but in a footnote (*BC,* 180–81, n. 56). My argument is not based on the footnote, as the reader may verify for himself. In proceeding as if it were, Father Eschmann avoids the true reason I give for my conclusion.

126. "Sicut autem ex modo visionis apparet diversus gradus gloriae in Beatis, ita ex eo quod videtur apparet gloria eadem: nam cujuslibet felicitas est ex hoc quod Dei substantiam videt, ut probatum est. Ideo ergo est quod omnes Beatos facit: non tamen ab eo omnes aequaliter beatitudinem capiunt" (*Summa contra gentes,* III, c. 58). "In quo etiam considerandum est quod quodammodo contrarius est ordo corporalium et spiritualium motuum. Omnium enim corporalium motuum est idem numero primum subjectum, fines vero diversi. Spiritualium vero motuum, scilicet intellectualium apprehensionum et voluntatum, sunt quidem diversa subjecta prima, finis vero numero idem" (ibid.).

127. IIaIIae, q. 8, a. 3.

128. "Hoc etiam tempore [Gregorii X, Thomas] scripsit etiam super Philosophiam, videlicet de Caelo, et de Generatione, sed non complevit; et similiter Politicam. Sed hos libros complevit Magister Petrus de Alvernia, fidelissimus discipulus ejus, Magister in Theologia, et magnus philosophus, et demum Episcopus Claromontensis" (*Hist. Ecclesiastic Rev. Ital. Script.,* t. XI, 1170. Apud Pierre Mandonnet, O.P., *Siger de Brabant* [Louvain, 1911], first part, p. 204).

129. Trans. H. Rackham, Loeb Classical Library.

130. It seems that besides this continuation of St. Thomas' commentary, Peter wrote his own commentary on the *Politics.* "Du commentaire des livres III–VIII de Pierre Auvergne, il faut bien distinguer ses Questions sur les livres I–V et VII de la Politique, qui se lisent dans le manuscrit lat. 16089 ff. 274–318 de la Biblio. Nat."

(Msgr. A. Pelzer, *Revue Néoscolastique* (1920): 219). Msgr M. Grabmann refers to this conclusion in *Die Werke des hl. Thomas von Aquin* (Munster, 1931), 86; see also *Mittelalterliches Geistesleben* (Munich, 1936), t. II, p. 230; P. Glorieux, *Répertoire des Maîtres en théologie de Paris au XIIIe siècle* (Paris, 1933), t. I, p. 415; E. Hocedez, *La vie et les oeuvres de Pierre d'Auvergne, Gregorianum* (1933): 23 and 29. If these *Questiones* are really a distinct commentary it would be interesting to compare the two. The whole matter is further complicated by the fact that there were two Peters of Auvergne, which raises a problem of the authenticity for the many works attributed to the first, excluding of course those mentioned by Ptolemy. Here, however, we are concerned merely with the internal value of the present commentary which was intended as a continuation of St. Thomas' own.

131. *Metaph.*, I, c. 2, 982b25.

132. For the reader who may have a particular interest in this principle of the prudential operation of the City as a whole, here are a few considerations which may be helpful. First we should recall what is laid down in *In I Ethicorum*, lectio l, already quoted above, n. 32. Let him note the lines: "*Habet nihilominus et ipsum totum aliquam operationem, quae non est propria alicujus partium, sed totius,* puta conflictus totius exercitus. Et tractus navis est operatio multitudinis trahentium navem. . . . Non autem ad eamdem scientiam pertinet considerare totum quod habet solam ordinis unitatem, et partes ipsius." Hence, if there were no operation proper to the whole, there would be no distinct science of Politics. Immediately following this passage, St. Thomas gives the divisions of moral philosophy. "Et inde est, quod *moralis philosophia,* in *tres* partes dividitur. Quarum *prima* considerat operationes unius hominis ordinatas ad finem, quae vocatur monastica. *Secunda* autem considerat *operationes multitudinis domesticae,* quae vocatur oeconomica. *Tertia* autem considerat *operationes multitudinis civilis,* qua vocatur politica." This should take care of the genus "operatio totius societatis." Now prudence is *recta ratio agibilium.* In regard to political society, there are two kinds of prudence: the one is called *regnativa,* that is, the prudence of the one who governs the community toward the common good; the other is called simply, *politica,* i.e. that prudence of the subjects governed, by which they, freely and in conformity with the government, direct their actions toward the common good (*In VI Ethicorum* lectio 7; IIaIIae, q. 50, aa. l, 2; *In III Sententiarum,* d. 33, q. 3, a. l, sol. 4). But prudence does not consist mainly in counsel and in judging what should be done, but in actually commanding what should be done (IIaIIae, q. 47, a. 8, c.). Hence, the *operatio prudentiae totius civitatis* will be the prudential operation of society as a whole, involving right reason both on the part of the governing power and of the governed. And to the degree that this is realized there is a chance that the community will enjoy what is, in Aristotelian terms, the practical felicity of the whole society. On *felicilas civitatis,* see St. Thomas' own commentary *In II Politicorum,* lectio 2.

133. IaIIae, q. 3, a. 5, ad 2.

134. ". . . Aliquid dicitur perfectum dupliciter: absolute, et secundum quid. Perfectio quidem beatitudinis absoluta est solius Dei: quia solus ipse tantum cognoscit se et amat quantum cognoscibilis est et amabilis (infinite enim cognoscit, et amat infinitam veritatem et bonitatem suam): et quantum ad hoc, ipsum summum bonum, quod

est beatitudinis objectum, et causa, non potest esse majus et minus: non enim est nisi unum summum bonum, quod est Deus. Secundum quid autem, idest secundum aliquas conditiones temporis, naturae et gratiae; et sic unus potest esse beatior alio secundum adeptionem hujus boni, et capacitatem uniuscujusque hominis: quia quanto homo magis est ejus capax, tanto magis participat ipsam, inquantum scilicet est melius dispositus et ordinatus ad ejus fruitionem" (*In Joann.*, c. 14, lectio l).

135. We must not forget that for God the created persons are themselves *operabilia,* and that they have their root in divine goodness. ". . . Quamvis possit dici quod intueatur ea [quae scil. facere potest] in sua potentia, quia nihil est quod ipse non possit, tamen accommodatius dicitur quod intuetur ea in sua bonitate, quae est finis omnium quae ab eo fiunt" (*Q.D. de veritate,* q. 2, a. 8, c.).

136. John 14:2.

137. *In Joann.*, c. 14, lectio 1.

138. Matth. 18:19–20.

139. "L'incommunicabilité des personnes dans l'acte de vision rompt-elle l'universalité de l'objet? Et l'amour que suscite cet objet, porte-il sur le bien universel comme tel, ou sur le bien pour son appropriation à la personne singulière? Et ce bien, est-il comme un bien commun inférieur dont la distribution entraîne, par voie de conséquence, une division de lui-même et une particularisation où il est dû à la partie comme telle et où il perd sa raison de communauté?" (*BC,* 64–65). Bread would be an example of this lowest kind of common good. Its distribution involves division, particularisation, reduction to proper goods. Yet, in the Blessed Sacrament, under the appearances of this most tenuous form of common good, is really present the highest common good. ". . . Bonum commune spirituale totius Ecclesiae continetur substantialiter in ipso Eucharistiae sacramento" (IIIa, q. 65, a. 3, ad 1). We are reminded of the *Lauda Sion*: "A sumente non concisus, non confractus, non divisus: integer accipitur. Sumit unus, sumunt mille: quantum isti, tantum ille: nec sumptus consumitur. . . . Fracto demum sacramento, ne vacilles, sed memento, tantum esse sub fragmento, quantum toto tegitur. Nulla rei sit scissura: signi tantum fit fratura, qua nec status nec statura signati minuitur." Again, we must weigh, in this connection, the words of St. Paul, 1 Cor. 10:17: *Quoniam unus panis, unum corpus multi sumus, omnes, qui de uno pane participamus.*

140. *Acta Apostolicae Sedis,* 20 July 1943, p. 222. Reference to St. Thomas: *Q.D. de veritate,* q. 29, a. 4, c. Italics mine.

141. In footnote, Father Eschmann here gives his reference as follows: 3 *Sent.*, d. 35, l. 4, sol. 1 ad 2; 4 *Sent.*, d. 49 1.1, sol. 3, ad 1.

142. *In IV Sententiarum,* d. 49, q. 1, a. 1, qa 3, sol. 3.

143. The text Father Eschmann here disposes of (*Q.D. de spiritualibus creaturis,* a. 8, ad 5) was quoted above, at n. 46.

144. IaIIae, q. 5, a. 6, c.

145. I should like it understood that my remarks on such employment of 'history' as this are not to be interpreted as a reflection on authentic historical research like that undertaken even on this continent, and whose quality could not be enhanced by any praise of mine.

146. IIaIIae, q. 26, a. 4, ad 1.

147. On connection with St. Thomas's doctrine on humility, it might be well to ponder the following words: "... Aliquis absque falsitate potest *se credere et pronuntiare omnibus viliorem* (Regula S. Benedicti), secundum defectus occultos quos in se recognoscit, et dona Dei quae in aliis latent. Unde Augustinus dicit, in libro *de Virginitate: Existimate aliquos in occulto superiores, quibus estis in manifesto meliores*" (IIaIIae, q. 161, a. 6, ad 1).

148. "... Secundum ea quae pertinent proprie ad propriam personam alicujus, plus debet exhibere dilectionis effectum parentibus quam extraneis; nisi forte in quantum in bono alicujus extranei penderet bonum commune, quod etiam sibi ipsi imponere quisque debet; ut cum aliquis seipsum periculo mortis exponit ad salvandum in bello ducem exercitus, vel in civitate principem civitatis, in quantum ex eis dependet salus totius communitatis. Sed secundum ea quae pertinent ad aliquid ratione alicujus adjuncti, utpote in quantum est civis vel miles, plus debet obedire rectori civitatis, vel duci, quam patri" (*Q.D. de caritate,* a. 9, ad 15).

149. John of St. Thomas, *Cursus theologicus* (ed. Vivès), t. vii, q. 23, p. 403.

150. "... Sicut ens est quoddam essentiale, et quoddam accidentale; ita et bonum quoddam essentiale, et quoddam accidentale; et eodem modo amittit aliquis bonitatem sicut esse substantiale et accidentale" (*Q.D. de veritate,* q. 21, a. l, ad 6). "... Sicut ens multiplicatur per substantiale et accidentale, sic bonitas multiplicatur; sed tamen inter utrumque differt. Quia aliquid dicitur ens esse absolute propter suum esse substantiale, sed propter esse accidentale non dicitur esse absolute: unde cum generatio sit motus ad esse; cum aliquis accipit esse substantiale, dicitur generari simpliciter; cum vero accipit esse accidentale, dicitur generari secundum quid; et similiter est de corruptione, per quam esse amittitur. De bono autem est e converso. Nam secundum substantialem bonitatem dicitur aliquid bonum secundum quid, secundum vero accidentalem dicitur aliquid bonum simpliciter; unde hominem injustum non dicimus bonum simpliciter, sed secundum quid, in quantum est homo; hominem vero justum dicimus simpliciter bonum. Cujus diversitatis ista est ratio. Nam unumquodque dicitur esse ens in quantum absolute consideratur; bonum vero, ut ex dictis, art. l, ad 6 argum., patet, secundum respectum ad alia. In seipso autem aliquid perficitur ut subsistat per essentialia principia; sed ut debito modo se habeat ad omnia quae sunt extra ipsum, non perficitur nisi mediantibus accidentibus superadditis essentiae; quia operationes quibus unum alteri conjungitur, ab essentia mediantibus virtutibus essentiae superadditis progrediuntur; unde absolute bonitatem non obtinet nisi secundum quod completum est secundum substantialia et secundum accidentalia principia. Quidquid autem creatura perfectionis habet ex essentialibus et accidentalibus principiis simul conjunctis, hoc totum Deus habet per unum suum esse simplex; simplex enim ejus essentia est ejus sapientia et justitia et fortitudo, et omnia hujusmodi, quae in nobis sunt essentiae superaddita. Et ideo ipsa absoluta bonitas in Deo idem est quod ejus essentia; in nobis autem consideratur secundum ea quae superadduntur essentiae. Et pro tanto bonitas completa vel absoluta in nobis et augetur et minuitur et totaliter aufertur, non autem in Deo; quamvis substantialis bonitas in nobis semper maneat" (ibid., a. 5, c.). For other references, see above, n. 86.

151. Spinoza, *Ethica,* pars IV, def. vii.—"Per virtutem et potentiam idem intelligo hoc est (*per Prop.* 7, p. III), virtus, quatenus ad hominem refertur, est ipsa hominis essentia seu natura, quatenus potestatem habet quaedam efficiendi, quae per solas ipsius naturae leges possunt intelligi" (ibid., def. viii.) "Causa autem, quae finalis dicitur, nihil est praeter ipsum humanum appetitum, quatenus is alicujus rei veluti principium seu causa primaria consideratur. Ex. gr. cum dicimus habitationem causam fuisse finalem hujus aut illius domus, nihil tum sane intelligimus aliud, quam quod homo ex eo, quod vitae domesticae commoda imaginatus est, appetitum habuit aedificandi domum. Quare habitatio, quatenus ut finalis causa consideratur, nihil est praeter hunc singularem appetitum, qui revera causa est efficiens, quae ut prima consideratur, quia homines suorum appetituum causas communiter ignorant" (ibid., praefatio).

152. I will quote again a passage from the series of articles on *The Theory of Democracy* by M. J. Adler and Father Walter Farrell: "The intentions of natural justice are selfish. They do not regard the good of another man *as such,* but only as a part of the community which must be preserved for one's own good. Now just as natural justice and natural love are selfish, so neither is heroic. Neither leads men to martyrdom. Though natural love is less selfish than justice, in that it involves some genuine forgetfulness of self, and though natural love, unlike justice, impels men to the generosity of sacrifice, it remains, nevertheless, on the plane of imperfect action, in which the agent always seeks to perfect himself as well as another, and in fact regards the other as an extension of self—as an *alter ego*" (*The Thomist* 4, no. 2 [April 1942]: 329–30). As I pointed out in my book (Appendice II, p. 129 ff.), these authors are at least consistent; they have the candor and courage to go the full distance. "With respect to the common good, it is *necessary to reject as false* [italics theirs] all the passages in which St. Thomas declares that the common good is supreme in the natural, temporal order; or, *if* this is not a fair interpretation of all those texts in which St. Thomas says that the common good takes precedence over the individual good because the good of the whole is greater than the good of its parts, *then* we must at least reject this false interpretation of what St. Thomas *seems* to say, even though it has prevailed among his commentators and followers to this day" (ibid., 336–37).

153. The original was quoted above in the text at n. 76.

154. IIaIIae, q. 26, a. 4, c. ". . . Unitate naturae nihil est magis unum quam nos; sed unitate affectus, cujus objectum est bonum, summe bonum debet esse magis nobis unum quam nos" (*Q.D. de caritate,* a. 9, ad 7).

155. IIaIIae, q. 64, a. 2.

156. *Acta Apostolicae Sedis,* March 1937, p. 79.

157. ". . . Homo non ordinatur ad communitatem politicam secundum se totum, et secundum omnia sua" (IaIIae, q. 21, a. 4, ad 3).

158. In a footnote, Father Eschmann here refers to: *Corpus Iuris Canonici,* C. 19, Q. 2, c. 1: Ed. Lips. Sec. (Friedberg), vol. I, col. 839 ff., and to his own paper: *Bonum commune melius est quam bonum unius. Eine Studie ueber den Wertvorrang des Personalen bei Thomas von Aquin, Mediaeval Studies* (Toronto) 6 (1944): 62–120. His specific reference is to pp. 100 ff., namely section IV: "Das 'Privileg des Heiligen Geistes' und der Thomistische Begriff des Personalen." He also refers to his rather incomplete

and disorderly *Glossary* (*Mediaeval Studies* 5 [1943]: 142 ff.) as a "complete collection of the Thomistic texts regarding the *dictum authenticum* of the relative primacy of the common good" (*DM*, 201). A much more complete group of texts will appear in a forthcoming issue of the *Laval théologique et philosophique.*

159. IaIIae, q. 90, a. 4, c

160. Ibid., a. 3, c.

161. Ibid., q. 91, a. 1, c.

162. Ibid., ad 3.

163. Ibid., a. 2, c.

164. IaIIae, q. 91, a. 3; q. 95, a. 2. It should be noted, however, that the conclusions "habent etiam aliquid vigoris ex lege naturali," whereas the mere determinations "ex sola lege humana vigorem habent" (ibid., q. 95, a. 2, c.). The reader will observe that the instance of a *lex publica* (which is of course a human law) given by the *Canon Urbani* belongs to the latter kind of law.

165. Ibid., q. 91, a. 4, ad 1.

166. Ibid., q. 96, a. 4, c.

167. Ibid.

168. Art. cit., p. 101.

169. "Considerandum est etiam quod, cum proprium alicujus bonum habet ordinem ad plura superiora, liberum est voleni ut ab ordine alicujus superiorum recedat et alterius ordinem non derelinquat, sive sit superior sive inferior" (*Summa contra gentes*, III, c. 109).

170. "Sicut miles, qui ordinatur sub rege et sub duce exercitus, potest voluntatem suam ordinare in bonum ducis et non regis, aut e converso. Sed si dux ab ordine regis recedat, bona erit voluntas militis recedentis a voluntate ducis et dirigentis voluntatem suam in regem, mala autem voluntas militis sequentis voluntatem ducis contra voluntatem regis: ordo enim inferioris principii dependet ab ordine superioris" (ibid.).

171. Thus we have our own doctrinal "Humanism," "Liberalism," "Naturalism," and even "Catholic Communism" has been suggested. Father Eschmann himself states that the term "personalism," is "in itself, no doubt, a bad one"; that it must be "purged of the connotations it has through its sources in modern philosophy" (*DM*, 208). On this subject, Cardinal Villeneuve says, in the Preface to *BC*, xi–xii: "Présentement, c'est le personnalisme qui est devenu à la mode. Des esprits très sincères le préconisent. On exalte la dignité de la personne humaine, on veut le respect de la personne, on écrit pour un ordre personnaliste, on travaille à créer une civilisation qui serait pour l'homme. . . . Tout cela est très bien, mais trop court, car la personne, l'homme, n'est pas sa fin à elle-même ni la fin de tout. Elle a Dieu pour fin, et à vouloir emprunter le langage des autres, même quand on paraît le corriger par l'envoûtement des meilleurs des adjectifs (n'est-on pas allé jusqu'à parler du 'matérialisme dialectique d'Aristote et de saint Thomas' pour désigner leur doctrine naturelle?), même si on n'exclut pas les sous-entrendus que suppose l'orthodoxie, on laisse sous-entrendre aussi la pensée des autres, une pensée naturaliste, athée, ne fût-ce que par son indifférence, radicalement humaniste, et on favorise le renveresement de la civilisation parce qu'on renverse le

langage and avec le langage la philosophie et la théologie. C'est contre quoi l'auteur s'élève. Il n'a pas tort. It est temps plus que jamais, en effet, de crier casse-cou. Et de vouloir que les sociétés ne se réorganisent pas en fonction de la personne individuelle, mais en fonction du bien commun, à ses divers degrés, c'est-à-dire, de la fin soveraine, c'est-à-dire en fonction de Dieu.

"L'auteur s'attaque ouvertement aux personnalistes, mais pour défendre vraiment la dignité de la personne humaine. Son étude insiste sur la grandeur de la personne sans flatter les personnes. Elle s'oppose à toute doctrine qui, sous prétexte de la glorifier, diminue et atrophie la personne humaine et la prive de ses biens les plus divins."

172. *Summa contra gentes,* III, c. 93 (*BC,* 188, n. 86).

173. IIIa, q. 16, a. 8, c. See Cajetan's commentary, n. 2.

174. *Sermo III, "Attendita a falsi prophetis . . . ," Opera Omnia,* ed. Fretté (Paris, Vivès), t. 32, p. 676.

175. On the personalist conception of marriage. Cardinal Villeneuve has this to say: "Ce n'est donc pas dans une *conception personnaliste du mariage,* ni dans un soi-disant *personnalisme chrétien et socialiste,* qui résultent l'une et l'autre de concessions spéculatives et éthiques à l'erreur, qu'on pourra trouver la solution aux problèmes que soulèvent de plus en plus tragiquement les déviations de la vérité. C'est toujours la vérité qui doit nous délivrer. Or, ces conceptions ne visent qu'à pousser jusqu'à l'exaspération la périlleuse solitude où se trouve plongée la personne, une fois qu'on la détache et qu'on l'isole, sous prétexte de l'exalter, de son appui naturel, le bien commun" (*BC,* xviii).—Referring to H. Doms' successful *Vom Sinn und Zweck der Ehe* (*Du sens et de la fin du mariage*) [Paris: Desclée De Brouwer, 1937]; *The Meaning of Marriage* [New York: Sheed and Ward, 1939), I ventured the unpopular opinion that it presents a "deeply perverse conception of marriage" (*BC,* 181, n. 58).

176. "De matrimonii finibus eorumque relatione et ordine his postremis annis nonnulla typis edita prodierunt, quae vel asserunt finem primarium matrimonii non esse prolis generationem, vel fines secundarios non esse fini primario subordinatos, sed ab eo independentes.

"Hisce in elucubrationes primarius conjugii finis alius ab aliis designantur, ut ex. gr.: conjugum per omnimodam vitae actionisque communionem complementum ac personalis perfectio; conjugum mutuus amor atque unio fovenda ac perficienda per psychicam et somaticam propriae personae traditionem; et hujusmodi alia plura.

". . . Proposito sibi dubio: 'An admitti possit quorundam recentiorem sententia, qui vel negant finem primarium matrimonii esse prolis generationem et educationem, vel docent fines secundarios fini primario non esse essentialiter subordinatos, sed esse aequae principales et independentes'; respondendum decreverunt: *Negative*" (*Acta Apostolicae Sedis,* April 20, 1944; NCWC, August 20, 1944, p. 103).

Nor is the following passage from the *Acta Tribunalium* of the *Sacra Romana Rota* to be overlooked: "Recentissimis his nostris temporibus auctores quidam, de finibus matrimonii disserentes, hoc 'mutuum adjutorium' alio modo explicant, inquantum scl. 'esse personale' conjugum auxilium et complementum accipit, atque contendunt, non secundarium sed primarium finem matrimonii esse hanc 'personae' conjugum *evolutionem atque perfectionem,* quam tamen non omnes eodem, sed alii sub alio respectu considerant atque urgent. Hi novatores in re matrimoniali a vera certaque doctrina recedunt, quin solida et probata argumenta pro suis opinionibus afferre valeant" (*Acta Apostolicae Sedis,* June 26, 1944; NCWC, October 20, 1944, p. 188).

177. St. Gregory, *Magni Moralium,* lib. 34, c. 21 (Migne, PL 76, col. 740, 1134 B).

178. *Sermo XVII in Cantica,* PL 183, col. 857, 1319 C.

179. Ia, q. 63, a. 2, c.

180. *Cursus theologicus* (ed. Vivès), t. IV, d. 23, a. 3, nn. 950–51 (*BC,* 169).

181. ". . . Naturam humanam assumptam a Dei Verbo in Persona Christi, secundum praedicta, Deus plus amat quam omnes angelos; et melior est, maxime ratione unionis. Sed loquendo de humana natura communiter, eam angelicae comparando, secundum ordinem ad gratiam et gloriam, aequalitas invenitur; cum eadem sit *mensura hominis et angeli,* ut dicitur Apoc. xxi; ita tamen quod quidam angeli quibusdam hominibus, et quidam homines quibusdam angelis, quantum ad hoc potiores inveniuntur. Sed quantum ad conditionem naturae, angelus est melior homine. Nec ideo naturam humanam assumpsit Deus, quia hominem absolute plus diligeret: sed quia plus indigebat. Sicut bonus paterfamilias aliquid pretiosius dat servo aegrotanti, quod non dat filio sano" (Ia, q. 20, a. 4, ad 2).

182. ". . . Nec enim nos dicimus peccatum superbiae in Angelo processisse ex judicio intendente consecutionem beatitudinis sine gratia, seu ex viribus propriis: hujusmodi enim consecutio non potest esse intenta sine errore, sed orta fuit ex judicio recusandi quodlibet bonum, et gloriam et gratiae, et beneficio alieno, et sine singularitate aliqua, sed communicando cum inferioribus: hoc enim maxime dedignantur superbi. Et quia daemones superbissimi fuerunt, ideo omnem communicationem cum aliis inferiorbus, etiam in gloria, et omnem modum habendi aliquid ex gratuito beneficio, et non ut debitum, etiam gloriam ipsam recusarunt: unde virtualiter voluerunt illam, si non esset ex gratia, sed ex propria virtute assequibilis. Et sic nullo modo ex relatis in argumento voluerunt positive, et formaliter beatitudinem, sed efficaciter voluerunt illam recusare, quia erat ex gratia, et communcabilis omnibus, virtualiter autem volebant illam, si sic non esset: efficaciter vero, et positive adhaerebant propriae excellentiae ut fini, quia ibi tales conditiones non inveniebant, sed erat propria, non communis, et non ex gratia speciali, sed solum jure creationis conveniens" (John of St. Thomas, op. cit., 953–54).

183. *De civitate Dei,* XII, c. 1

184. Wisdom 2:24.

185. John 8:44.

186. Apoc. 20:7.

187. 1 Peter 5:8.

188. I readily admit that of all those who dare to go into print, I have possibly the least skill as a writer. However, I cannot allow my shortcomings to prevent me from taking up so great a cause. For, as has been well said: If a thing is worth doing at all, it is worth doing ill.

189. "*Adam non est seductus, sed mulier.* Seductio autem duplex est, sc. in universali, et in particulari eligibili, quae est ignorantia electionis. Quicumque ergo peccat,

seducitur ignorantia electionis in particulari eligibili. *Mulier autem fuit seducta*, ignorantia in universali, quando sc. credidit quod serpens dixit; sed vir non credidit hoc, sed deceptus fuit in particulari, sc. quod gerundus esset mos uxori, et cum ea comedere deberet, et inexpertus divinae severitatis credidit quod facile ei remitteretur" (*In I ad Tim.*, c. 2, lectio 2).

190. John of St. Thomas, op. cit., 954.

191. My Adversary says he "would never have come out with this judgment had not Father Baisnée written his article" (*DM*, 187, n. 9). His reference is to Father Jules A. Baisnée's "Two Catholic Critiques of Personalism," which appeared in *The Modern Schoolman* 22 (Jan., 1945): 59–75. On page 185, Father Eschmann writes: "Father Baisnée reveals himself to be much impressed by the weight of the authorities which, according to him, Professor De Koninck's anti-personalist position commands." Father Baisnée had said, "Cardinal Villeneuve, Archbishop of Quebec, added the weight of his authority to this condemnation of the new theory in which he saw a real danger of revival of Pelagianism [art. cit., 50]. . . . Is there lurking in the movement of Personalism an opposite [of Totalitarianism] but equally serious danger of fostering 'by loose thinking which goes to evil to find good' what Cardinal Villeneuve does not hesitate to call 'a revival of the polycephalous monster of Pelagianism'?" (ibid., 74; *BC*, xxii). On the same page 185 Father Eschmann exclaims: "*Pacem, amici!* Would it not be better for us to stop short on the road of censuring and adding-up authorities, before it is too late and Catholic scholarship is once more a laughing stock?" As many will remember, when I originally presented my paper on the common good at the annual meeting of the *Académie canadienne saint Thomas d'Aquin*, it was generally believed my position was not in agreement with the Cardinal's opinion. Yet, on the dignity of the person and the function of society, His Eminence had never said more nor less than what is repeated in the preface.

192. "Ainsi veut-on détruire une proposition *per se nota* résultant de la seule notification du bien commun" (BC, 71).

193. The "Enemy" is, of course, God.

194. C. S. Lewis, *The Screwtape Letters* (Toronto: S.J. Reginal Saunders, 1945), 92.

The Dialectic of Limits as Critique of Reason

1945

εἶναι γαρ ἕν σοφόν
—Heraclitus

In order to know natures distinctly, we need a number of means of knowing, that is, concepts, intelligible species, proportional to the multiplicity of natures themselves. The means of attaining knowledge of the circle is distinct from the means of attaining knowledge of the polygon. It is true, however, that we can grasp in a single concept objects different by definition—but this sort of reduction to unity involves indetermination, without confusion. Thus in the common genus "figure" we do not grasp either the circle or the polygon in the sense of knowing what properly constitutes them.

But we also know that this multiplication of our concepts, this scattering, arises fundamentally from the proper structure of our intelligence and from the nature of the multiple objects in themselves. Our means of knowing are at the lowest degree of intensive universality.

> Summum autem hujus universalitatis est in Deo, qui per unum, scilicet per essentiam suam, omnia cognoscit: infimum autem in intellectu humano, qui ad unumquodque intelligibile indiget specie intelligibili propria et ei coaequata.

> Non est igitur per formas universaliores apud substantias superiores imperfectior cognitio, sicut apud nos. Per similitudinem enim animalis, per quam cognoscimus aliquid in genere tantum, imperfectiorem cognitionem habemus quam per similitudinem hominis, per quam cognoscimus speciem completam: cognoscere enim aliquid secundum genus tantum, est cognoscere imperfecte et quasi in potentia, cognoscere autem in specie est cognoscere perfecte et in actu. Intellectus autem noster, quia infimum gradum tenet in substantiis intellectualibus, adeo particulatas similitudines requirit quod unicuique cognoscibili proprio oportet respondere propriam similitudinem in ipso: unde per simulitudinem animalis non cognoscit rationale, et per consequens nec hominem, nisi secundum quid. Similitudo autem intelligibilis quae est in substantia

> separata, est universalioris virtutis, ad plura repraesentanda sufficiens. Et ideo non facit imperfectiorem cognitionem, sed perfectiorem: est enim universalis virtute, ad modum formae agentis in causa universali, quae quanto fuerit universalior, tanto ad plura se extendit et efficacius producit. Per similitudinem igitur unam cognoscit et animal et differentias animalis: aut etiam universaliori modo et contractiori, secundum ordinem substantiarum praedictarum.[1]

Add to this that our most distinct intelligible species are so poor that we need to make a return to the senses in order to attain the singular.

> [I]ntelligentiae superiores habent formas magis universales: hoc tamen observato, quod in infimis angelis sunt formae adhuc universales, in tantum quod per unam formam possunt cognoscere omnia individua unius speciei, ita quod illa species sit propria uniuscujusque particularium secundum diversos respectus ejus ad particularia sicut essentia divina efficitur propria similitudo singulorum secundum diversos respectus. Sed intellectus humanus, qui est ultimus in ordine substantiarum intellectualium, habet formas in tantum particulatas quod non potest per unam speciem nisi unum quid cognoscere: et ideo similitudo speciei existens in intellectu humano non sufficit ad cognoscenda plura singularia; et propter hoc intellectui adjuncti sunt sensus, quibus singularia accipiat.[2]

Our intellectual knowledge being abstractive, we know that we cannot have a purely natural science that would be at once speculative and practical. The universal of speculative science and the singular to be made or done are, for us, in contrary directions. In this respect too, there is a finitude quite characteristic of human reason.

But we will be told that we know these limitations. Thus in a certain way our intellect surmounts its own finitude. It knows that, our knowledge being by origin and resolution empirical, it is the lowest level of universality; it knows that its concepts are always inferior to objects;[3] it knows that it imposes, even on the objects which are proportioned to it, limitations which arise from our mode of knowing. Doubtless this limitation is not that which mathematizing idealism imposes, but we must recognize that it affects the very structure of our intelligible universe.

Note too that at the beginning of our learning there is only confused knowledge.[4] More profoundly, it is common and permanent to every finite intelligence. The more that intelligence is raised above the multiplicity of intelligible species which represent objects most distinctly, the more it departs from distinction and retreats into a certain indetermination. The confusion of knowledge by fewer concepts brings to light the limitation of the more distinct means.

This limiting distinction that characterizes the finite intelligence is also the reason for the changeable nature of its acts of thought. For the actual and distinct consideration of the totality of its intelligible universe can be accomplished only by acts which succeed one another by the before and after of duration. In going from one thought to another, every created intelligence fabricates time.[5] This dispersion in successive duration is a reflection of the extraneity and irreducibility of species in the treasury of intelligence. In short, the finite intelligence, proceeding in its proper manner, can experience the plenitude of its intelligible universe only in a discreet time constituted by the succession of acts of thought.

If successive duration finds its cause in the imperfection of the intelligence which knows by multiple means, it is also the sign of the freedom of intelligence—it is free of determined and limited species. Going from one species to another, thanks to this movement,[6] intelligence can convert and reestablish in an at least dynamic unity the multiplicity of its isolated species. This dynamic unity furnishes it with a substitute for the simultaneous view of the totality of its intelligible universe.

It is an incontestable fact that our intellect tends naturally to see one nature as the limit of another. This comes about most easily in mathematics. We are not content to see the elements—point, line, surface, volume—in their absolute and irreducible natures. We think we grasp them better when, besides, we can define them as limits; when, knowing quite well that we can never attain it without contradiction, we proceed nonetheless resolutely *as if* we wished to generate these elements one from the other in their proper and abstract nature.

This mathematical model can, to a degree, be extended to physical natures where we can conceive or artificially interpolate a comparable order, under some aspect, to that of infinite converging series. Thus we can in a way bring forth the reason for a degradation of *intellectus*. In sum, the whole of the *De divinis nominibus* employs this method. This work in particular presents insurmountable difficulties when one tries to interpret it in a directly real sense. Moreover, in order to legitimize such an interpretation one must ignore the many warnings of the author.

We define a circle as the limit of a regular inscribed polygon whose sides increase infinitely in number. In this way we try to see the circle, in that which distinguishes it from the polygon, by means of the intelligible likeness of the polygon. Thanks to the movement impressed on natures, we force ourselves to go around the divisive structure of our intelligence. Does that mean, though, that the knowledge of beings in their absolute nature is, as such, imperfect? That natures are radically identical? That their differences are only phenomenal modes? That their diversity is only a reflection of a single nature too rich in its identity to be expressed all at once by a fragmentary intelligence? That would be to ruin intelligence and its object. The identity of different natures could not be the point of this tentative *reductio,* for if, *per impossibile,* intelligence could attain the end to which it appears to approach nearer and nearer, the effort of intelligence would be at its term frustrated by the mutual destruction of the natures it wanted to penetrate more profoundly in their very distinction. If, *per impossibile,* it could attain the limit point of the line or the limit point of the polygon, it would confront an unextended extension or a one-sided figure with many sides.

This effort of the intelligence toward unification is, at bottom, only an attempt to reduce the multiplicity, not of the natures conceived, but of the means of knowing them. And that is its only point: to arrive at a higher view, more unified, consequently, more penetrating of the diversity of natures; to tend toward this unity and this penetration that simple coordination, however perfect, could never give us. In a sense, this view is only a limit that one never ceases approaching but never really attains. What we can do in this sense is only a shadow of a mode of knowing at the least superhuman. The view to which we can approach is only a tendency, an effort, an attempt at assimilation. Once more, it is only a shadow, but it is truly a shadow of intelligence properly so called.

To the degree that objects known by us are in a finite understanding, our concepts constitute an a priori structure that keeps us on this side of objects themselves. By objects themselves we here mean the object as known by the divine intelligence. But, again, we know, in a certain measure, this limitation: intelligence dominates its structure a priori, although the judgment that it makes on itself remains purely extrinsically critical and negative and can only pronounce its insufficiency. By contrast, using the method of limits, reason seeks to surmount the division in our knowledge, to surpass the finitude of our concepts always inferior to objects, and all that finitude entails of composition in our judgment and of formal discourse in our science; our intelligence finds itself instituting a positive and constructive critique of its own limitations. It tends, in a certain way, to repress reason, that is, it tends toward the condition of intelligence properly so called.[7] What characterizes reason is the diffusion of its objects: there is need of formal discourse in order to attain their scientific coordination. Therefore it only imitates intelligence in a twisted manner; it does not surmount, even in a tentative fashion, the division of our concepts. On the contrary, in its critical movement toward a limit, our intelligence tends toward the universality of intelligences. Properly speaking, it tries to make itself similar to, so far as this is possible, the deiform intellect of separated substance,[8] even to imitate in its fashion the supreme Wisdom which, in a unique intelligible species, contains the plenitude of all intellectual knowledge.

The movement and the infinite which underlie this mode of knowing are not considered as objects, as natures, which are very obscure in any case; even the succession of distinct acts involves obscurity. We use movement and the infinite as pure means. It is by them that the intelligence tries to build a bridge between natures, or, more exactly and profoundly, to abolish the divisions which weaken our means of knowing. We accord thus to movement and the infinite in question a power which transcends in a way our finite understanding, a power of dissolving our concepts and of arriving at a tentative more profound unification. Still one notices that this power is, radically, that of intelligence itself using its means.

Manifestly, the movement that we thus impose on natures, the dynamism which pushes them, not only to interlock one to another but, despite their differences and immobility, to converge toward identity, is only a dialectical dynamism, a product of reason and grounded not in objects as such but in the object in its *esse cognitum.* So too the infinite in question is considered only with respect to a limit that only reason can envisage as such.

The projection of motion into natures is thus only a shadow of the interior convergence of our concepts: these under the attraction of the one infinitely beyond the concept of being, the one of exemplary Wisdom, tend by means of the infinite and the movement of reason to fuse themselves in more simple means of knowing of a more intensive universality. Thus we attempt to suppress the gaps that separate our concepts, to erase their irrational heterogeneity in order to grasp more profoundly and in a higher fashion the rational heterogeneity of objects, of natures.

How can the interpolation of the infinite serve reason in its tendency toward perfect rationalization—toward *intellectus?* Because reason can use it to attempt to fill in the gaps, to make dense the distance which separates concepts. Thanks to this density (which is never given save in potency, which is never actual), intelligence attempts to pass from one object to another without changing concepts; to try to see the different in the same. It forces itself to pass from one to two by interpolating all the convergent series it can conceive; it tends toward a perfect continuity from one to two, toward that "continuum cujus partes ad unum terminum communem copulantur." But if this attempt could succeed, intelligence would be involved in a contradiction: in the identity of one and two, it would have converted rational heterogeneity of natures into the irrational continuity of the continuum. In this manner, the rational in itself would be rationalized for us by a reduction to the irrational in itself.[9]

Let us try to see now the liberty toward which intelligence tends by the application of the method of limits. Thanks to its capacity to reflect on itself, our intelligence can multiply intelligible relations to infinity.

> [I]n nobis relationes intelligibiles in infinitum multiplicantur; quia alio actu intelligit homo lapidem, et alio actu intelligit se intelligere lapidem, et alio etiam intelligit se hoc intelligere; et sic in infinitum multiplicantur actus intelligendi, et per consequens relationes intellectae. Sed hoc in Deo non habet locum, quia uno actu tantum omnia intelligit.[10]

Intelligence can use no matter what object in order to produce this multiplicity. It can even concentrate on its act of understanding, building on it an infinity of understandings of understanding. The infinite in question is a purely rational infinite: the intelligence opens onto it by the free multiplication of its acts. It is to itself that it owes this faculty of redoubling the same thing to infinity. The case of the judgment of identity is significant.

> Nam non potest intelligi relatio nisi inter duo extrema. Sicut cum dicitur aliquid esse idem sibiipsi. Tunc enim intellectus utitur eo quod est unum secundum rem, ut duobus. Alias ejusdem ad seipsum relationem designare non posset. Unde patet, quod si relatio semper requirit duo extrema, et in hujusmodi relationibus non sunt duo extrema secundum rem sed secundum intellectum solum, relatio identitatis non erit relatio realis, sed rationis tantum, secundum quod aliquid dicitur idem simpliciter. Secus autem est, quando aliqua duo dicuntur esse idem vel genere vet specie. Si enim identitatis relatio esset res aliqua praeter illud quod dicitur idem, res etiam, quae relatio est, cum sit idem sibi, pari ratione haberet aliam relationem, quae sibi esset idem, et sic in infinitum. Non est autem possibile in rebus in infinitum procedere. Sed in his quae sunt secundum intellectum nihil prohibet. Nam cum intellectus reflectatur super suum actum, intelligit se intelligere. Et hoc ipsum potest etiam intelligere, et sic in infinitum.[11]

In the proposition *A is A,* the subject and the predicate are rationally different. The intelligence thus effects the multiple, exteriority, and otherness of the same. The relation of identity of A to A can be related to itself to infinity.[12] In doing this, it produces an order of time and position which are purely rational. It can compare these relations among themselves, thus making arise all the species of relation: equality, similarity, priority, etc. as well as the contraries of these, all purely rational. Notice that throughout this proliferation, it is always the same object that we deal with; it is its identity that assures the rational purity of the system in question. It is the same A in itself that we encounter at every point in the system. Between any two terms, however close they are, we can interpose as many relation-terms as we like and render the system as dense as we like. In all this proliferation, however varied, the object in itself is never separated from itself, no matter what term in the system is susceptible of an identical proliferation identical to any other. Each of the terms of the ensemble can be elaborated in an ensemble in every point identical with that of which it is a part. Thus what is said of the whole can be said in turn of no matter what part. In each member of the ensemble we rejoin the same ensemble, etc.

Because in this construction of a purely rational universe the intelligence depends only on the object it gives itself and builds the whole system by a pure reprise of the same, of which it is itself the cause, it depends on no given plurality or any otherness as such nor on any actual relation; that it never has to do with anything new that it itself does not produce; it can rejoin the whole system in the same. Because it is itself principle and author of this universe, one can say that intelligence finds itself entirely at home, that it is very free there. The reflection of intelligence on itself being a sort of lived identity, the universe in question is like a projection of the presence of intelligence to itself.

In the method of limits, it is by a similar free construction that intelligence can surpass the heterogeneity of given natures. This heterogeneity is envisaged as a matter to be rationalized by the application of form—the purely rational universe—that we have described. The rationalization will consist in fabricating the formally diverse with itself. It will use only objects that it produces itself by the same act. It is by a pure repetition of the same operation that $1 + 1/2 + 1/4$ leads to 2, and that we tend to fill up in their turn the gaps left by the commensurable numbers with the incommensurables.[13] Intelligence thus uses its own lived identity to fill up the abysses which separate formally distinct objects, or, more exactly, to surmount the division of its means of knowing. It tries to see everything without changing concept. Isn't this the ideal of mathematical rigor that one pursues in wanting to demonstrate that 2 and 2 make 4?

There is the fact of the multiplicity of our means of knowing, of the succession in which we have to engage to consider them actually, of the time we fabricate in making the tour of our intelligible universe. In order to remove ourselves from these necessities, this dispersion, we have recourse to singularly paradoxical means. We use division to conquer division, plurification to conquer the multiple; we engage in movement and project time, but this time with as many things as the intelligence has produced by fixation on the same where it never ceases to be at home in a lived identity. Moreover, there always remains this infinity that our intelligence can name, can define, but that it cannot get through because it is not there to be gotten through.[14] In sum, there is only the fact of being merely human or of not being divine which prevents our attaining the term that we pursue in tending to the limit.

Poincaré has defined the point of this method quite well as it pertains to mathematics. In *La science et l'hypothèse,* he expressed the root of the matter. "Why therefore does this judgment impose itself on us with an irresistible evidence? It is that it is only the affirmation of the power of mind capable of conceiving the indefinite repetition of the same act if that act is once possible. The mind has this power of a direct intuition and the experience can be for it only an occasion to serve it and become conscious."

This method cannot be directly applied to virtual quantity. That was the error of the mathematician Nicolas of Cusa. One can compare its legitimate employment in philosophy to the use that we make of the concept of movement applied to God in the *De divinis nominibus.*[15] One will also notice that this method presupposes distinct knowledge according to the natural mode of knowing. By the natural mode I mean that opposed to the logical or dialectical.

The excellence of this mode of knowing, even its necessity in order to tend toward a more profound sapiential view of things, is beyond question. But does it not imply that there is in man purely as man a mode of knowing higher than metaphysics? It is true that it tends to reduce to absolute unity the pluralism of the being that is the object of metaphysics, that it tends to dissipate whatever remains in it of the undetermined,[16] but this dialectical method never really escapes the limits of the object of metaphysics. It is always, in the final analysis, metaphysics which makes the transcendental critique of our mode of knowing, it is metaphysical wisdom that uses this dialectic in order to judge itself in a more radical manner, to live in a manner at once more critical and truly constructive, its distance from the first Wisdom. The tendency to know, as much as is humanly possible, the divine principle of its subject obliges metaphysics to use this dialectical mode. It would not be wisdom if it did not always seek theology or take on as divine a mode as is possible for man.

The method of Hegel—"the absolute, unique, supreme, infinite force that no object can resist"—is it anything other than this mode of knowledge abusively carried to the limit? And when Hegel identifies it with "the tendency of reason to find itself, to recognize itself in everything" is it not more exactly a matter of the tendency in virtue of which intelligence seeks to devour its concepts which impede its drive toward a more intensive intuition, effectively to rejoin that which is most divine in it, to live more in conformity to its principles,[17] more conformably to that intelligence which quite naturally it remembers?

NOTES

1. *II Summa contra gentes,* c. 98. See too *In II Sententiarum,* d. 3, q. 3, a. 2; *Q.D. de veritate,* q. 8, a. 10; *I Summa contra gentes,* cc. 50–54; Ia, q. 14, a. 6; q. 55, a. 3; *In librum de causis,* lectio 10.

2. *Quodlibet VII,* a. 3, c.

3. "Sed quia primum principium nostrae cognitionis est sensus, oportet ad sensum quodammodo resolvere omnia de quibus judicamus; unde Philosophus dicit in III Caeli et Mundi, quod complementum artis et naturae est res sensibilis et visibilis, ex qua debemus de aliis judicare; et similiter dicit in VI *Ethic.* (cap. viii, in fine) quod *sensus est extremi sicut intellectus principiorum* extrema appellans illa in qua fit resolutio judicantis" (*Q.D. de veritate,* q. 12, a. 3, ad 2).

4. "Et sic esse intentionale semper est inferius, licet aliunde in ratione spiritualis entitatis possit aliquando conceptus superare obiectum. Et quando dicimus obiectum esse principalius et perfectius, loquimur de objecto primario et formali conceptus; nam obiectum materiale et secundarium habet se accessorie, nec est necesse quod sit perfectius, cum pro illo non directe et per se substituat conceptus" (John of Saint Thomas, *Cursus philosophicus* [ed. Reiser], t. I, p. 696a4–14).

5. *In II Sententiarum,* d. 3, q. 3, a. 4; *In III Sententarium,* d. 14, a. 2, qa. 4; *Q.D. de veritate,* q. 8, a. 14; *Quodlibet VII,* a. 2; *I Summa contra gentes,* c. 55; Ia, q. 12, a. 10; q. 58, a. 2; q. 85, a. 4.

6. Since it is a question of a succession of acts, we are not using the term 'movement' in the strict sense of *actus entis in potentia in quantum hujusmodi.*

7. *Q.D. de veritate,* q. 15, a. 1, c.; *In de trinitate,* q. 6, a. 1, ad 3.

8. ". . . ratione cujus [angelos] intellectuum habere dicuntur" (*Q.D. de veritate,* q. 15, a. 1, c.).

9. In Emile Meyerson, as in Parmenides and Anaxagoras, the problem of the noetic one and many and the problem of the natural one and many are confused. They do not distinguish what is more knowable to us from what is more knowable in itself (Aristotle, *Physics* I, 1) or rationality in itself from rationality for us. Meyerson, if I have correctly understood him, does not distinguish the reduction of the singular to the universal, the coordination of laws in ever more universal principles, the relations of causality, the connection of concepts (in the sense of *rationes*) which make up the object of science, from the reduction of the many which arises from the empirical nature of our knowledge. The "irrational" elements which resist our effort of identification are nothing else than the distinctions of the heterogeneity of objects, therefore, of that which is most rational in the world. Even this rationality is relatively irrational *for us,* as the most knowable in itself is the least knowable for us. Meyerson is right in saying that if these irrational gaps in the world did not exist—and one could say the same of the mathematical world—all our knowledge would be an immense tautology; all of mathematics, the expression of all the theorems which fill so many volumes, would be, as Poincaré says, merely complicated ways of saying A = A. Nonetheless, Meyerson's work remains one of the most important modern contributions, if not the most important, to the philosophy of the sciences.

10. Ia, q. 28, a. 4, ad 2.

11. *In V Metaphysicorum,* lectio 11, n. 912.

12. On the expression "to infinity" see Cajetan, Ia, q. 85, a. 6, n. 6.

13. The number 2 in play is clearly the predicamental number, *a* 2, and not the numbering number or abstract *two* ones.

14. IIIa, q. 10, a. 3.

15. Chapter 9. Above all, see the commentary of St. Thomas, lectio 4.

16. John of St. Thomas, *Cursus philosophicus,* 499a37–500a47.

17. "Substantiis autem separatis, quae sunt principia intellectus humani, et ad quae intellectus humanus se habet ut imperfectum ad perfectum, non conjungitur homo nisi per intellectum" (*In I Metaphysicorum,* lectio 1).

Notes on Marxism

1945

I. Calculus and Dialectic

In the *Anti-Dühring* as well as in *The Dialectics of Nature*, Engels recognized the fundamentally dialectical character of the calculus.[1] "[T]he mathematics of variable magnitudes, of which the most important part is the infinitesimal calculus, is essentially nothing other than the application of dialectics to mathematical questions."[2] "Moreover, I don't think that one can give the calculus as an example of the sort of contradiction that dialectic taken in the Hegelian or Marxist sense demands, that is, a contradiction objectively existing in things and phenomena themselves. . . ."[3] J. B. S. Haldane alludes to this.[4] Let's examine the question in more detail in order to assure ourselves of the direction and degree of the concession that the Marxist seems compelled to make today.

The problem is not without interest since Engels saw in it, as well as in the case of infinity and movement, a glaring example of a contradiction in the Marxist sense. But if the Marxist on the one hand calls calculus dialectical because of this contradiction, and if on the other hand we must now recognize that we do not have there the sort of contradiction that Marxist dialectic demands—at least not at the level indicated by Engels—how *does* the Marxist understand the dialectical character of the calculus? Should he say that the calculus is no longer dialectical and that this claim must now be consigned to history? To come to Engels' aid by saying that in this manner he was only basing himself on the state of mathematics in his time would be, I am persuaded, unjust to him.

Here is a relevant passage that I take from the *Anti-Dühring*:

> I distinguish x and y: that is, I suppose that x and y are infinitely small and that they disappear in relation to any magnitude no matter how small, however little one posits it as really existing, in such wise that of x and y there exists only their reciprocal relation without, so to say, any material foundation, a quantitative relation without any quantity. The expression dy/dx, that is, the relation of two differentials of x and of y is therefore equal to 0/0, but this 0/0 is posited as the expression of y/x. I note in passing that this relation between two disappeared magnitudes,

> fixing the moment of their disappearance, implies a contradiction; but this contradiction should not trouble us any more than it troubled mathematicians for more than two hundred years.[5]

This interpretation calls into question the notion of limit. But, in terms of the method of limits, it could be a question of a contradiction in the example given by Engels only if he supposes that the limit is truly attained. But that is contrary to the very definition of limit: the successive values of the variable x must approach a fixed given number in such a way that the difference x – a ends by becoming and remaining a value less than any given number, e, no matter how small. It is essential to the definition of the derivative that dx, that is x – a, differs from o. In order for a contradiction to appear in the example given, the magnitudes in question would have to be determinately equal to o; that they not only be magnitudes on the way to disappearing and becoming equal to o, but that they have entirely disappeared and are actually equal to o.

We know that the first interpreters of the calculus had a false understanding of the infinitely small and returned, in effect, to the ancient error of Antiphon. One need only consider the infinitely small as a static entity for it to become as contradictory as a square circle. In fact, it is contradictory only if one takes away its dynamic character. As infinitely small, it has no definite magnitude. No doubt it is something negative with respect to a definite magnitude from which it departs. As Engels himself would say: it disappears with reference to any magnitude. But it remains in the state of "disappearing," and what has disappeared always refers to a given magnitude from which it departs. Since it is never equal to o, the infinitely small retains at the same time its positive character.

But then can one say that "of x and of y there subsists only their reciprocal relation, without so to say any material foundation, a quantitative relation without any quantity"? That is what would be if, *per impossibile,* one could attain the limit; if dx/dy = o/o expressed an accomplished equality. Only on that condition could one speak of a contradiction. But in fact there is no contradiction since it is a matter of only a tendency toward equality.

Still let us allow that Engels appears to have grasped the profound difference between a given magnitude and a magnitude in the process of varying, in this case, one in the process of disappearing. The latter can be defined only in an extrinsic manner, thanks to the given magnitude of the constant limit.

Isn't this the case with any motion? It is neither determinately act nor determinately potency. One denies the proper nature of motion as soon as one seeks to reduce it to simple act or simple potency. One denies it just as much in seeking to reduce it simultaneously to both: in composing it from that which determinately is and that which determinately is not. In the one case, as in the other, one has a purely static conception of motion. That is what comes from wanting to introduce only the immobile into motion; that is to deny that motion is fundamentally other than immobility and to want to express the whole of motion in terms of absolute fixity.

It is the same with the infinitely small: the very fact of wishing to reduce it to one or the other of the terms we need to define is already to deny it. Such a reduction would make us say: the infinitely small both is and is not perfectly equal to 0. Or it both becomes and does not become equal to 0 since it already is: one has and has not attained the limit. That would be contradictory.

Even if we were disposed to accept a contradiction, we could not do it in the present case since it arises from a too obvious mathematical error. One could use a too obvious escape-hatch: since for the Marxist, movement is an evident contradiction, why should the present contradiction give him pause? But since it is only a question of a contradiction arising from a false understanding of the limit, I am convinced the Marxist would not accept this escape-hatch.

But let it be clear. I have no wish to exclude from the notion of limit every relation to contradiction. I would then be denying its dialectical character. But this contradiction plays only an extrinsic and negative role; it does not enter as an accomplished thing. Tending toward a limit, we get closer and closer to what would be contradictory if, *per impossibile*, one could adequately reach that limit. But the contradiction which *would be* will never be.

Provided that the identity in question is only the identity of becoming and not an achieved identity, I can sustain the identity of the differences, the identity of contraries without being either Marxist or Hegelian. I can also then accept contradiction: a contradiction in becoming. It is only the identity of differences or of achieved contraries that goes against the principle of contradiction. The example that Engels cites is an example of contradiction in becoming—let us coin the expression "dynamic contradiction." But, I repeat, such a contradiction is not contradictory, and consequently cannot be cited as an example of a true, that is achieved, contradiction, which dialectic understood in the Hegelian or Marxist sense requires.

It remains to be seen whether Engels wants to speak of a contradiction other than that which would be found at the limit if, *per impossibile,* one could adequately attain the limit. On that I don't think any doubt is possible. One can see this in the passage I cited above. One can see it, too, in a passage where he speaks of contradiction in the real world:

> It is possible that this proposition have for reasonable people who are sane enough, the same evident value as this one: what is straight cannot be curved, what is curved cannot be straight. Still the differential calculus, despite the protests of the same reason, posits in certain circumstances the straight and the curved as identical and thereby achieves the results which sane reason, refractory to this absurdity, would never attain, the identity of the straight and curved.[6]

I cite as well this passage from *Dialectics of Nature*: "*Straight and curved* in the differential calculus are in last resort put as equal. . . ."[7] And later he cries, "O metaphysics!" But it is only on condition of identifying straight and curved that one could have an accomplished contradiction and the exclamation would have a foundation. He well knows that he is putting in play here the principle of contradiction, that is, it is impossible to be and not to be at the same time and in the same respect. Did he not already say with respect to identity and difference: "The dialectical relation is already seen in the differential calculus, where dx is infinitely small, but yet is effective and performs everything."[8] But that the infinitely small were "effective and performs everything" would astonish only if it were perfectly equal to o.

But the passage that seems truly astonishing to me is the paragraph on asymptotes.[9] It contains quite correct remarks which contradict openly the position he wants to adopt. Take the first phrase: "Geometry begins with the discovery that straight and curved are absolute opposites, that straight is absolutely inexpressible in curved, and curved in straight, that the two are incommensurable." This must be understood. There is a *tendency* to surmount this opposition, there is a tendency toward identity, we can speak of an identity in becoming, but the opposition is never removed, the identity never accomplished. When he adds, "Yet even the calculation of the circle is only possible by expressing its periphery in straight lines," we must recognize that its expression in terms of the straight will never be adequate, that the difference remains absolutely irreducible. Straight and curved differ by definition. One can define the one as the limit of the other, but the definition of straight as the limit of the circumference of the circle whose radius increases infinitely is not a definition of the straight, but of the straight as the limit that the curved can never reach. It is only in the case of differences which are in the last analysis irreducible that it is possible to define the one as the limit of the other.

He goes on. "For curves with asymptotes, however, straight becomes totally merged in curved and curved in straight. . . ." But after having used the expression "totally merged," he adds this precision that contradicts the idea he seems to wish to illustrate: "just as much as the notion of parallelism: the lines are not parallel, they continually approach one another and yet never meet; the arm of the curve becomes more and more straight, *without ever becoming entirely so,* just as in analytical geometry the straight line is regarded as a curve of the first order with an infinitely small curvature. The x of the logarithmic curve may become ever so large, *y can never = o.*" I have emphasized the expressions which openly contradict the first half of the paragraph. Let us add that "a curve of the first order with an infinitely small curvature" is not a straight line, although we can say that it is on the way to becoming a straight line. The straight is always only the limit.

Does my insistence seem superfluous? J. B. S. Haldane, in fact, settles the matter in saying: "This was of course written before 'rigorous' proofs based on the theory of limits were introduced into most books on the calculus. Engels is quite correct concerning the calculus taught in his day."[10] And in his own work *The Marxist Philosophy,* he writes, "It is important to remember that [Engels] was historically correct." But as I said above, I don't think this expedient does Engels justice. Haldane adds, "It is however worth pointing out that even the so-called rigorous proofs ultimately rest on a logical basis which is not entirely free from contradiction." This reservation is not very reassuring since it always remains possible that one day the underlying contradiction would be recognized as resulting from an error.

One cannot do justice to Engels except by clearly distinguishing the partial truth that he has seen; he is right to introduce the idea of contradiction into every tendency toward a limit. But this contradiction that I have called dynamic, although it suffices to render the calculus dialectical in the classical sense, can never be given as an example of the sort of contradiction that the Marxist dialectic claims.

II. Two Sorts of Contradiction

Sometimes the Marxist, like the Hegelian, considers contradiction to be something impossible; sometimes he embraces it as something accomplished in the real world, as objectively existing, and so to speak in the flesh and blood of things and phenomena themselves. Thus he declares motion and the infinite, perfectly incontestable things, to be evident contradictions. We will keep it consequently on the terrain of the real.

What is the criterion for distinguishing the two sorts of contradiction? How is it defined? Doubtless, in the final analysis one should be possible for if it were impossible it could not be; the other would be impossible, for example, that motion is identical at every point with the immobile. Motion includes one and excludes the other.

But if, in the last analysis, one of the two contradictions should be characterized as impossible, how is "impossible" to be defined? I dare putting the question to the contemporary Marxist, since I know that Marx, Engels, and Lenin did not hesitate to face up to this kind of difficulty. The possible is always defined as that which does not involve a contradiction. But the Marxist holds that some contradiction is possible, for example, that the straight should be at the same time and in the same respect non-straight; or again that "simple mechanical change of place can only be accomplished because a body at one and the same moment of time is and is not in this place."[11] But then how to define the impossible? There would be a vicious circle that the Marxist does not accept.

For the possible contradiction to be truly contradiction, the terms the contradiction encloses would have to be opposed in such a way that their identity would be contradictory. But then, what does "contradictory" mean here? If, in this context, the term "contradictory" does not come down to meaning "impossible," what does it mean?

When the Marxist declares that motion is contradictory, must he not judge that to be in a single place and not to be there at the same time and in the same respect are contradictory? If a body could be in a place and at the same time and in the same respect not be in the place it is in, would he not have to say that, since it is possible, it is not contradictory? And wouldn't it follow that the possible contradiction is only on condition of being impossible?

III. Motion and Contradiction

Knowing how the contemporary Marxist holds Aristotle in esteem, I dare ask how he refutes the analysis of motion and the infinite in book III of the *Physics* and book XI (K) of the *Metaphysics*. The Marxist should consider it fundamentally erroneous. That motion and the infinite are contradictions would mean that, for Aristotle, they are impossible.

It is true that for Aristotle motion is a certain mixture of being and non-being. And yet, for him, motion has nothing of the contradictory. This is because there are two points to consider here: (a) the being and non-being in question are already divisions of being as such. The first is being in act; the non-being in question is nothing other than possible being, being in potency, which is non-being in comparison to being in act and which is quite different from the non-being opposed to being as such divided into act and potency, being and relative non-being. Motion would be contradictory if one confused the two sorts of non-being, the absolute and relative. And that would mean that motion is impossible, that it is not and could not be. (b) Besides, the mixture of being and non-being that is motion is not a mixture of that which is determinately in act with that which is in potency. Thus, the house under construction, in becoming, would be at once a perfectly achieved house and one purely in potency. That is how we would have to conceive motion in order for it to be a contradiction.

Since the Marxist takes motion for a fact at once absolutely evident and an evident contradiction, he should be able to say that motion is a possible contradiction, for if it is not possible, it could not be. In short, it is necessary that motion be a contradiction which is not contradictory in the sense of impossible. We thus rejoin the point made above in section II.

Every contradiction implies some non-being. But every non-being does not imply a contradiction. If the non-being which implies the "possible contradiction" is the relative non-being of which I spoke earlier, one removes from possible contradiction its element of contradiction. On the other hand, if the non-being which possible contradiction implies is the non-being identical with the impossible, possible contradiction is impossible. In short, it is only on condition of being impossible that possible contradiction would be possible.

IV. Quantity and Quality

With regard now to quantity which is transformed into quality: In the example given by Engels and reproduced by Stalin in *Le materialisme dialectique et le materialisme historique,*[12] we notice that the quantitative change of the quality of a thing (for example, the increase and decrease in the temperature of water) can provoke the passage of this thing from one qualitative state to another (from the liquid to the gaseous state or frozen state). I do not see how from this proposition, stating this incontestable fact, one can infer (it is said, in fact, "consequently") the radically different proposition: "Quantity is transformed into quality." I do not think that the passage from the one proposition to the other is immediately evident. Therefore some intermediary proposition must be understood to justify the inference.

But if one looks at the inference closely he will see that he should suppose, at the outset, the quantitative change of quality (temperature, for example) is in every way identical to that same quality, the one not differing from the other save in name. But that entails a double difficulty. In effect, (a) to have even a semblance of a proper inference, we must suppose some difference between the quality and the quantitative change; (b) on the other hand, if we suppose the quantitative change is in every way identical with the quality of which it is the quantitative change, the inference can establish nothing; it is no longer an inference.

Besides, if we want to use the example given above, the proposition "the quantity is transformed into quality" should be made concrete as follows: when water passes from a liquid to a solid state, *the quantitative change is transformed into a solid state.*

Note further that the example given does not even permit us to say that the quality that is subject to quantitative change is transformed into another quality. In fact, when water passes from the liquid to the solid state, we do not say that the temperature is transformed into a quality other than temperature. From the point of view of temperature, there is only a change of degree, a quantitative change of the quality. It is the qualitative *state* that has changed. Engels says, moreover, that there is a modification of the state of cohesion and that the water is transformed into ice. It is neither the diminution of the temperature nor the temperature which passes from the liquid to the solid state.

Perhaps it will be said that I should not grant the proposition that "the quantity is transformed into quality" because I have not admitted possible contradiction. To that I reply that, even so, for the demands of argument, I will grant this contradiction, the more so that I should admit at the outset a contradiction (call it A) which renders the contradiction called possible impossible (call it B). In fact, I should admit that the quantitative change of quality and the quality itself will be at once the same and different in the same respect. But, if I admit this contradiction A, I render contradiction B impossible, since the quantitative change being already both the same and different in the same respect as the quality, the transformation of the quantitative into quality becomes impossible.

NOTES

1. These notes were written in the summer of 1943, at the request of Constantin Oumanski, then ambassador of the Soviet Union to Mexico. They represent some of the difficulties we have with the doctrine of dialectical materialism. I offer them to contemporary Marxists, not as a challenge, but in the hope of receiving a response that will help in the objective presentation of their doctrine.

2. Friedrich Engels, *Anti-Dühring*, trans. Bracke (Paris: Alfred Costes, 1931), vol. 1, ch. 13, p. 208.

3. Ibid., ch. 12, p. 182.

4. See J. B. S. Haldane, *The Marxist Philosophy and the Sciences* (New York: Random House, 1939), 56.

5. Engels, *Anti-Dühring*, vol. 1, ch. 13, pp. 212–13.

6. Ibid., ch. 12, pp. 180–81.

7. The English translation of Dutt, *Dialectics of Nature* (New York: International Publishers, 1940), 200.

8. Ibid., 198.

9. Ibid., 199.

10. Ibid., 200, n. 1.

11. Engels, *Anti-Dühring*, vol. 1, ch. 12, p. 182.

12. In *Histoire de Parti communiste* (Moscow, 1939), 101–2. [Joseph Stalin, in *History of the Communist Party* (Moscow, 1939), ch. 4, section 2. The same section was issued separately by International Publishers, New York, under the title *Historical and Dialectical Materialism*.]

This Is a Hard Saying

1945

Abscondes eos in abscondito faciei tuae.
—Psalm 30

I

A contemporary author who has many readers and is of great influence recently wrote that he could not understand how some of the most intelligent people of his acquaintance could still believe the most afflicting foolishness they are taught by the Catholic Church. We will not be so simplistic as to reject out of hand this invective. The Good Lord has told us "it pleased God by the foolishness of our preaching to save those who believe." And that "the doctrine of the cross is foolishness to those who perish." "But the foolish things of the world God has chosen to put to shame the strong." "But the sensual man does not perceive the things that are of the Spirit of God, for it is foolishness to him and he cannot understand."[1]

Some philosophers have thought that the relation of properly divine truth to natural truth can be compared to a series converging at its limit. But this comparison tends to confuse the incomparable otherness of the two truths. Properly divine truth so far exceeds natural truth that nothing of the latter can give any presentiment of the former. Indeed it is written that "the things God has prepared for those who love Him have not entered into the heart of man."[2]

Others have held that divine truths are so different that natural reason can teach the contrary of the faith. Against this extreme we say, with St. Thomas, that if it is absolutely impossible for natural reason to arrive at knowledge of truths which are of pure divine faith, it is equally impossible for reason to call them impossible. If natural reason can determinately know that there are truths that it cannot conceive, it cannot say determinately what those truths are. If, for example, reason cannot say that the Trinity is impossible, it cannot say determinately that it is possible. We cannot call possible that whose impossibility we cannot see. That is the limit point of philosophical knowledge.

Let us not then seek to weaken the force of the Apostle's words. A bad understanding of the doctrine of analogy could destroy its profound meaning. If natural theology knows God *sub ratione entis,* it in no way attains Him as part of the subject of metaphysics—it knows Him only as the absolutely extrinsic principle of its subject. It cannot know this principle in terms of what it properly is, that is, in that which constitutes it in its very otherness. If God were contained within the limits of that being which makes up the subject of metaphysics, faith would only serve to reinforce in a way the *ratio entis* and convert it into the *ratio deitatis.* Or again, the latter would only make us know distinctly what the former makes known to us confusedly. Knowledge of God according to His deity would then be at least in the direct line of philosophical wisdom, it would be like a limit toward which a series converges.

That is why the wisdom of this world becomes pure foolishness as soon as it wants to judge properly divine wisdom. On the contrary, judged by the wisdom of the world, divine wisdom itself becomes, according to the strong saying of the Apostle, folly. Philosophical wisdom, applying itself to things which are of pure divine faith, can only lead to error. It is in that sense that the words of Denis the Areopagite "all human knowledge is error in comparison to divine knowledge," are perfectly formal and not simple hyperbole. Philosophy can aid theology under the title of *ancilla.* Philosophical truth can influence a theological conclusion only in virtue of its being lifted up by the sapiential judgment of theology. St. Thomas says that sacred science makes use of the sciences not out of defect or insufficiency, but because of the weakness of our minds which, moving from things known by natural reason, can be more easily led by the hand to things which are above reason and the object of this sacred science.

The truths to which we must adhere by faith are so far above man purely as man, they are of an otherness so profound and obscure, that the degree of our faith can be judged by the inclinations or repugnances we can feel in giving them our firm assent. "One of the most signal favors," says St. John of the Cross,

> that the Savior gives to a soul during this life—though it is fleeting, not lasting—is to grant it a knowledge so clear and a sentiment so elevated of his divinity, that the soul grasps and sees in a very clear way the impossibility of fully having here below that knowledge and sentiment. When a soul bases itself on its own knowledge, or on its tastes and sen-

> timents, to go to God, when it does not see that such means are without value and without proportion to such an end, it easily goes astray, or is stopped on the way, because it does not blindly attach itself to faith alone which is the true guide.

There is no mystery of our faith where properly divine ways so shine forth as in that of the blessed Eucharist. It is with respect to this doctrine that the Gospel reports, "Many of his disciples, having heard him, said, 'This saying is hard, and who can bear it?' Jesus knowing in himself that his disciples murmured on this subject, said to them, 'Does this scandalize you?' From that moment, many of his disciples went away and walked no more with him."[3] Let us consider for a few moments the care with which God has taken to make us adhere to the great mystery of His Eucharist by faith alone.

II

David Hume, in his *Enquiry Concerning Human Understanding,* first of all attacks the notion of causality. It is by denying the possibility of demonstrating the existence of a cause by its effect that he denies the possibility of knowing God by natural reason. He takes away from natural wisdom what it can have of the divine; therefore he denies natural wisdom. As Aristotle said, "Wisdom (we mean here the wisdom man can acquire by reason alone) is a science that would be most worthy of God to have, and which treats of divine things. But wisdom alone has this double character: God appears to be indeed a cause and principle of all things, and such a science God alone, or at least God principally, can have" [*Metaphysics,* I, 2].

The second part of the *Enquiry* (section 10) is directed against the Catholic faith. Hume attacks miracles as motives of credibility. But this attack is, at first glance, rather curiously made. It opens with a disdainful charge against the doctrine of the *real presence:* "a doctrine so little worthy of a serious refutation." Should one attribute this to stupidity or astuteness? He cites the Eucharist as an example of the miracles to which the Apostles witness—"who were eye-witnesses to those miracles of our Savior"—and by which Christ proved his divine mission "by which he proved his divine mission." But, he goes on, the doctrine of the real presence contradicts the senses. "It contradicts sense, though both the scriptures and tradition, on which it is supposed to be built, carry not much evidence with them as sense; when they are considered merely as external evidences, and are not brought home to everyone's breast, by the immediate operation of the Holy Spirit" [*Humes's Enquiries*, ed. Selby-Bigge (Oxford, 1902), p. 109].

Clearly he is here guilty of a double confusion. He supposes that we assimilate the certitude of the miracle as extrinsic motive of credibility to the certitude of faith. But, what is much more insidious, he destroys the certitude of the extrinsic motive, even renders it absurd, by citing, whether by ignorance or design, the miracle of transubstantiation as exemplar of the miracles by which the truth of the Christian religion is proved. He seeks to ignore that transubstantiation is precisely an absolutely invisible miracle, that this miracle is in no way a sign of the truth of revelation and a motive of credibility. Indeed we adhere to it by divine faith alone. This miracle is in effect that which the Apostles neither saw nor touched nor tasted. They simply heard the words of Christ as we would have heard them. They have reported them to us. And like them we believe in this miracle because Christ said it.

Notice the shrewdness, conscious or unconscious, of Hume's procedure. Not only does he destroy the type of miracles that are extrinsic motives of credibility by citing a miracle which belongs to another kind, but he destroys, with the same blow, the most hidden sacrament, the most profound, the greatest of the whole Christian life. He attacks the faith in its purest form, that by which we adhere not only to that which is invisible like the Most Sacred Trinity, but firmly as well that which contradicts the principle of the origin of all our knowledge, the senses. He chose the instance where the divine truth is most obviously beyond the knowledge of reason founded on the senses. In the Blessed Eucharist, in fact, although the senses do not err with respect to their proper object—they see truly the accidents of bread and wine—nevertheless, it is because of sense that intelligence, without faith, errs with regard to the hidden substances of those accidents.

Visus, tactus, gustus in te fallitur,
Sed auditus solus tuto creditur.[4]

Et si sensus deficit,
Ad firmandum cor sincerum,
Sola fides sufficit.[5]

It is in the faith in this sacrament that God asks of us the most complete abnegation of that which is profoundly human—I mean the loss of our proper judgment on the substance of the object that is most proportioned to our intelligence.

But there is more. God chooses to hide the flesh and blood of the divine Savior, not under a nutritional substance produced by the operations of nature alone; He has chosen the appearances of substances produced by the application of human art, bread and wine.

The complete hiddenness of this *sacrum secretum* pertains to the perfection of faith, says St. Thomas. Since God wishes us to participate in that divinity that no creature can know by its proper means—*no one has seen God*[6]—isn't it fitting that He ask of us a total faith in this Sacrament of the Way, the Truth, and the Life, a universal faith where reason is deprived of its reliance on the senses? Since our faith bears not only on the divinity of Christ but also on His humanity by which He is assimilated to us, is it not fitting for the very perfection of that faith that in the Sacrament that contains substantially the very principle of all sanctification and which, in this perfect obscurity, announces in most appropriate manner our union with Him in the light of the future life, Christ shows us His flesh and blood in an invisible manner?

In cruce latebat solo deitas,
At hic latet simul et humanitas.[7]

On the cross, where the redemptive passion was accomplished, only divinity was hidden. The senses could grasp the humanity of Jesus—a crushed humanity, it is true, but at least visible. Whereas in the Sacrament, the humanity too is hidden under the appearances of a different sensible substance. *Sacramentum regis abscondere bonum est.*[8] It is good, says Scripture, to hide the sacrament of the King, of the King who is made known to us in the total abnegation of the Passion. It is good that the *semetipsum exinanivit—He has made himself nothing*[9]—be shared by the senses in this sacrifice that we make in His memory. It is good, it is wise that the Light of the perfect Pasch be fully manifest to us in its present obscurity, in its deep darkness. *Et nox illiminatio mea.*[10] *Nox nocti indicat scientiam.*[11]

It is then by a truly unheard-of mercy that God has deigned to meet us in the perfect night and that, to lift us up to Himself, He has overwhelmed our insufficiency, He has asked us in our act of faith for an abnegation analogous to that of His Son. "This is a hard saying. Who can listen to it?"[12]

But the hardness of the saying for human ears, is not that a reason to adhere to it with a yet firmer faith? Is it not a wonderful manifestation of the otherness of the divine truth from now on made 'ours'? In place of being scandalized along with the murmuring disciples then we have on the contrary every reason to cry out with St. Peter, "Lord, you have the words of eternal life!" The more the word in which you ask me to believe is other than ours, the more you—the Word of Life—draw me to you as you are in yourself. "May he kiss me with the kisses of his mouth."[13] Let Him say to me the unheard-of word, the word I hear without understanding. Lord, you are the Word of life. You are the Word that no human word can express. The human word makes vain the cross of Christ.[14] Speak to me the words that *you* have formed.

"Do you also wish to go away?" Simon Peter answered Him: "Lord, to whom shall we go?" Is it not the most admirable mercy of all that, abandoned by all, we can now go to Him only in the abandonment of the mystery of faith where hidden in a perfectly appropriated silence is He whose name is the Word.

NOTES

1. 1 Cor. 1 and 2 passim.
2. 1 Cor. 2:9.
3. John 6:60–61, 66.
4. "Sight, touch, and taste here fail, / Hearing only can be believed" (from the *Adoro te devote*).
5. "Though senses fail to see this truth, / To confirm the sincere heart / Faith alone suffices" (from the *Pange Lingua*).
6. John 1:18.
7. "On the cross divinity alone was hid, / While here is hid humanity as well" (from the *Adoro te devote*).
8. Tob. 12:7.
9. Phil. 2:7.
10. Ps. 138:3.
11. Ps. 18:3.
12. John 6:61.
13. Canticle 1:1.
14. 1 Cor. 1:17.

Review of *Between Heaven and Earth*

1946

During a recent stay in the United States I was astonished at the publicity being given by Catholic periodicals and bookshops to a work entitled *Between Heaven and Earth.* Let it be said at once that it is not from a Catholic pen. From one point of view there is no reason to be surprised at the favorable reception it has received even among the devout, coming as it does from the author of the *Song of Bernadette,* the novel many found irreproachable and moving. Franz Werfel had written so well of Our Lady that his readers might have expected the same fidelity in *Between Heaven and Earth.* But having examined this work, I am obliged to say that it calls for the most serious reservations. In fact, it teaches errors clearly and directly contrary to the two most fundamental truths of our religion. "In our faith," says St. Thomas, "two things are principally to be believed: the mystery of the Trinity, and the mystery of the Incarnation." Surely Catholic readers should be put on their guard against the pernicious errors contained in this book; and it will also be failing in charity toward these outside the fold not to expose so false an understanding of our faith.

Here are several propositions cited from the English text: "The second person of the Trinity is the social aspect in the Deity. *Verbum caro factum.* This not only applies to the Son but, in another sense, to the world also. All created things are expressed words. The Son is that essence of God which does not contemplate itself, like the Father, but has to do exclusively with the world, as the first and uncreated and yet incarnate Word above all words" (p. 144).

In this brief passage are implicit several grave errors which are also closely interrelated. In truth, it is independently of all creation and independently of the Incarnation that the Word is the Son of God, consubstantial with the Father, equal to the Father. He knows Himself as the Father knows Himself. *Genitum, non factum.* In God He is begotten and not made. The Word was *made* flesh. The Word that is exclusively concerned with the world is the incarnate Word as incarnate—not the Word as such. The begotten Word is concerned with creation only secondarily.[1] Similarly, it is the incarnate Word only as considered in His humanity who is inferior to the Father.

Aequalis Patri secundum divinitatem: minor Patre secundum humanitatem.[2] To identify the Word as such with the incarnate Word is to render the Word as such inferior to the Father, to lower His divinity and to make creation and the Incarnation necessary acts.

Nor are we drawing inferences which the author himself fails to state. Here are his plain words: ". . . The mystery of Incarnation . . . is an intrapersonal act within the Deity, having nothing to do with creation" (p. 143). The author affirms that, since life is a communication of self, God cannot help but turn outward; self-contemplation may not suffice Him. So that: "Every motion directed outward from within arises out of insatiability" (pp. 145–46).

Mr. Werfel's mistakes are not always coherent. Having lowered the dignity of the Son of God, that is, of God Himself, he next speaks of "the Son, who is the Father, the Father who is the Son" (p. 183). Such enormities are not only contrary to theology; they are directly contrary to formally revealed truth.

On the motive of the Incarnation, Mr. Werfel teaches an error comparable to that of Ludwig Feuerbach in *Das Wesen des Christenthums,* namely that the Incarnation is necessary to God Himself. Were it not for the Incarnation, he thinks, the creature would hold an advantage over God which God could not permit. "Since the Deity is *Integritas* and *Incorruptibilitas* personified, humanity has the advantage of one experience over Him: suffering and death" (pp. 144–45). This experience, the fruit of sin, is that of wretched and fallen nature, it is true. "And yet, suffering and death are the deepest experience of created beings; and it is precisely this deepest experience that the creator does not possess" (ibid.). It is this, says Werfel, that reveals to us one of the most mysterious reasons for the Incarnation of God. "The Creator cannot suffer His created being to have any advantage over Him even if it be pain and decay." Hence, "Something like a longing for debasement and twilight awakens in the Absolute Light." "The Lord descends so that man may not be superior to Him even in that which is negative." In a word, according to Werfel, God did not humble Himself solely to snatch us from our misery and save the world, *propter nostram salutem;* He humbled Himself in order to enrich Himself with an experience in which otherwise man would have an advantage over Him: "He descends not only to save the world, but *in His very own interest,* to taste of death" (ibid., italics ours).

Werfel has not understood that it is sheer goodness, that it is the most pure and disinterested liberality, in short, infinite Mercy, which is at the base of all the works of God—*miserationes ejus super omnia opera ejus*—that God has humbled Himself for us: that from this God derives not the slightest profit. The words just quoted from *Between Heaven and Earth* are among the most frightful and blasphemous that have ever been written. We must believe that Werfel did not know the import of what he was saying. It is not our task, fortunately, to judge the intentions of the author; our concern is simply with the sense of his words as attacking God in His condition of extreme humiliation. How could we hear without shuddering that, if *God emptied himself, taking the form of a servant, and humbled himself, becoming obedient unto death, even to the death of the cross,*[3] He did it not out of pure mercy, to save us from the misery in which we find ourselves, and to assimilate us to Himself, but rather in virtue of a Self-interest essential to His divinity, so that man might have no advantage over Him, so that man might not be His superior. According to Werfel, if God must taste suffering and death, the responsibility is not entirely ours, since God also empties Himself "in His very own interest." "The mystery of the *Passion,* or the sacrificial death . . . is suffered for the sake of the Creator no less than for the sake of the creatures" (p. 131).

Let us hearken rather to Saint Paul: *Quoniam propter* vos *egenus factus est, cum esset dives, ut illius inopia* vos *divites essetis.*[4] For *you,* being rich He became poor, that through His poverty you might be rich. Are we to shamelessly protest to our God that we are not the only ones enriched by His poverty? That His becoming humbled was not through mercy alone? That fundamentally He profited more by it than we? It is an odious blasphemy. Truth forces us to cry from the house-tops to all readers of Mr. Werfel that no one could be more wrong. We have no right to permit the illusion that the author of the *Song of Bernadette* knows revealed doctrine on the most fundamental points of our faith. *Fides ex auditu.* Faith comes from preaching heard. Even they who are well disposed to believe these revealed truths, must yet be made to hear them. And these truths are hard. *Durus est hic sermo.*

One should not conclude from this criticism that *Between Heaven and Earth* contains only errors. But I make bold to say that the author goes astray whenever he takes up a question that is in the least profound. It is regrettable that Werfel should have thought himself competent in these matters. He appears to be the victim of his talent of stating things pleasingly. The gravest errors and the most vain conceptions can be phrased in a language correct and elegant.

These condemnations and reservations will displease some; but broad-mindedness has its limits; and there are things which God does not allow. *Omne verbum otiosum, quod locuti fuerint homines, reddent rationem de eo in die judicii. Ex verbis enim tuis justificaberis, et ex verbis tuis condemnaberis.*[5] We cannot allow the monstrous errors of *Between Heaven and Earth* to be spread on the strength of the charm and piety of the *Song of Bernadette.* Doctrines so fundamental must be defended in season and out of season.

NOTES

This note is the English version of a radio-talk delivered on the Laval University Hour of April 1945. The original text appeared at the time in the Quebec *Semaine Religieuse.* Several American priests have asked us to render it available to the English-speaking public, since the book in question is still being favorably reviewed by Catholic periodicals and on sale in Catholic bookshops.

1. Cf. St. Thomas, Ia, q. 34, a. 3: *Utrum in nomine Verbi importetur respectus ad creaturam; Q.D. de veritate,* q. 4, a. 5: *Utrum Verbum relationem ad creaturam importet.*
2. Symbolon *Quicumque,* St. Athanasius.
3. Phil. 2:7–8.
4. 2 Cor. 8:9.
5. Matt. 12:36–37.

Concept, Process, and Reality

1946

It is taken for granted that Thomistic doctrine can neither assimilate, nor account for, one of the most fundamental ideas of philosophy, which modern thought, since Nicholas of Cusa, has set in bold relief: the idea of process extended to otherwise irreducible natures. Most modern philosophers would agree with the late Ernst Cassirer's statement of our problem:

> Aristotle's logic is unexcelled in the precise working out of contradictions, in setting up the categories by which the classes of being are distinguished. But it is unable to overcome this opposition between the various classes of being; it does not press on to their real point of unification. Hence it remains caught in the empirical and the finite; it is unable to rise to a truly speculative interpretation of the universe. The physical universe of Aristotle is dominated by the opposition between "the straight" and "the curved"; motion in straight lines and motion in circles are for him essentially and radically distinct. But the transition to the infinitely large and the infinitely small shows that this is a matter not of an absolute but of a relative distinction. The circle with an infinite radius coincides with the straight line; the infinitely small arc is indistinguishable from its chord.[1]

If the unification referred to by the author were understood as accomplished identity, there could be no compromise. For, at infinity, straight would have to be, in the same respect, non-straight. We are not concerned, at present, with the philosophers who accept this contradiction. We merely wish to point out that the irrevocable opposition between the two conceptions illustrated by the passage quoted from Cassirer arises only when we cease to understand "at infinity" as the limit of unending process. While the convergent series is fixed once and for all by a definite law which holds for all possible values of the series, a smallest or a greatest possible value is impossible; nor is the limit itself a possible value. The very law precludes this possibility; for what it guarantees is convergence toward, not attainment of, the limit. Nor could the law be fixed and definite as it is if the variable and the limit were not irreducibly distinct, or if there were a smallest possible difference to keep them apart. At first glance, this qualification might be interpreted as a rejection of the very idea of process associated with formally distinct natures; yet I venture the opinion that the very idea of process depends upon it.

No one will deny that the definition of a term as the limit of a process is a decisive achievement, however well this term may already be known in itself. Not only does it bring out an otherwise unknown relatedness of formally distinct natures: it also implies a peculiar unity in the very *mode of knowing* the one and the other. In fact, it is precisely because we naturally seek a more unified mode of knowing that, when faced with natures related as variable and limit, we are urged to "press on to their real point of unification." But the problem is: does the tendency toward unification in the mode of knowing arise from an actual identity of the objects known?

The problem of the One and the Many is usually confined to the manner in which things in themselves are one and many. Yet there is also a question of a One and Many with regard to the cognitive *means* by which we reach *what* we know. The latter (we shall call it the noetic as opposed to the natural problem) is amply treated by St. Thomas who, in this connection, draws from Platonic, and more particularly from Neo-Platonic sources. His teaching on this subject is as follows.[2] For each object distinctly known we require a distinct means of knowing. Thus, the concept by which we reach the object "circle" is other than that by which we attain "triangle." It is true that both objects may be known simultaneously by some common concept such as that of figure, but the genus "figure" cannot represent them distinctly. Whenever, by means of one concept, we actually consider many objects, we inevitably do so at the expense of distinction. In fact, distinct knowledge requires in us a multitude of cognitive means directly proportioned to the multitude of objects we know. This dispersion of our means of knowing is due to the empirical nature of our mind. Any finite intellect, knowing things in its own mode, requires a manifold of intelligible species, but the number of species, the extent to which the intellect is broken up and scattered about within itself, will be in proportion to its specific degree of perfection. Thus, if our mind were of a more exalted nature, a single concept such as figure might well represent simultaneously the several irreducible kinds of figure with even sharper distinction than that attainable by separate concepts used in succession. Indeed, the Divine Intellect knows all things by means of the single intelligible species which is Its indivisible Essence.—The general concept by which distinct objects are known in confusion only, is called universal in predication ("secundum praedicationem"), whereas the intelligible species which represents distinct objects in their very distinction is said to be of universal power ("universalis virtute," akin to Cassirer's "concrete universality").[3]

Now let us see how we might apply this doctrine to the problem of reduction by process. Why should we attempt any reduction of formally distinct natures at all? Why not rest in their essential and radical distinction?

Whatever is first and more easily known may be called "more rational to us," whereas that which is known only by further application may be termed "less rational to us." Yet, what is relatively irrational to us may be more rational in itself, such as that proper and formal distinction of things which requires the formation, through experience, construction or inference, of new and distinct concepts to be used in time-producing succession. Hence, the formal distinction of natures, requiring as it does restricted and separate means of knowing, takes on the appearance of an irrational gap between natures. Thus the irreducible givenness of objects, the distinction between straight and curve, polygon and circle, one and two, continuum and discrete, is conceived as a break, as an obstruction, to full understanding. And so it is, if by full understanding we mean a simultaneous, undivided and yet more penetrating view of such natures. However, this irrationality may be tentatively cleared by the insertion of convergent series, for, whenever we can define a notion as the limit of a variable containing it in inchoation, as it were, we somehow overcome the givenness of that notion: it is as if, from the notion of "curve," we were moving toward, and about to reach, that of "straight" in its very difference, without dependence upon the distinct concept of "straight." Could this process be carried through to the end, our mind would be freed from the meshes of its conceptual network. The mere desire to grasp more thoroughly even the objects we already know, would rouse us to seek them in the light of process wherever the interpolation of "infinity"[4] and movement could afford a "nexus"—to use Cusanus' term.

Yet if, *per impossibile,* this tentative rationalization could be carried through to identity, the universe, as the author of *Identité et réalité*[5] insisted, would collapse into the absolute irrationality of contradiction. It is precisely the irreducible distinctions we have called "irrational gaps" which ensure, at the same time, the rationality of natures.[6] Otherwise, as Henri Poincaré has said, all of mathematics—where we find the first, direct and most complete illustration of process—would turn out to be but a roundabout way of saying that A is A.[7]

When viewed in this perspective, Bergson's distinction between analysis and intuition ("l'analyse opère sur l'immobile, tandis que l'intuition se place dans la mobilité ou, ce qui revient au même, dans la durée")[8] becomes quite acceptable. Now, as Hegel had pointed out in this connection, dialectical and speculative reason (understood in the Hegelian sense) presuppose the distinctions of understanding.[9] Or, in other words, what has been called the "Platonic mode" must follow upon the "Aristotelian mode." It is true, then, that "*Philosopher consiste à invertir la direction habituelle du travail de la pensée.*"[10]

Let us suppose for one moment that this procedure could be extended to all objects (the basic postulate of mathematism): its ultimate limit then would be a single means of knowing, a unique intelligible species, a numerical, indivisible "one" in the Platonic sense. Above and free from the multiplicity of finite concepts, this absolute idea would be infinite, all-comprehensive. If the rapprochement between such a limit and the noetic one were believed irrelevant, we might quote from Cassirer:

> Die Idealität der Mathematik hebt den Geist zu seiner höchsten Höhe und bringt ihn zu seiner eigentlichen Vollendung:—sie beseitigt die Schranke, die die mittelalterliche Anschauung zwischen der Natur und dem Geist einerseits, zwischen dem menschlichen und dem göttlichen Intellekt andererseits aufgerichtet hatte.[11]

And Hermann Weyl, discussing the problem of being and possibility, of continuum and discrete, has expressed his

> conviction that the origin and the reconciliation of this divergence can lie only in God. The attempt presented by realism to elevate the object to the dignity of absolute being was doomed to fail from the start; as was also the opposite attempt of idealism to endow the subject with the same high independence. In mathematics also, partly through the dependence on philosophy, the inclination toward the absolute which is evidently deeply rooted in man has asserted itself.[12]

It is held that Aristotelian logic, resting on the principle of contradiction which excludes a middle third, is a mere logic of the finite and must therefore be essentially invalid where it deals with "the contemplation of the infinite";[13] that "it demands a new mathematical logic based on the coincidence of opposites."[14] Now, coincidence of formally distinct objects does indeed contradict the excluded middle when interpreted as formal identity. However, if we distinguish the One and Many of things from the One and Many of cognitive means, of intelligible species, the formal identity of the concept of "straight" and of the concept of "curved" does not conflict with the excluded middle: a single concept representing simultaneously and indivisibly two or more irreducible natures implies no contradiction, since the known is in the knower according to the mode of the knower. Of course, if the excluded middle were interpreted to mean that any knowledge of distinct natures requires distinct means of knowing, it would be quite false: if by "the opposition between 'the straight' and 'the curved'" were meant the distinction between the concept by which we know "the straight" and that by which we know "the curved," we would unhesitatingly agree "that this is a matter not of an absolute but of a relative distinction"—relative to a given type of intellect. In the identity of a single concept distinctly representing different objects in their difference, not in a contradictory identity of the objects themselves, the ideal of rigor would indeed be achieved.[15]

Whether or not our mind can ever reach such rigor is another matter. But there is no doubt that we tend toward it. Defining a nature as the limit of an unending process we at least approach noetic identity; to use Hermann Weyl's expression: our mind "is open toward the infinite," toward that absolute "which 'is not of this world,' and of which the eye of our consciousness perceives but reflected gleams."[16] However, in consequence of the empirical nature of our mind we can never attain to it by the method we are bound to employ—i.e. the tentative reduction toward identity of formally distinct objects—since "at infinity" both object and concept would be destroyed. For this reason the tendency toward a limit can be no more than the symbolic expression of the identity found in that perfect concept which we have termed a universal in power.

If by the logic of contradiction we mean the method of considering an object as the limit of a process which, at infinity, would actually involve contradiction; if this logic receives its denomination from what *would* be, and not from what is believed to be or to be *possible*, I fail to see wherein it contradicts or escapes the logic based on the principle of contradiction. To recognize that the process of a series converging toward a limit is fixed once and for all by a definite law does not imply that we must accept the existence of an ultimate value identical with the limit; that there *is* an arc identical with its chord. To be valid and universal the law does not have to apply to what is impossible. We should not be led to believe that the present dialectical method proceeds from what is quite the opposite of the principle of contradiction, simply because nothing short of identity could arrest the process, while at the same time this impossible term must be kept in mind[17] if we are not to sacrifice the nature of the process as converging toward a limit.

When we attend to a merely noetic "point of unification" the pursuance of rationalization is no longer seen as leading toward a term which implies contradiction, toward destruction of the universe of specific natures:[18] even if it is impossible for us to attain the noetic identity we pursue, such identity does not in itself involve contradiction. Compared to the unity we approach by imitation, our method is but a toy. Yet, even a toy is a likeness. Perhaps the allegory of the Cave would be more fitting.

NOTES

1. "Giovanni Pico della Mirandola," in *Journal of the History of Ideas* 3, no. 3 (June 1942): 322–23. Permission to quote has kindly been granted by the publishers.

2. E.g., *II Summa contra gentes*, cc. 98–101; *Super librum de causis*, lectio 10.

3. Ernst Cassirer, *Substance and Function* (Chicago: Open Court, 1923), ch. 1, p. 19.

4. As A. J. Maréchal, S.J., reminds us in his *Précis d'histoire de la philosophie moderne:* "La position de Leibniz devant l'idée d'infini a ceci de particulier qu'il ne rejette pas simplement cette idée dans la zone de l'irrationnel ou du métarationnel (comme les Anciens), mais lui fait une place positive dans le cadre de la raison, sans toutefois supposer l'infini connaissable en lui-même par une idée intuitive, claire et distincte (à la manière des cartésiens). C'est du côté du fini, que l'infini se laisse pressentir et jusqu'à un certain point qualifier par nous; ainsi capté dans un geste d'approche plutôt que dans une vue, il se révèle le fond constitutif et le principe d'intelligibilité du fini" ([Louvain: Museum Lessianum, 1933], vol. 1, ch. 7, n. 103, p. 128).

5. Emile Meyerson, more particularly in his essay on *La notion de l'identique* in *Essais*, preface by Louis de Broglie (Paris: Vrin, 1936), 187–208.

6. Louis de Broglie, distinguished physicist and close friend of Meyerson, summarizes this view as follows: "En montrant cette tendance instinctive de notre raison, le grand philosophe en soulignait hardiment le caractère paradoxal, car l'effort identificateur de la raison, s'il pouvait complètement réussir, aboutirait à l'abolition de toute diversité et de toute hétérogénéité, c'est-à-dire à une espèce de négation du monde même qu'il cherche à expliquer. Et si la raison parvient néanmoins à s'échapper de ce cercle vicieux et à constituer une science qui incontestablement progresse, c'est, pense-t-il qu'elle laisse se glisser dans nos constructions théoriques quelques éléments irrationnnels dont l'introduction plus ou moins subreptice permet à l'ensemble de nos identifications successives de ne pas constituer seulement une immense tautologie. . . . Cette situation se trouve résumée par une phrase de M. Paul Valéry, phrase sans doute inspirée par la lecture même des ouvrages de M. Meyerson: 'L'esprit humain est absurde par ce qu'il cherche; il est grand par ce qu il trouve'. Mais, comme en définitive l'univers ne peut pas se réduire à une vaste tautologie, nous devons forcément nous heurter cà et là dans notre description scientifique de la nature à des éléments 'irrationnels' qui résistent à nos tentatives d'identification, l'effort jamais lassé de la raison humaine s'acharnant à circonscrire ces éléments et à en réduire le domaine" (*Matière et lumière* [Paris: Albin Michel, 1937], 318–20).

7. Henri Poincaré, *La Science et l'Hypothèse* (Paris: Flammarion, 1932), 9–10.

8. Henri Bergson, *La pensée et le mouvant* (Paris: Alcan, 1934), 228.

9. Hegel, *Logic* (Encycl.), trans. W. Wallace, ch. 6. The viewpoint of understanding is also called 'Metaphysics.' Dialectical materialists now use the latter term in this sense. See, e.g., Joseph Stalin's *Dialectical and Historical Materialism* (New York: International Publishers, 1940), 7 ff.

10. Henri Bergson, op. cit., 241. Because Bergson believed that movement, becoming and duration in nature are a true instance of the object of what he called intuition, as opposed to the "concepts fixes" of analysis, the interpretation we put upon this most fundamental idea of his philosophy might seem far-fetched. Yet, immediately following the line just quoted we read: "Cette inversion n'a jamais été pratiquée d'une manière méthodique; mais une histoire approfondie de la pensée humaine montrerait que nous lui devons ce qui s'est fait de plus grand dans les sciences, tout aussi bien que ce qu'il y a de viable en métaphysique. La plus puissante des méthodes d'investigation dont l'esprit humain dispose, l'analyse infinitésimale, est née de cette inversion même. La mathématique moderne est précisément un effort pour substituer au *tout fait* ce qui *se fait,* pour suivre la génération des grandeurs, pour saisir le mouvement, non plus du dehors et dans son résultat étalé, mais du dedans et dans sa tendance à changer, enfin pour adopter la continuité mobile du dessin des choses. Il est vrai qu'elle s'en tient au dessin, n'étant que la science des grandeurs. . . . Il est donc naturel que la métaphysique adopte, pour l'étendre à toutes les qualités, c'est-à-dire à la réalité en général, l'idée génératrice de notre mathématique. Elle ne s'acheminera nullement par là à la mathématique universelle, cette chimère de la philosophie moderne. Bien au contraire, à mesure qu'elle fera plus de chemin, elle rencontrera des objets plus intraduisibles en symboles. Mais elle aura du moins commencé par prendre

contact avec la continuité et la mobilité du réel là où ce contact est le plus merveilleusement utilisable. Elle se sera contemplée dans un miroir qui lui renvoie une image très rétrécie sans doute, mais très lumineuse aussi, d'elle-même" (op. cit., 241–43). What Bergson here calls "métaphysique" is akin to Hegel's speculative reason, and not to understanding.

11. Ernst Cassirer, *Individuum und Kosmos in der Philosophie der Renaissance* (Leipzig: Teubner, 1927), 171.

12. Herman Weyl, *The Open World* (New Haven: Yale University Press, 1932), 68. Permission to quote has kindly been granted by the publishers.

13. "... Jetzt aber erfolgt eine weit radikalere Scheidung: denn die Aristotelische Logik, die auf dem Satz des ausgeschlossenen Dritten beruht, erweist sich eben damit für Cusanus als eine bloss Logik des Endlichen, muss also immer und notwendig versagen, wo es sich um die Anschauung des Unendlichen handelt" (Cassirer, op. cit., 12).

14. "... Daher beruft sich Cusanus, um Sinn und Ziel der *visio intellectualis* zu bezeichnen, nicht sowohl auf die mystische Form der passiven Kontemplation, als er sich vielmehr auf die Mathematik beruft. Sie wird ihm zum eigentlichen, zum einzig wahrhaften und "präzisen" Symbol des spekulativen Denkens und der spekulativen Zusammenschau der Gegensätze. "Nihil certi habemus in nostra scientia nisi nostram mathematicam": wo die Sprache der Mathematik versagt, da gibt es für den menschlichen Geist überhaupt nichts Fassbares und nichts Erkennbares mehr. Wenn die Gotteslehre des Cusanus sich daher von der scholastischen Logik, von der Logik der Gattungsbegriffe, die unter dem Satz des Widerspruchs und des ausgeschlossenen Dritten steht, lossagt, so fordert sie dagegen einen neuen Typus der mathematischen Logik, die die Koinzidenz der Gegensätze nicht ausschliesst, sondern die eben diese Koinzidenz selbst, eben dieses Zusammenfallen des Absolut-Grössten und des Absolut-Kleinsten, als ständiges Prinzip und als notwendiges Vehikel der fortschreitenden Erkenntnis braucht" (ibid., 15).

15. On the subject of mathematical rigor we refer to J. Pierpont's "Mathematical Rigor Past and Present," in *Bulletin of the American Mathematical Society* 34 (1928). The author's conclusion is: "Personally we do not believe that absolute rigor will ever be attained and if a time arrives when this is thought to be the case, it will be a sign that the race of mathematicians has declined" (p. 23).

16. Op. cit., 84, 72.

17. "Ainsi il est indispensable de maintenir fermement, dans l'analyse de la pensée, cette notion de l'identique strict, en dépit de ce que cet identique demeure entièrement inaccessible, que le concept paraisse contradictoire en son essence, et que, bien entendu, en sa pureté ce concept ne se rencontre nulle part ni dans la pensée commune, ni dans le savoir physique, dans son sens le plus large, ni dans le savoir philosophique, puisque ce que l'on trouve c'est toujours et partout l'identique partiel" (Meyerson, *Essais*, 197).

18. "... Aber eben aus dieser Zerstörung erwächst jetzt um so dringlicher die Aufgabe, die Gesamtordnung des Seins und Geschehens aus der eigenen Kraft und mit den eigenen Mitteln des Intellekts wieder aufzubauen. Der Intellekt muss lernen, sich ohne

sinnliche Hilfe und Stütze in seinem eigenen Medium, im freien Aether des Gedankens zu bewegen, um kraft dieser Bewegung der Sinnlichkeit Herr zu werden und sie zu sich emporzuziehen. Dami kehrt sich, verglichen mit der Aristotelisch-scholastischen Physik, die Ordnung der Probleme um. Was für diese als Anfangspunkt galt, das wird jetzt zum End- und Zielpunkte der kosmologischen Betrachtung" (Cassirer, op. cit., 188). See also, in connection with the subject of this paper, Alexandre Koyré's sharp presentation of Descartes in *Entretiens sur Descartes* (New York: Brentano's, 1944), as well as Jean Wahl, *Du rôle de l'idée de l'instant dans la philosophie de Descartes* (Paris: Alcan, 1920).

Index

CHARLES DE KONINCK

was on the faculty of Québec's Université de Laval and was Director of Laval's philosophy faculty from 1939 to 1956. He determined the course of philosophy at Laval and in much of French Canada through his publications and his connections with the Roman Catholic Church. He lectured frequently in the United States, as well as in Latin America, Europe, and Canada.

Printed in the USA
CPSIA information can be obtained
at www.ICGtesting.com
CBHW050013090724
11321CB00003B/16

9 780268 025977